P9-DCG-820

THE WAY WE THINK

THE WAY WE THINK

Conceptual Blending and the Mind's Hidden Complexities

GILLES FAUCONNIER
MARK TURNER

A Member of the Perseus Books Group

Copyright © 2002 by Gilles Fauconnier and Mark Turner

Published by Basic Books,
A Member of the Perseus Books Group

All rights reserved. Printed in the United States of America. No part of this book may be reproduced in any manner whatsoever without written permission except in the case of brief quotations embodied in critical articles and reviews. For information, address Basic Books, 10 East 53rd Street, New York, NY 10022-5299.

Designed by Trish Wilkinson
Set in 11-point AGaramond by the Perseus Books Group

Fauconnier, Gilles.
The way we think : conceptual blending and the mind's hidden complexities / Gilles Fauconnier, Mark Turner.
p. cm.
Includes bibliographical references and index.
ISBN 0-465-08785-X (alk. paper)
1. Concepts. 2. Thought and thinking. I. Turner, Mark, 1954– II. Title.

BF433 .F38 2002
153.4—dc21 2001052925

02 03 04 05 / 10 9 8 7 6 5 4 3 2 1

PREFACE

FIFTY THOUSAND YEARS AGO, more or less, during the Upper Paleolithic Age, our ancestors began the most spectacular advance in human history. Before that age, human beings were a negligible group of large mammals. After, the human mind was able to take over the world. What happened?

The archeological record suggests that during the Upper Paleolithic, humans developed an unprecedented ability to innovate. They acquired a modern human imagination, which gave them the ability to invent new concepts and to assemble new and dynamic mental patterns. The results of this change were awesome: Human beings developed art, science, religion, culture, sophisticated tools, and language. How could we have invented these things?

In this book, we focus on *conceptual blending*, a great mental capacity that, in its most advanced "double-scope" form, gave our ancestors superiority and, for better and for worse, made us what we are today. We investigate the principles of conceptual blending, its fascinating dynamics, and its crucial role in how we think and live.

Conceptual blending operates largely behind the scenes. We are not consciously aware of its hidden complexities, any more than we are consciously aware of the complexities of perception involved in, for example, seeing a blue cup. Almost invisibly to consciousness, conceptual blending choreographs vast networks of conceptual meaning, yielding cognitive products that, at the conscious level, appear simple. The way we think is not the way we think we think. Everyday thought seems straightforward, but even our simplest thinking is astonishingly complex.

The products of conceptual blending are ubiquitous. Students of rhetoric, literature, painting, and scientific invention have noticed many specific products of blending, each one of which, in isolation, seemed remarkable at the time, in its strange and arresting way. These scholars, ranging from Aristotle to Freud, took these specific instances to be exceptional, marginal eruptions of meaning, curious and suggestive. But none of them focused on the general mental capacity of blending or, as far as we can tell, even recognized that there is such a mental capacity. Attentive to the specific attraction—the painting, the poem, the dream, the scientific insight—they did not look for what all these bits and pieces have in common. The spectacular trees masked the forest.

Our own work started with just such curious and suggestive examples. But by making precise their underlying principles, we began to get glimpses of an entire forest behind the trees. We discovered that the same cognitive operation—conceptual blending—plays a decisive role in human thought and action and yields a boundless diversity of visible manifestations.

This was an exciting but also shocking discovery, running as it does against much conventional wisdom. We had certainly not set out to prove anything of the sort. Rather, like Aristotle and Freud, and others less illustrious in this tradition, we began by looking at striking and, we thought, exotic examples of creativity, such as analogical counterfactuals, poetic metaphors, and chimeras like talking donkeys. By 1993, we had amassed overwhelming evidence from many more fields—grammar, mathematics, inferencing, computer interfaces, action, and design. This launched a general research program into the nature of conceptual blending as a basic mental operation, its structural and dynamic principles, and the constraints that govern it.

Coming from a different angle and with very different kinds of data, several "creativity theorists" were speculating on the existence of a general mental capacity—called "cognitive fluidity" by Stephen Mithen—that brings together elements of different domains. Mithen and others linked the availability of this capacity to the explosion of creativity in tool-making, painting, and religious practice, dated by archaeologists to roughly 50,000 years ago.

In this book, we argue that conceptual blending underlies and makes possible all these diverse human accomplishments, that it is responsible for the origins of language, art, religion, science, and other singular human feats, and that it is as indispensable for basic everyday thought as it is for artistic and scientific abilities. Above all, it is our goal to do what has not been done before: to explain the principles and mechanisms of conceptual blending.

ACKNOWLEDGMENTS

THE COLLABORATIVE RESEARCH PRESENTED in this book began in 1992–1993 at the University of California, San Diego. Gilles Fauconnier was a member of the department of cognitive science, and Mark Turner was on leave from the University of Maryland as a visiting scholar in the departments of cognitive science and linguistics. Mark Turner is grateful to the John Simon Guggenheim Memorial Foundation for fellowship support provided during that year and to the departments of cognitive science and linguistics at the University of California, San Diego, for their hospitality. He is additionally grateful to the Center for Advanced Study in the Behavioral Sciences for a fellowship year during 1994–1995 and the Institute for Advanced Study for a fellowship year during 1996–1997, periods in which this collaboration was continued.

The writing of this book was conducted principally during 1999–2000, on the campus of Stanford University. Gilles Fauconnier is grateful for fellowship support to the John Simon Guggenheim Memorial Foundation, the American Philosophical Society, the Center for Advanced Study in the Behavioral Sciences, where he was a Fellow, and the John D. and Catherine T. MacArthur Foundation, which provided grant number 95–32005–0 to the Center. During 1999–2000, Mark Turner was a visiting scholar in the department of linguistics and the Center for the Study of Language and Information at Stanford University. He is grateful to them for their support and hospitality. He is additionally grateful to the University of Maryland for the many assignments to research leave it has granted during this collaboration.

For exceptional overall support, encouragement, and the sharing of many insights, we are deeply indebted to Tina Fauconnier and Megan Whalen Turner. Many friends and colleagues have helped us along the way and have made significant contributions of their own to the conceptual blending research program. They include Per Auge Brandt, Nannette Brenner, Marco Casonato, Seana Coulson, Margaret Freeman, Joseph Grady, Jennifer Harding, Masako Hiraga, Doug Hofstadter, Ed Hutchins, Michael Israel, George Lakoff, Nili Mandelblit, Shweta Narayan, Todd Oakley, Esther Pascual, Adrian Robert, Tim Rohrer, Vera Tobin, and Bob Williams.

We jointly thank Jeff Elman and the Center for Research in Language at the University of California, San Diego, for technical support. We thank Wildlife

Education, Limited, for permission to reprint the image of the dinosaur evolving into a bird from *Dinosaurs*, an issue of *Zoobooks*. We thank Michelle Moore of Wildlife Education, Limited, and Michael Gosnell of the UCSD Resources Center for their assistance. We thank the Education Excellence Partnership of the Business Roundtable for permission to reprint the image of the Bypass advertisement.

We are especially grateful to William Frucht, our editor at Basic Books, for his guidance and insight.

CONTENTS

OVERVIEW BY CHAPTER

PART ONE: THE NETWORK MODEL

principles behind the possibilities for conceptual blending, and we must grapple with that entire system to explain any one of its products.

Much of that system concerns conceptual compression. Compression in blending networks operates on a surprisingly small set of relations rooted in fundamental human neurobiology and shared social experience. These vital relations, which include Cause-Effect, Change, Time, Identity, Intentionality, Representation, and Part-Whole, not only apply across mental spaces but also define essential topology within mental spaces. Blending, it turns out, is an instrument of compression *par excellence.* One of the overarching goals of compression through blending is to achieve "human scale" in the blended space, where a great deal of conscious manipulation takes place.

Compression and decompression of vital relations can produce spectacular examples, such as the scientific explanation of the biological evolution of the American pronghorn and the cultural notions of metempsychosis and reincarnation expressed poetically in William Butler Yeats's "Fergus and the Druid."

In an elaborate typology of networks, four kinds stand out on a continuum of complexity: *simplex, mirror, single-scope,* and *double-scope.* At the high end of the continuum of blending complexity, double-scope networks blend inputs with different (and often clashing) organizing frames to produce creative emergent frame structure in a blended space. Double-scope blending is what we typically find in scientific, artistic, and literary discoveries and inventions. Indeed, double-scope creativity is perhaps the most striking characteristic of our species.

Conceptual integration creates mental products that often seem completely different from one another. This apparent dissimilarity misled previous thinkers into assuming that these products must arise from different mental capacities,

THE WAY WE THINK

Part One

THE NETWORK MODEL

One

THE AGE OF FORM AND THE AGE OF IMAGINATION

> It is far more useful to view computational science as part of the problem, rather than the solution. The problem is understanding how humans can have invented explicit, algorithmically driven machines when our brains do not operate in this way. The solution, if it ever comes, will be found by looking inside ourselves.
>
> —*Merlin Donald*

WE LIVE IN THE age of the triumph of form. In mathematics, physics, music, the arts, and the social sciences, human knowledge and its progress seem to have been reduced in startling and powerful ways to a matter of essential formal structures and their transformations. The magic of computers is the speedy manipulation of 1s and 0s. If they just get faster at it, we hear, they might replace us. . . . Life in all its richness and complexity is said to be fundamentally explainable as combinations and recombinations of a finite genetic code. The axiomatic method rules, not only in mathematics but also in economics, linguistics, sometimes even music. The heroes of this age have been Gottlob Frege, David Hilbert, Werner Heisenberg, John von Neumann, Alan Turing, Noam Chomsky, Norbert Wiener, Jacques Monod, Igor Stravinsky, Claude Lévi-Strauss, Herbert Simon.

The practical products of this triumph are now part of our daily life and culture. We eat genetically engineered corn; we announce births and send wedding congratulations and buy guns on the Internet. We buy groceries by having our credit cards scanned. Our taxes are determined by formulas invented by demographers and economists. We clone sheep. Serialist composers choose their notes according to mathematical principles.

ACHILLES AND HIS ARMOR

All of these wonders come from systematic manipulation of forms. By the magic of such transformations, the picture of your newborn baby becomes a long string of 1s and 0s. They are transmitted electronically over thousands of miles and turned back into the same picture on the other end. The powerful and deeply meaningful image appears therefore to be the same as a bunch of 1s and 0s. Form carries meaning with no loss. A picture is worth a thousand 1s and 0s, and vice versa.

A college student enrolled in economics, once a branch of ethics, will now spend considerable time manipulating formulas. If she studies language, once firmly the province of humanists and philologists, she will learn formal algorithms. If she hopes to become a psychologist, she must become adept at constructing computational models. The manipulation of form is so powerful and useful that school is now often seen as largely a matter of learning how to do such manipulation.

Formal approaches lead us not only to reconceive hard problems but also to ask new questions previously inconceivable or inexpressible. Systematic study by Zelig Harris, Noam Chomsky, and their students revealed that linguistic form is astonishingly complex and difficult to account for, thereby compelling psychologists to abandon simple associative modes of explanation. The Bourbaki group and others, by the same kind of systematic analysis, revealed how shaky the foundations of mathematics had been for centuries. Most impressive, Kurt Gödel, by recasting mathematical questions into purely formal schemes such as Gödel numbering, showed inherent limits on proofs within axiomatic systems, thereby using form to analyze itself.

Claude Lévi-Strauss showed how ostensibly different myths shared meaning in virtue of having a shared structure. Vladimir Propp gave formal structures applicable to all Russian folktales. Roman Jakobson and others used as their primary method of literary analysis the investigation of formal relationships among the sounds, rhythms, and orthography of the work. Abstract expressionism came to see the height of meaning as carried by the intersections and juxtapositions of form. Many of these efforts were controversial, but our century has seen enormous energy devoted to the discovery and manipulation of meaning through systematic analysis of form.

These approaches could lead us to think that scientific knowledge is only a matter of finding deep hidden forms behind ostensible forms. On the other hand, common sense tells us that form is not substance: The blueprint is not the house, the recipe is not the dish, the computer simulation of weather does not rain on us. When Patroclos donned the armor of Achilles to battle the Trojans, what the Trojans first saw was the spectacular armor, and they naturally assumed it was Achilles, and were terrified, and so the armor by itself looked as if it was turning the battle. But it didn't take long for the Trojans to discover that it was just Achilles's armor, not Achilles himself, and then they had no pity. In our century, we often look at form the way the Trojans looked at the armor, and indeed, the armor is indispensable—without it even Achilles would fail. The gods may put considerable effort into making superior armor for the mortals, but they take the power of the warrior for granted. Clearly the miracles accomplished by the armor depend on the invisible warrior inside.

Like the Trojans, we in the twenty-first century have come to realize that the miracles of form harness the unconscious and usually invisible powers of human beings to construct meaning. Form is the armor, but meaning is the Achilles that makes the armor so formidable. Form does not present meaning but instead picks out regularities that run throughout meanings. Form prompts meaning and must be suited to its task, just as the armor of Achilles had to be made to his size and abilities. But having the armor is never having Achilles; having the form—and indeed even the intricate transformations of forms (all those 1s and 0s)—is never having the meaning to which the form has been suited.

The famous computer program "Eliza" cleverly delivers canned responses on the basis of superficial word matches to questions and statements made by a real human being. For example, "Tell me more" is a catch-all production of the machine that easily fits into almost any real conversation. People who encounter Eliza are amazed to find that they cannot help feeling they are taking part in a rich human conversation. Even when they know the program's tricks, they cannot suppress the urge to feel that Eliza is manipulating meanings and that the meanings are causing the expressions it produces. Even when they know that Eliza is an empty suit of armor, they cannot help feeling that they are standing before the flesh-and-blood Achilles.

When we see a picture of the newborn baby, we cannot suppress our feeling that we are seeing a baby. In fact, the two-dimensional arrangement of colors in the photograph has almost nothing in common with a baby, and it takes a brain evolved over three billion years and trained through several months of early life to construct the identity between the picture and the baby. Because the brain does this instantly and unconsciously, we take the construction of meaning for granted. Or rather, we tend to take the meaning as emanating from its formal representation, the picture, when in fact it is being actively constructed by staggeringly complex mental operations in the brain of the viewer.

The illusion that meaning is transmitted when we send the digitized picture over the Internet is possible only because there is a brain on each end to handle the construction of meaning. This illusion takes nothing away from the technological feat of transmitting the picture—just as the Trojans took nothing away from the divine technological feat of constructing Achilles' armor—but the picture still needs the human brain just as the armor still needs the human warrior.

Achilles got the best armor, made by Hephaestos, because he was the best warrior, and to be useful at all, the armor must be suited to the warrior. Just so, as we argue in this book, human beings have the most elaborate forms (language, math, music, art) because they have the most effective abilities for the construction of meaning. The forms are especially impressive because they have been suited to the meanings they prompt, but on their own the forms are hollow. In particular, meaning is not another kind of form. Inside the armor is not more armor.

What is in the armor is not a thing at all but a potential force that, no matter the circumstances, can be unleashed dynamically and imaginatively upon the Trojans to lethal effect. Just so, what is behind form is not a thing at all but rather the human power to construct meanings. It, too, no matter the circumstances, can be unleashed dynamically and imaginatively to make sense.

The theme of this book is what the form approaches have assumed as given: the operations of identity, integration, and imagination. These operations—basic, mysterious, powerful, complex, and mostly unconscious—are at the heart of even the simplest possible meaning. We will show that they are the key to the invention of meaning and that the value of even the simplest forms lies in the complex emergent dynamics they trigger in the imaginative mind. We will argue that these basic operations are more generally the key to both everyday meaning and exceptional human creativity. Surprisingly—but, as it turns out, crucially—even the most basic forms, the chestnuts of the form approaches, are prompts for massive imaginative integration.

In investigating identity, integration, and imagination, we will return repeatedly to certain themes:

- *Identity.* The recognition of identity, sameness, equivalence, A = A, which is taken for granted in form approaches, is in fact a spectacular product of complex, imaginative, unconscious work. Identity and opposition, sameness and difference, are apprehensible in consciousness and so have provided a natural beginning place for form approaches. But identity and opposition are finished products provided to consciousness after elaborate work; they are not primitive starting points, cognitively, neurobiologically, or evolutionarily.
- *Integration.* Finding identities and oppositions is part of a much more complicated process of conceptual integration, which has elaborate structural and dynamic properties and operational constraints, but which typically goes entirely unnoticed since it works fast in the backstage of cognition.
- *Imagination.* Identity and integration cannot account for meaning and its development without the third *I* of the human mind—imagination. Even in the absence of external stimulus, the brain can run imaginative simulations. Some of these are obvious: fictional stories, what-if scenarios, dreams, erotic fantasies. But the imaginative processes we detect in these seemingly exceptional cases are in fact always at work in even the simplest construction of meaning. The products of conceptual blending are always imaginative and creative.

Identity, integration, and imagination—the mind's three *I*'s—are the subject of this book.

CHINKS IN THE ARMOR

The spectacular success of form approaches in many domains, combined with the Eliza effect, which leads us to see forms as carrying far more meaning than they actually do, naturally encouraged people to develop these approaches as far as they would go in fields like artificial intelligence, linguistics, cybernetics, and psychology. Yet, invariably, form ran up against the mysteries of meaning. What looked simplest—seeing a line, picking up a cup, telling the difference between *in* and *out*, combining a noun and an adjective, making the analogy between your mommy and your mommy's mommy—turned out to be diabolically hard to model. Learning and development had looked like unfortunate primitive aspects of evolutionary systems, which the more powerful and precise instruments of formal manipulation would simply leap over. But evolution turned out to be far more powerful than the logicians could conceive. Human babies, who looked incompetent and bumbling, doomed to long and tedious processes of learning even that a shoe is a shoe, turned out to be incomparably more capable than anything form approaches could offer, on paper or silicon. It was natural to think that if form approaches could handle apparently hard things like chess and the Goldbach conjecture, it would be child's play for them to account for much more rudimentary things like child speech or navigating a new room or seeing a simple analogy. But not so.

As more and more effort and money were devoted to solving these problems, researchers developed not the expected solutions but a deep respect for their intractability. Problems that were supposed to take a few years at most—machine translation, machine vision, machine locomotion—became entire fields. Although brute-force statistics has often provided improved performance, many take the view that in such cases the form approaches have not improved our understanding of the conceptual processes at work.

In fact, the situation is graver than this. Phenomena that were once not even perceived as problems at all have come to be regarded as central, extremely difficult questions in cognitive neuroscience. What could be simpler than recognizing that a tree is a tree? Yet when we look at works in cognitive neuroscience, we find this recognition problem listed under "conceptual categorization," already regarded as a higher-order problem, beyond the already difficult feat of "perceptual categorization." Apparently simpler still would be the simple recognition of a single entity, as when we look at a cup of coffee and perceive the cup of coffee. As neuroscience has shown, the many aspects of the cup of coffee—the color of the cup, the shape of the opening, the topology of the handle, the smell of the coffee, the texture of the surface of the cup, the dividing line between the coffee and the cup, the taste of the coffee, the heavy feel of the cup in the hand, the reaching for the cup, and so on and on—are apprehended and processed differently in anatomically different locations, and there is no single site in the brain where these various apprehensions are brought together. How can the coffee cup, so obviously a single thing for us at the conscious level, be so many different things and operations for the neuroscientist looking at the unconscious level? Somehow, the combination of three billion years of evolution and several months of early training have resulted in the apprehension of unities in consciousness, but neuroscience does not know the details of that unification. How we apprehend one thing *as one thing* has come to be regarded as a central problem of cognitive neuroscience, called the "binding problem." We do not ask ourselves how we can see one thing *as one thing* because we assume that the unity comes from the thing itself, not from our mental work, just as we assume that the meaning of the picture is in the picture rather than in our interpretation of its form. The generalized Eliza effect leads us to think the form is causing our perception of unity, but it is not. We see the coffee cup as one thing because our brains and bodies work to give it that status. We divide the world up into entities at human scale so that we can manipulate them in human lives, and this division of the world is an imaginative achievement. Frogs and bats, for example, divide the world up in ways quite different from our own.

These chinks in the armor of form show us that elements of mental life that look like primitives for formal analysis turn out to be higher-order products of imaginative work. The next step in the study of mind is the scientific study of the nature and mechanisms of the imagination. Having investigated form with an array of instruments, we are now turning to the investigation of the fundamental nature of meaning on which form relies. Our own research, developed in this book, will focus on a wide array of ostensibly quite different phenomena in the construction of meaning, in a number of different fields—art, mathematics, grammar, literature, counterfactuals, cartoons, and so on. Like the neuroscientist considering the perception of the coffee cup, we will show that these apparently simple mental events are the outcome of great imaginative work at the cognitive level.

BACK TO ARISTOTLE

The view we take here—that form approaches are a special kind of capacity, as useful to the imaginative human being as armor is to the great warrior—has a long and honorable tradition. Aristotle, for example, in surveying the scope of human knowledge, including botany, the generation of species, and ethics, gave a sharp and influential analysis of special areas of human knowledge in which precise formal operations can be of some help. In particular, he noticed that there are certain patterns of language that preserve or change meanings in systematic ways that depend on parts of speech but not on the specific nouns or adjectives we pick. These language patterns are of course the famous syllogisms of the type:

- All men are mortal.
- Socrates is a man.
- Therefore, Socrates is mortal.

Here, nothing depends on *Socrates* or *man* or *mortal*. Aristotle's syllogism is a formal, truth-preserving manipulation of meaning that we could also code as "All As are B, C is an A, therefore C is B." Aristotle's observation and systematization are the seed for all the approaches that we have been referring to as "form approaches." Their power lies in the reliability of symbolic or mechanical manipulations that preserve truth no matter how involved the manipulation. Neither Aristotle nor any of his successors in the analysis of form considered this a general solution to the problem of knowledge. Clearly, however, scientific and mathematical progress was accompanied by ever more formal sophistication. With the explosive development of the form approaches at the beginning of the twentieth century, advanced by thinkers like Bertrand Russell and David Hilbert, the prevailing view is still that correlation of meaning and form is highly desirable but not found in so-called messy, soft, fuzzy, everyday, nontechnical natural systems like language. For example, Russell, a man known for expressing what he confidently viewed as truths about nearly every human sphere, including sex, war, and religion, nonetheless had a stark view of formal mathematics: "Mathematics may be defined as the subject in which we never know what we are talking about, nor whether what we say is true." One fundamental goal of these approaches, then, is to construct artificial languages that have rigorous and reliable form-meaning correlations. The pursuit of this goal brought great success in mathematics as well as physics, chemistry, logic, and, later, computer science.

The same goal became important in philosophy and the social sciences, but the successes were not as clear. In these fields, pursuit of the goal again brought out great and often unperceived complexity of problems, but not effective solutions. For example, Rudolph Carnap's herculean efforts to develop inductive logic were remarkably helpful in highlighting unexpected complexities of reasoning, but they did not lead to an all-unifying logic.

The development of formal systems to leverage human invention and insight has been a painful, centuries-long process. Some forms assist meaning construction much more effectively than others. As Morris Kline writes, "The advance in algebra that proved far more significant for its development and for analysis than the technical progress of the sixteenth century was the introduction of better symbolism. Indeed, this step made possible a science of algebra." It is a commonplace that no one who wants to learn differential and integral calculus will try to learn it through Newton's nearly impenetrable notation; the notation developed by Leibniz is incomparably more perspicuous. Once the appropriate forms are invented, they are easily learnable. Schoolchildren everywhere have little trouble learning to manipulate simple equations like $x + 7 = 15$ or $x = 15 - 7$ or $x = 8$ or $8 = x$, but developing this notation took the efforts of many mathematicians over centuries in many different cultures—Greek, Roman, Hindu, Arabic, and others. In the twelfth century, the Hindu mathematician Bhaskara said, "The root of the root of the quotient of the greater irrational divided by the lesser one being increased by one; the sum being squared and multiplied by the smaller irrational quantity is the sum of the two surd roots." This we would now express in the form of an equation, using the much more systematically manageable set of formal symbols shown below. This equation by itself looks no less opaque than Bhaskara's description, but the notation immediately connects it to a large system of such equations in ways that make it easy to manipulate.

$$\sqrt{\left(\sqrt{\frac{n}{k}}+1\right)^{2} k} = \sqrt{k} + \sqrt{n}$$

We can see the struggle involved in developing this formalism by looking at moments of partial progress. For example, Kline says of Hindu notation:

> There was no symbol for addition; a dot over the subtrahend indicated subtraction; other operations were called for by key words or abbreviations; thus *ka* from the word *karana* called for the square root of what followed. For the unknowns, when more than one was involved, they had words that denoted colors. The first one was called the unknown and the remaining ones black, blue, yellow, and so forth. The initial letter of each word was also used as a symbol. This symbolism, though not extensive, was enough to classify Hindu algebra as almost symbolic and certainly more so than Diophantus' syncopated algebra. Problems and solutions were written in this quasi-symbolic style.

Similarly,

> Cardan . . . wrote $x^2 = 4x + 32$ as *qdratu aeqtur* 4 *rebus p*:32.

Historically, the development of armor was a long and effortful process, involving the discovery of metals, the invention of mining and refining, and the evolution of all of the techniques and tools of the smith. Just so, the development of formal systems is an admirable tradition in the expansion of human knowledge, and one that cultures do not get for free.

In the evolutionary descent of our species, in the history of a science, and in the developmental history of an individual person, systems of form and systems of meaning construction intertwine, so that it is not possible to view them as separable. As Kline points out, the advance in algebra in the sixteenth century was simultaneously conceptual and formal, each aspect being necessary for the other. Formal systems are not the same kind of thing as meaning systems, nor are they small translation modules that sit on top of meaning systems to encode and decode work that is done independently by the meaning systems. Like the warrior and the armor, meaning systems and formal systems are inseparable. They co-evolve in the species, the culture, and the individual.

Just as we have emphasized that the Eliza phenomenon involves seeing more in form than is there, so we emphasize that form is not an ancillary or illusory aspect of the human mind. Much of our effort in this book will go toward unraveling the complex ways in which forms prompt largely unconscious and unnoticed constructions of the imagination.

THE MIND'S THREE *I*'S—IDENTITY, INTEGRATION, IMAGINATION

As we noted earlier, the binding problem—the problem of how we can perceptually apprehend *one integrated thing*—has its counterpart in neurally inspired computational modeling. Psychologists and cognitive scientists at the University of California–San Diego in the early 1980s developed the theory and implementations of parallel distributed processing (PDP), a remarkably successful approach to modeling cognitive phenomena. PDP was widely acclaimed as a major advance in the understanding of cognition, and its merits were contrasted with shortcomings of the traditional symbolic approach, which used logic-like computer programming languages to try to represent cognitive phenomena. But, startlingly for anyone who thinks identity is simple or primitive, the major challenge for this new kind of modeling turned out to be capturing identities and linking roles to values. For example, Zeus as a bull and Zeus as a god and Zeus as a swan are the same, and in turn, the Cloud-Gatherer (a role) is the "same" as Zeus (its value). But the sameness of the god, the bull, and the swan is not a matter of resemblance and shared features. Even now this problem is by no means resolved, and the exceptionally complex and technical solutions that have been proposed for it look nothing like an intuitive representation of identity. Paul Smolensky's approach uses tensor products, and Lokendra Shastri's depends on temporal synchrony. In short, connectionist modeling, like neuroscience, has come to recognize that identity, sameness, and difference, far from being easy primitives, are the major and perhaps least tractable problems involved in modeling the mind.

A related area of research that has undergone tremendous development is the study of analogy. Here, too, what initially seemed easy and primitive, the explicit characterization of sameness, turned out to be extraordinarily complex. Matching and aligning the elements of two domains, finding the common schematic structure that motivates an analogy between them, are now recognized as formidable feats of imaginative work to which the current state of computational modeling cannot do justice. Yet the ability to perceive everyday analogies, like the ability to perceive everyday identities, is completely taken for granted by human beings at the conscious level. It seems like no work at all. In the common view, taking cube roots is hard but finding the door out of a room is no work at all. In fact, extracting cube roots is extremely easy to model computationally, but present-day robots waste a lot of time trying to get out of rooms, and often fail. Understanding the room you are in by comparing it with rooms you already know is an everyday analogy. We find such an analogy trivial because the complex cognitive processes that provide the solution run outside of consciousness (and because "everybody can do it!"). Only the "obvious" solution to this analogy comes into consciousness, and quietly at that. Because we have no awareness of the imaginative work we have done, we hardly even recognize that there was a problem to be solved.

Why didn't the form approaches run up earlier against these extremely difficult problems of identity, sameness, and difference? The quick answer is that human beings who ran the procedures handled the problems unconsciously, so that no one noticed the difficulty. Consider, for example, a logical formula like $\forall x, p(x) \Rightarrow q(x)$. This logical form sets up a schema, according to which anything with property p has property q. A human being who understands the formula can then use it to discover particular truths by instantiating the properties p and q for a specific thing or individual. How does the human being know that the same individual who has property p also has property q? He knows it because the identical letter x has been used in the formula. But that formal identity itself is not itself a binding; it is only a prompt for real binding to occur in the mind of the interpreter of the form. What the real binding allows a real brain to do is to apply the general schema behind the logical formula to particular things and individuals and to keep track of when they count as the same and when they count as different. "Choose a point in the plane such that $x = 1$" asks us to lump together, for the purpose of the direction, an entire set of points as equivalent. By binding all these points, we create an integrated object: the line. "Choose a point in the plane such that $x = -1$" asks us to do the same, and although it uses the same x, it's a different line. The lumping together of points as the "same" is a mental achievement that creates an integrated object.

Formal approaches, in prompting these integrations, take their cue from human perceptual and conceptual systems. We are disposed to construct objects and preserve identities, so that although we hold and move and see and feel "the wine bottle" in many different situations, we effortlessly and unconsciously bind together all these events as involving a single wine bottle. Conversely, we are equally able to use the very same perceptual evidence to distinguish "two" "wine bottles," to all appearances identical and yet not the "same" object.

"I was born in 1954" prompts us to bind an infant in 1954 with an adult living many decades later as the "same," despite the manifest and pervasive differences. "Chaucer's London bore no resemblance to the London of today" allows us to construct and keep separate two cities that we know to be the "same" from another perspective. And "If I were you, I would wear a black dress" prompts us to bind the "I" and the "you" with respect to some aspects but not others. This marvelous capacity for binding turns out to depend on very sophisticated cognitive and neurobiological processes.

In form approaches, identity is taken for granted; analogy, by contrast, is typically not even recognized. How can this be? The answer is that analogy is smuggled in as part of the formal system through a number of back doors. As an instance of the smuggling in of analogy, consider the relationship between "Paul loves Mary" and "John kicks Joe." They share not a single word, so at least at the most obvious level, they have no identical parts. Yet we recognize instantly that they are similar in *form*. In production system approaches such as generative grammar, this similarity falls out not from an explicit analogical mapping between the two specific sentences but from their sharing a common part of a syntactic derivation. In such theories, there are typically hidden layers of structure, so that what looks like an analogy at one level is treated as a superficial by-product of structural identity at a deeper level. Analogical mapping *per se* is not part of the theoretical apparatus; nor is it viewed as part of the child's learning apparatus. So, paradoxically, although the child may be equipped with vast analogical capacities in all kinds of domains, the view of formal linguistics has been that the learning of grammar does not involve analogical mapping. Rather, to learn the grammar is to induce a production system (the formal grammar) on the basis of innate *a priori* constraints (the universal grammar). Perceived analogy will be a by-product of that system, not one of its theoretical concepts nor, surprisingly, a means for the child to apprehend that system.

In form approaches, as we noted, analogies are replaced by structural identities at hidden levels. But because the form approaches take identities for granted, the apprehension of the structural identities is not seen as posing any problem. Analogy thus seems to be dispensable. In fact, the form approaches have been forced to smuggle in some analogy, even if unwittingly, in the guise of formal manipulation, but have suffered from not being able to bring in yet more analogy.

A powerful and, at first, highly promising feature of form approaches such as generative grammar was the possibility of postulating successive invisible levels of form (such as deep structure, or logical form) behind the superficial appearances. Mysteries of formal organization at one level would thus be explained in terms of regularities at a higher one. This technique is what we described earlier as looking for more armor inside the armor. In itself it is not as absurd as it sounds—a warrior could have additional protection under his topsuit of armor, and hidden layers of form are a plausible explanatory technique. The absurdity would come from assuming that the *only* thing that can lie behind a form is yet another form.

Analogy has traditionally been viewed as a powerful engine of discovery, for the scientist, the mathematician, the artist, and the child. In the age of form, however, it fell into disrepute. Analogy seemed to have none of the precision found in axiomatic systems, rule-based production systems, or algorithmic systems. When these new and powerful systems came to be viewed as the incarnation of scientific thinking, analogy was contemptuously reduced to the status of fuzzy thinking and mere intuition. The absence of formal mechanisms for analogy was mistakenly equated with a supposed absence of analogy itself as a fundamental cognitive operation. At the high point of the popularity of rule-based systems, analogy had lost status as an important scientific topic and was ridiculed as a method of discovery and explanation. But toward the end of the 1970s, analogy and its disreputable companions—metonymy, mental images, narrative thinking, and, most unpalatable of all to the formally minded, affect and metaphor—made a roaring comeback.

There were many convergent reasons for this comeback. First, analogy came to be seriously studied by psychologists whose methodologies included both clinical experimentation and computer modeling. The results left little doubt that analogy, as a cognitive operation, was intricate, powerful, and fundamental. New modeling techniques, most notably connectionist systems, provided both better and more realistic models of analogical thinking and more precise insights into its real complexities. Analogy became respectable again as a phenomenon, exactly because it could now be modeled along formal lines. But as the limits of the formal line became apparent, it was recognized that analogy posed a formidable challenge to both the modeler and the experimental psychologist. Mental images made a comeback for similar reasons: Researchers like Roger Shepard and Stephen Kosslyn developed clever experimental techniques for investigating visual perception, visual imagination, and their relationship. Mental images were suddenly viewed as respectable scientific phenomena with surprising complexity. The same story began to be repeated: What was easiest for human beings to accomplish with no thought at all turned out to be far harder to model than chess and other seemingly difficult mental tasks.

Linguists and philosophers made a powerful case for the centrality of metaphor in human cognition, and, again, clever methodologies were invented—for investigating metaphoric thought in very young children, for discovering regularities in metaphoric expression across families of languages, for teasing out complexities of metaphoric comprehension by human subjects, and for analyzing the role of metaphor in both sign language and nonverbal communication systems like gesture. Of course, traditional lines of inquiry before this century had often accepted, even gloated over, the powerful role of metaphor in scientific discovery, artistic creativity, and childhood learning, but that acceptance was entirely canceled during the ascendancy of form approaches.

What analytic philosophers gloated over now was the complete exclusion of figurative thought from "core meaning." Core meaning is, as the formally minded philosopher sees it, the part of meaning that can be characterized formally and truth-conditionally. Therefore, goes the logic, it must be the only important and fundamental part of meaning. Inevitably, these analytic approaches were blind to the imaginative operations of meaning construction that work at lightning speed, below the horizon of consciousness, and leave few formal traces of their complex dynamics.

As we continue to see, work in a number of fields is converging toward the rehabilitation of imagination as a fundamental scientific topic, since it is the central engine of meaning behind the most ordinary mental events. The mind is not a Cyclops; it has more than one *I;* it has three—identity, integration, and imagination—and they all work inextricably together. Their complex interaction and their mechanisms are the subject of this book.

We will focus especially on the nature of integration, and we will see it at work as a basic mental operation in language, art, action, planning, reason, choice, judgment, decision, humor, mathematics, science, magic and ritual, and the simplest mental events in everyday life. Because conceptual integration presents so many different appearances in different domains, its unity as a general capacity had been missed. Now, however, the new disposition of cognitive scientists to find connections across fields has revived interest in the basic mental powers underlying dramatically different products in different walks of life.

Two

THE TIP OF THE ICEBERG

> How can two ideas be merged to produce a new structure, which shows the influence of both ancestor ideas without being a mere "cut-and-paste" combination?
>
> —*Margaret Boden*

COMMON SENSE SUGGESTS THAT people in different disciplines have different ways of thinking, that the adult and the child do not think alike, that the mind of the genius differs from that of the average person, and that automatic thinking, of the sort we do when reading a simple sentence, is far beneath the imaginative thinking that goes on during the writing of a poem. These commonsense distinctions are unassailable, yet there exist general operations for the construction of meaning that cut across all these levels and make them possible. These are the kinds of powerful but for the most part invisible operations we will be interested in.

Commonalities across these divisions have been widely recognized. Analogical thinking, we saw, has lately become a hot topic in cognitive science. Conceptual framing has been shown to arise very early in the infant and to operate in every social and conceptual domain. Metaphoric thinking, regarded in the commonsense view as a special instrument of art and rhetoric, operates at every level of cognition and shows uniform structural and dynamic principles, regardless of whether it is spectacular and noticeable or conventional and unremarkable. Aristotle says both that metaphor is "the hallmark of genius" and that "all people carry on their conversations with metaphors." He is not offering a paradox but instead recognizing the distinction between the existence of the general cognitive operation and the different levels of skill with which it is used by different people. The various schemes of form and meaning studied by rhetoricians can be used by the skilled orator, the everyday conversationalist, and the child. Similarly, modern language science has shown that there are universal cognitive abilities underlying all human languages and shared by the adult and the child. A further demonstration of commonality is the complexity of commonplace reasoning, discovered when researchers in artificial intelligence unexpectedly encountered extreme difficulty in their attempts to model it explicitly. This extraordinary complexity, previously associated only with highly expert thought, turns out to cut across thinking at all levels and all ages.

It might seem strange that the systematicity and intricacy of some of our most basic and common mental abilities could go unrecognized for so long. Perhaps the forming of these important mechanisms early in life makes them invisible to consciousness. Even more interestingly, it may be part of the evolutionary adaptiveness of these mechanisms that they should be invisible to consciousness, just as the backstage labor involved in putting on a play works best if it is unnoticed. Whatever the reason, we ignore these common operations in everyday life and seem reluctant to investigate them even as objects of scientific inquiry. Even after training, the mind seems to have only feeble abilities to represent to itself consciously what the unconscious mind does easily. This limit presents a difficulty to professional cognitive scientists, but it may be a desirable feature in the evolution of the species. One reason for the limit is that the operations we are talking about occur at lightning speed, presumably because they involve distributed spreading activation in the nervous system, and conscious attention would interrupt that flow.

These basic mental operations are highly imaginative and produce our conscious awareness of identity, sameness, and difference. Framing, analogy, metaphor, grammar, and commonsense reasoning all play a role in this unconscious production of apparently simple recognitions, and they cut across divisions of discipline, age, social level, and degree of expertise. Conceptual integration, which we also call *conceptual blending*, is another basic mental operation, highly imaginative but crucial to even the simplest kinds of thought. It shows the expected properties of speed and invisibility. Our goal in this chapter is to convey a feel for how conceptual blending works, by walking through a few easy examples in which the blending is hard to miss. What is important in these examples is not so much their content as their manifestation of the process. It is crucial not to be misled by their exceptional appearance. We have chosen them as opening displays exactly because they are striking; but, for the most part, blending is an invisible, unconscious activity involved in every aspect of human life.

THE IRON LADY AND THE RUST BELT

In the early 1990s, British Prime Minister Margaret Thatcher—known as the Iron Lady—had great popularity among certain factions in the United States. It was common to encounter claims that what the United States needed was a Margaret Thatcher. The response we are interested in is "But Margaret Thatcher would never get elected here because the labor unions can't stand her."

Thinking about this requires bringing Margaret Thatcher together with U.S. electoral politics. We must imagine Thatcher running for president in the United States and develop enough structure to see the relevant barriers to her being elected. Crucially, the point of this reasoning has nothing to do with the objective fact that it is impossible for Margaret Thatcher to be elected, since in the real world she is already head of another state, she is not a citizen of the United States, and she has no apparent interest in running. The speaker's point, right or wrong, is that the United States and Great Britain, despite their obvious similarities, are quite different in their cultural and political institutions and will not choose the same kinds of leaders. This point is made by setting up a situation (the "blend") that has some characteristics of Great Britain, some characteristics of the United States, and some properties of its own. For example, in the imaginative blended scenario, Thatcher, who is running for the office of president of the United States, is already hated by U.S. labor unions, but not by virtue of any experience connected with the United States or its unions. Rather, the hatred of Thatcher is projected into the imaginative blended scenario from the original British history in which (quite different) British labor unions hate her because she was head of state and dealt with them harshly. In the historical situation, Thatcher had to be elected before she could earn the labor unions' hatred. In the blend, they already hate her and therefore will prevent her election, but not because of anything she has previously done to them.

After we have understood all this, the analogy looks obvious: The British prime minister corresponds to the U.S. president; the labor unions correspond to the labor unions; the United Kingdom corresponds to the United States; the British voters correspond to the U.S. voters. What could be simpler? But in fact these correspondences are imaginative achievements, as we can see by considering that in slightly different circumstances we would construe all these counterparts as strong oppositions. The British parliamentary system is, from one perspective, almost nothing like the American union of states with its electoral college. Entire books have been written on the radical differences between labor unions in the United States and those in Britain. In a different context, such as one in which a particular vice-president is viewed as having all the real power behind a figurehead president, the queen of England might be the natural, obvious, immediate counterpart of the president of the United States, and Prime Minister Margaret Thatcher the equivalent of the vice-president. After a blend has been constructed, the correspondences—the identities, the similarities, the analogies—seem to be objectively part of what we are considering, not something we have constructed mentally. Just as we feel that we see the coffee cup for the simple reason that there is a coffee cup to be seen, so we feel that we see the analogy because there is an analogy to be seen—that is, to be perceived directly and immediately with no effort. But analogy theorists and modelers have discovered, to their dismay, that finding matches is an almost intractable problem, even when, after the fact, the matches look as if they are straightforward. Nobody knows how people do it. The unconscious mind gives it, seemingly for free, to the thinking person.

In fact, finding correspondences that look as if they are objectively there requires the construction of new imaginative meaning that is indisputably not "there." Mere correspondences are based on inventive constructions. For example, neither the conceptual frame of the U.S. presidential election nor the history of Britain contains any prime minister campaigning in Michigan. That structure is not "there" in the analogues. But the blend has this novel invention, a Margaret Thatcher campaigning in Illinois and Michigan and hated by the U.S. labor unions. In that blend, Margaret Thatcher is defeated, an outcome not in any way contained in the two analogues. Because the imaginative blended scenario is connected to the real U.S. situation on the one hand and the real British situation on the other, inferences that arise through the creation of new meaning in the blend can project back to the two real situations, yielding the all-important conclusion that the speaker is asking us to build a *disanalogy* between the United States and Britain. The import of this disanalogy might be that although the United States may need a certain kind of leader, the intricacies of U.S. electoral politics make it impossible for that kind of leader to be elected. Or the import might be that the United States is lucky to have labor unions sufficiently vigilant to ensure that what happened to the British unions will never happen to them.

To set up and use this blend, we need to do much more than match two analogues, which is already an awesome task. Somehow we have to invent a scenario that draws from the two analogues but ends up containing more. We have to be able to run that scenario as an integrated unit, even though it corresponds to no prior reality or experience. Somehow, the dynamics of this imaginary scenario are automatic, even though it has never been run before. The blend ends up making possible a set of "matches" that seem obvious to us, even though we might never previously have matched "retired British Prime Minister" with "American presidential hopeful." You can't fully match the analogues without constructing that imaginative blended scenario, because what counts as a good match depends on whether the match gives you what you need for the blend. A little matching helps the blend run, and running the blend helps us find matches.

Finding the matches, however staggeringly impressive, is relatively minor when compared with the creation of new meaning in the blend. We cannot run the blend in just any way, but must somehow run it in the way that is relevant to the purpose at hand. For example, once Margaret Thatcher's bid is stopped by the labor unions, she does not then endorse another candidate, although this would be expected of an actual presidential hopeful. Somehow, by working inside the blend, we must be able to locate inferences that apply outside the blend. The fantastic notion of Thatcher's defeat in a U.S. election translates into a quite practical comment on real U.S. politics.

Yet the fantastic aspects of the blend do not seem to stop anyone from using it for everyday reasoning. It does not matter that this blend is remote from any possible scenario. Its very impossibility, in fact, seems to make the reasoning more vivid and compelling. We will see that blends may or may not have features of impossibility or fantasy. Many blends are not only possible but also so compelling that they come to represent, mentally, a new reality, in culture, action, and science.

THE SKIING WAITER

Conceptual integration is indispensable not only for intellectual work, as in "The Iron Lady and the Rust Belt," but also for learning everyday patterns of bodily action. When the ski instructor helps us learn how to propel ourselves on skis by inviting us to pretend that we are "pushing off" while roller skating, it may look as if we are simply to incorporate a known action pattern—pushing off—into skiing, but not so: Performing the exact action involved in roller skating would make us fall over. Rather, we must selectively combine the action of pushing off with the action of skiing and develop in the blend a new emergent pattern, known (not coincidentally) as "skating." Similarly, one of us had a ski instructor who prompted him to stand properly and face in the right direction as he raced downhill by inviting him to imagine that he was a waiter in a Parisian café carrying a tray with champagne and croissants on it and taking care not to spill them. This might seem like a simple execution of a known pattern of bodily action—carrying a tray—in the context of skiing, but again not so: When we carry a tray, we create equilibrium by exerting force against the weight of the tray, but in skiing, there is no tray, no glassware, no weight. What counts are direction of gaze, position of the body, and overall motion. The resulting integrated action in skiing is not the simple sum of carrying a tray while moving downhill on skis.

The creation of blends is guided by cognitive pressures and principles, but in the case of skiing it is also guided by real-world affordances, including biophysics and physics. Most motions that the skier can imagine are impossible or undesirable to execute. But within the conceptual blend prompted by the instructor, and under the conditions afforded by the environment, the desired motion will be emergent.

The instructor is astutely using a hidden analogy between a small aspect of the waiter's motion and the desired skiing position. Independent of the blend, however, this analogy would make little sense. The instructor is not suggesting that a good skier moves "just like" a competent waiter. It's only within the blend—when the novice tries to carry the tray mentally while skiing physically—that the intended structure (the improved body position) emerges. In this case, as in all others, we still have the construction of "matches" between "waiter" and "skier," but the function of these matches is not analogical reasoning: We are not drawing inferences from the domain of waiting on tables and projecting them to the domain of skiing. Rather, the point is the integration of motion. Once the right motion has emerged through integration and the novice has mastered it, the links to croissants and champagne can be abandoned. The skier need not forever think about carrying trays in order to perform adequately.

Both the Skiing Waiter and Iron Lady blends depend upon a widely recognized psychological and neurobiological property that we have not mentioned yet: The brain is a highly connected and interconnected organ, but the activations of those connections are constantly shifting. The great neurobiologist Sir Charles Sherrington, in his Gifford Lectures titled *Man on His Nature*, described the brain as "an enchanted loom where millions of flashing shuttles weave a dissolving pattern, always a meaningful pattern though never an abiding one; a shifting harmony of subpatterns." Activation makes certain patterns available for use at certain times, but it does not come for free. The fact that two neurons are connected in the brain does not necessarily mean they will be coactivated. The matching that we have talked about in the Iron Lady and Skiing Waiter examples is actually a powerful way to bind elements to each other and activate them. What counts as a "natural" match will depend absolutely on what is currently activated in the brain. Some of these activations come from real-world forces that impinge upon us, others from what people say to us, others from our purposes, others from bodily states like weariness or arousal, and many others from internal configurations of our brains acquired through personal biography, culture, and, ultimately, from biological evolution. But much of the shifting activation is the work of the imagination striving to find appropriate integrations. In the Iron Lady case, the activation of a political frame made various matches more available: prime minister to president, British voters to U.S. voters, and so on. Words themselves are part of activation patterns, so when the same word is appropriate for two elements, we can prompt someone to match them by using the same word for both. The language makes matching the British "labor unions" and the U.S. "labor unions" look trivial: The same expression picks them both out, despite their radical differences. But there is no single expression for both "ski poles" and "waiter's tray," so the ski instructor has to direct the novice explicitly to make the connection. Once activated, however, this binding is very strong and will feed the blend that becomes a new integrated motion.

THE GENIE IN THE COMPUTER

In the seemingly quite different realm of technological design, computer interfaces are prompts to activate, bind, and blend at the level of both conceptual structure and bodily action. The most successful interface is the "desktop," in which the computer user moves icons around on a simulated desktop, gives alphanumeric commands, and makes selections by pointing at options on menus. This interface was successful because novices could immediately use it at a rudimentary level by recruiting from their existing knowledge of office work, interpersonal commands, pointing, and choosing from lists. These domains of knowledge are "inputs" to the imaginative invention of a blended scenario that serves as the basis for integrated performance. Once this blend is achieved, it delivers an amazing number of multiple bindings across quite different elements—bindings that seem, in retrospect, entirely obvious. A configuration of continuous pixels on the screen is bound to the concept "folder," no matter where that configuration occurs on the screen. Folders have identities, which are preserved. The label at the bottom of the folder in one view of the desktop corresponds to a set of words in a menu in another view. Pushing a button twice corresponds to opening. Pushing a button once when an arrow on the screen is superimposed on a folder corresponds to "lifting into view." Of course, within the technological device that makes the blend possible—namely, the computer interface—no ordinary lifting, moving, or opening is happening at all, only variations in the illumination of a finite number of pixels on the screen. The conceptual blend is not the screen: The blend is an imaginative mental creation that lets us use the computer hardware and software effectively. In the conceptual blend, there is indeed lifting, moving, and opening, imported not from the technological device at hand, which is only a medium, but from our mental conception of the work we do on a real desktop.

Of course, the generalized Eliza effect makes it seem as if the desktop interface carries all of this meaning. In fact, the desktop interface is like the baby photo: evidently an effective tool to be used by our imaginations, but very thin and simple relative to them. The imaginative work we do when we use the desktop interface is part of backstage cognition, invisible to us and taken for granted.

Once learned, the entire activity of using the desktop interface is coherent and integrated. It is not hampered by its obvious literal falsities: There is no actual desk, no set of folders, no putting of objects into folders, no shuffling of objects from one folder to another, no putting of objects into the trash. The desktop interface is an excellent example of conceptual integration because the activity of manipulating it can be done only in the blend, and would make no sense if the blend were not hooked up to the inputs.

The user of the interface manipulates an integrated structure that derives some of its properties from different inputs—office work, commands, menus. But however much the interface takes from the inputs, it has considerable emergent structure of its own: Pointing and clicking buttons is not at all part of traditional office work or choosing from lists of words on paper; having little two-dimensional squares disappear under other little squares is not part of giving commands or of putting sheets of paper into folders; and dragging icons with the mouse is not part of moving objects on a desktop, ordering a meal, giving standard symbolic commands, or, *a fortiori*, using the machine language.

The user manipulates this computer interface not by means of an elaborate conscious analogy but, rather, as an integrated form with its own coherent structure and properties. From an "objective" point of view, this activity is totally novel—it shares very few physical characteristics with moving real folders, and it is novel even for the traditional computer user who has issued commands exclusively from a keyboard rather than from a mouse. Yet the whole point of the desktop interface is that the integrated activity is immediately accessible and congenial. The reason, of course, is that a felicitous blend has been achieved—a blend that naturally inherits, in partial fashion, the right conceptual structure from several inputs and then cultivates it into a fuller activity under pressure and constraints from reality and background knowledge.

The desktop also nicely illustrates the nonarbitrary nature of blending: Not just any discordant combination can be projected to the blend. Some discordant structure is irrelevant because it has no bad consequences—for example, the trashcan and the folders both sit on the desktop—but other discordant structure is objectionable. Dragging the icon for a floppy disk to the trash as a command to eject the disk from the drive is notoriously disturbing to users. The inference from the domain of working at a desk that one loses everything that is put into the trash and the inference from the domain of computer use that one cannot recover what one has deleted interfere with the intended inference that the trashcan is a one-way chute between two worlds—the desktop interface and your actual desk.

Another point illustrated by the example above is that inputs to blends are themselves often blends, often with an elaborate conceptual history. The domain of computer use has as inputs, among others, the domain of computer operation and the domain of interpersonal command and performance. It is common to conceive of the deletion of files as an operation of complete destruction performed by the system at the command of the user. In actual computer operation, however, the files are not permanently erased by that command and can often be recovered. The user's sense of "deletion" is already a blend of computer operation and human activity. More generally, by means of blending, keyboard manipulation is already conceived of as a blend of typing and high-level action and interaction, thus providing appropriate partial structure to later blends like desktops with icons. The existence of a good blend can make possible the development of a better blend. Conceptual structure contains many entrenched products of previous conceptual integration.

CRAZY NUMBERS

In the Iron Lady example, integration happens on-line and quickly and looks unremarkable. In the Computer Desktop example, there has been laborious design to develop an efficient blend involving novel computer hardware, but once it is developed, users can work with it quickly, automatically, and productively.

In other cases, the conceptual work of integration can take years or even centuries. This is often the case in scientific discovery. In Chapter 13, we will analyze in some detail the invention of complex numbers. The mathematical domain of complex numbers was fully accepted only in the nineteenth century. It turns out to be a blend of two much more familiar inputs: two-dimensional space and numbers. In this blend, complex numbers have all the usual properties of numbers (they can be added, multiplied, and so on), but they also have properties of vectors in two-dimensional space (magnitudes, angles, coordinates). This blend is a well-integrated structure, with no inconsistencies, with important properties, and with impressive new mathematical power. It has elegant emergent structure of its own: Numbers now have angles, multiplying numbers is now an operation involving addition of angles, and negative numbers have square roots.

All accounts of scientific discovery acknowledge the crucial importance of analogy, but analogy is not enough: Historically, the analogy between imaginary numbers and points in space was well known by the end of the seventeenth century, and moreover the formal manipulations yielded by that analogy were fully recognized. But the analogy alone did not produce an integrated notion of complex number, and so it was not embraced as part of number theory. As late as the end of the eighteenth century, outstanding mathematicians such as Euler thought that such numbers were harmless but impossible.

HOW SAFE IS SAFE?

The Iron Lady, the Skiing Waiter, the Computer Desktop, and Complex Numbers are representative cases in superficially very different aspects of human endeavor. They all display the imaginative complexity of activation, matching, and the construction of meaning. Complex blending is always at work in any human thought or action but is often hard to see. The meanings that we take most for granted are those where the complexity is best hidden.

Even very simple constructions in language depend upon complex blending. It is natural to think that adjectives assign fixed properties to nouns, such that "The cow is brown" assigns the fixed property *brown* to *cow*. By the same token, there should be a fixed property associated with the adjective "safe" that is assigned to any noun it modifies. Yet consider the following unremarkable uses of "safe" in the context of a child playing at the beach with a shovel: "The child is safe," "The beach is safe," "The shovel is safe." There is no fixed property that "safe" assigns to *child, beach,* and *shovel.* The first statement means that the child will not be harmed, but so do the second and third—they do not mean that the beach or the shovel will not be harmed (although they could in some other context). "Safe" does not assign a property but, rather, prompts us to evoke scenarios of danger appropriate for the noun and the context. We worry about whether the child will be harmed by being on the beach or by using the shovel. Technically, the word "safe" evokes an abstract frame of *danger* with roles like victim, location, and instrument. Modifying the noun with the adjective prompts us to integrate that abstract frame of *danger* and the specific situation of the child on the beach into a counterfactual event of *harm* to the child. We build a specific imaginary scenario of *harm* in which *child, beach,* and *shovel* are assigned to roles in the *danger* frame. Instead of assigning a simple property, the adjective is prompting us to blend a frame of *danger* with the specific situation of the child on the beach with a shovel. This blend is the imaginary scenario in which the child is harmed. The word "safe" implies a disanalogy between this counterfactual blend and the real situation, with respect to the entity designated by the noun. If the shovel is safe, it is because in the counterfactual blend it is sharp enough to cause injury but in the real situation it is too dull to cut.

We can create many different blends out of the same inputs. The process is the same in all of them, but the results are different. In "The shovel is safe," the child is the victim in the blend if we are concerned about the shovel's injuring the child, but the shovel is the victim in the blend if we are concerned about the child's breaking the shovel. Furthermore, any number of roles can be recruited for the *danger* input. In the imaginary blend for "The jewels are safe," the jewels are neither victim nor instrument; they are *possessions* and their *owner* is the victim. If we ship the jewels in packaging, then in the imaginary blend for "The packaging is safe," the jewels are the *victim*, external forces are the *cause of harm*, and the packaging is the *barrier to external forces*. Other examples showing the variety of possible roles would be "Drive at a safe speed," "Have a safe trip," "This is a safe bet," and "He stayed a safe distance away."

The noun-adjective combination can prompt even more elaborate blends, involving several roles, as in "The beach is shark-safe" versus "The beach is child-safe." In the context of buying fish at a supermarket, the label on a can of tuna can report that "This tuna is dolphin-safe," meaning that the tuna was caught using methods that prevent accidents from happening to dolphins. This blend looks more unusual, but it is put together according to the same dynamic principles as the blends we assemble for unremarkable phrases like "safe beach" or "safe trip."

"The beach is safe" shows that the "matches" are not achieved independent of the blend, and that there is nothing simple about "matching." The beach in the real situation is matched to the role "doer of harm" in the harm scenario because we have achieved an imaginary blend that counts as counterfactual to the real situation. That match, however, is a match between a role in a frame and a specific element that is in fact *not* an instance of the role. The real "safe beach" is not a "doer of harm." That's the point of the utterance. The role "doer of harm" in the harm input is matched to a beach in the counterfactual blend that *is* imaginatively a doer of harm. And the beach in the real situation is matched to the beach in the counterfactual blend because they are opposites in the way that matters for this situation: One is a doer of harm, and the other is not.

"Safe" is not an exceptional adjective with special semantic properties that set it apart from ordinary adjectives. Rather, it turns out that the principles of integration suggested above are needed quite generally. Even color adjectives, which at first blush look as if they must assign fixed features, turn out to require noncompositional conceptual integration. "Red pencil" can be taken to mean a pencil whose wood has been painted red on the outside, a pencil that leaves a red mark (the lead is red, or the chemical in the pencil reacts with the paper to produce red, or . . .), a pencil used to record the activities of a team dressed in red, a pencil smeared with lipstick, or a pencil used only for recording deficits. Theories of semantics typically prefer to work with examples like "black bird" or "brown cow" since these examples are supposed to be the principal examples of compositionality of meaning, but, as we will show later, even these examples illustrate complicated processes of conceptual integration.

THE IMAGE CLUB

While dogs, cats, horses, and other familiar species presumably must do perceptual binding of the sort needed to see a single dog, cat, or horse, human beings are exceptionally adept at integrating two extraordinarily different inputs to create new emergent structures, which result in new tools, new technologies, and new ways of thinking. The archeologist Stephen Mithen, who has provided independent archeological evidence for such a capacity, argues that it is quite recent in human evolution and responsible for not only the sudden proliferation of novel tools but also the invention of art, religion, and science. It is easy to point with pride to the invention of complex numbers, but the strikingly human capacity for two-sided blending is equally powerful throughout human domains of every description. For Mithen, blending, which he calls "cognitive fluidity," is what made possible the invention of racism. In a section of *The Prehistory of the Mind* titled "Racist Attitudes as a Product of Cognitive Fluidity," he writes: "Physical objects can be manipulated at will for whatever purpose one desires. Cognitive fluidity creates the possibility that people will be thought of in the same manner. . . . There is no compulsion to do this, simply the potential for it to happen. And unfortunately that potential has been repeatedly realized throughout the course of human history." Similarly, mass killing of certain groups of people can be blended with ordinary bureaucratic frames to produce a blended concept of genocide as a bureaucratic operation. Because the projection to the blend is only partial, people who could not bring themselves to operate in the frame of genocide may find themselves operating comfortably in the blend. Documentaries such as Claude Lanzmann's *Shoah* reveal in great detail how bureaucrats in Nazi Germany could talk and think about the enormous killing machine they served as an ordinary transfer of goods and merchandise. The blend has the two inputs of genocide and bureaucracy, but the bureaucrat can recruit a third input, war, to supply the frame according to which self-imposed ignorance is a virtue for a citizen and may be necessary for the security of the nation.

Blending imaginatively transforms our most fundamental human realities, the parts of our lives most deeply felt and most clearly consequential. Meaning goes far beyond word play. Meaning matters, in ways that have relevance for the individual, the social group, and the descent of the species. Human sexual practices are perhaps the epitome of meaningful behavior because they constitute a deeply felt intersection of mental, social, and biological life. It is remarkable how different they are from the sexual behaviors of the most closely related species. This realization has been central to theories of the unconscious such as Freudian psychoanalysis, but curiously, it is almost taboo inside cognitive science. Even though modern cognitive science emphasizes the embodiment of the mind, philosophy in the flesh, it deprives itself of sexuality as a source of data and as a laboratory of analysis. Yet the role of meaning construction and imagination in the elaboration of human sexual practices is phenomenal and has direct, real-world social consequences. From the *Odyssey* to *Ulysses*, with *Othello* in between and *Lolita* after, the world's literatures explore the febrile and exquisite sophistications of mental sexual fantasies and their grave consequences in reality. This fundamental theme in literature—the connection between the mental apprehension of sex and the historical patterns of war, rape, suicide, alliances—merely reflects our everyday reality. These practices, which intertwine psychology, biology, and social life—through which we, as individuals and as cultures, define ourselves—are unique to our species. We believe this pervasive aspect of human life has the richness and complexity it does because of the imaginative processes of blending.

The latest inventions and twists in cultural sexual practices are reported routinely in newspapers as if they were mere curiosities. Under the title "A Plain School Uniform as the Latest Aphrodisiac," a *New York Times* article described "several hundred" bordellos (called "image clubs") in Tokyo in which the rooms are made to look like schoolrooms, complete with blackboards, and the prostitutes, chosen for their youthful looks, dress in high-school uniforms and try to act like apprehensive teenagers, while the customer takes the role of a teacher. Because sexual issues and sexual fantasy are very familiar, even if not talked about, we may find such examples mundane, but the imaginative construction of emergent meaning in this instance is astounding. The inputs to the blend are the scenario with an imaginary high-school student and the real situation involving both the man in the "image club" and the prostitute (who, in the specific case reported in the *Times*, is actually twenty-six years old). But the blend has a teacher and a high-school student.

Since neither the customer nor the prostitute is deluded, why should this make-believe have any power or attraction at all? The answer is that while the customer can of course have sex with a prostitute, he can't have sex with a high-school student, except in the blend, which he can inhabit mentally without losing his knowledge of the actual situation. These are mental-space phenomena, about which we have more to say later. Mental spaces can exist routinely alongside incompatible mental spaces. When we look in the refrigerator and see that there is no milk to be had, we must simultaneously have the mental space with no milk in the refrigerator and the counterfactual mental space with the milk in the refrigerator. The customer in the image club similarly has at the same time the mental space with the experienced and trained prostitute, the mental space with the imaginary and unattainable high-school student, and the blended mental space with the woman in the club as the attainable innocent high-school student. The high-school student is projected to the blend from the imaginary input, while the actual sexual act that takes place is imported from the material reality linked to the mental space with the prostitute. The blend has the essential new structure: sex with the high-school student.

Far from just mixing the features in free-for-all fashion from two situations—the classroom and the bordello—blending demands systematic matches between the inputs and selective projection to the blend according to a number of constraints that we will discuss in depth in this book. The teacher's privileges and responsibilities in the classroom do not, for the most part, project to the blend: The customer is not supposed to demand that the prostitute learn how to factor polynomials. Many other projections are equally inappropriate. Just as in the Iron Lady example we had to match U.S. and British political domains, here we must match classroom and house of ill-repute. The matching is not obvious and preconstructed. It is driven by the intended blend, not by any obvious analogy between the schoolhouse and the whorehouse. Also, as in both the Iron Lady and Skiing Waiter examples, there will be only partial projection from the inputs, but the resulting blend must have integrated action and meaning, on the one hand, and enough disintegration that the participants can connect it to both of the inputs, on the other. In the Iron Lady example, we do not want to get lost in an escapist fantasy about an imaginary life for Margaret Thatcher, forgetting that the point is to make projections back to the real U.S. political situation. In the Skiing Waiter example, we do not want the skier to start believing that he has the job of delivering food to other people on the slopes. And in the Image Club case, the customer is not supposed to turn himself in to the police for having assaulted a high-school student, but he is supposed to pay her, in keeping with the prostitution input.

Identity and analogy theory typically focus on compatibilities between mental spaces simultaneously connected, but blending is equally driven by incompatibilities. Often the point of the blend is not to obscure incompatibilities but, in a fashion, to have at once something and its opposite. Consider, for example, the sexual response of the female in each of the two input spaces and in the blend.

There are many possibilities, but one standard assumption would be that in the prostitution input, the workaday prostitute has no real passionate response, and in the input with the imaginary high-school student, there is no response at all because the man only desires but takes no action. In the blend, however, the woman will have a spectacular, never-to-be-forgotten ecstatic experience. These oppositions are not suppressed; on the contrary, activating them all simultaneously is part of the purpose of the network of spaces, and the participants must keep these spaces distinct. The human capacity to construct and connect strikingly different mental spaces is what makes such sexual fantasies and practices possible to begin with. Needless to say, the capacity is much more general. The *raison d'être* of mental spaces is to juggle representations that, in the real world, are incompatible with each other. This mental juggling gives rise, among other things, to phenomena that logicians and philosophers of language call "opacity," "counterfactual reasoning," and "presupposition projection."

In fact, there is even more complex blending going on in the Image Club case. What actually happens in the blend may vary considerably depending on the specific projections from the two inputs, fine-tuned by the participants in the moment. Also to be considered is the practice the prostitute goes through to be able to sustain the act well but not too well.

Whether distasteful or attractive, the schoolroom blend, as an object of contemplation, discussion, or experience, does not leave us indifferent. But conceptual blending of just this sort typically operates in ways that do not make us self-conscious. The owner of a Dodge Viper sports car told *Parade* magazine, "My Viper is my Sharon Stone. It's the sexiest vehicle on the road." Apparently he felt no hesitation in committing himself to this blend of sexuality and motoring. In our culture, the general version of this blend is pervasive and supported by efforts of corporations and advertisers. Sexual blends pervade the culture, but this fact neither causes chronic embarrassment nor keeps the citizenry in a permanent sexual frenzy.

GRADUATION

As we shall see throughout this book, one of the central benefits of conceptual blending is its ability to provide compressions to human scale of diffuse arrays of events. Graduation is an example everyone knows. Going to college involves many semesters of registration, attendance at courses, listening to lectures, completing the courses, and moving on to other walks of life. A ceremony of graduation is typically a compression achieved by blending this diffuse array with the more general schema of a special event with speakers in a limited time, such as a presidential inauguration. The graduation ceremony is a blend of "going to class" and attending the special event. It compresses, into two or three hours, four years of being a student: You hear a distinguished person who conveys wisdom and knowledge; you have your family at your side; you see all your college friends going with you through the same process; and the process culminates in a transition. Amazingly enough, the graduation ceremony, which is already a strong compression, contains compressions of itself as well as compressions of those compressions. The thirty seconds of the actual conferral of the individual degree has the student rising, taking the stage, engaging in a transitional moment that includes a very compressed conversation with the university official, and departing, transformed into a graduate. These thirty seconds are then compressed in turn into a single moment, the movement of the tassel from one side of the mortarboard to the other. And the entire event is compressed into an abiding material anchor that you take with you and hang on the wall: your diploma.

CHAPTER 2
ZOOM OUT

COUNTERFACTUALS

We began this chapter with an everyday example of thinking about presidential politics.

Question:

- Is there really anything deep going on in such examples?

Our answer:

Philosophers of language would call the Iron Lady example a "counterfactual," because it is contrary to fact. Counterfactuals include *reductio ad absurdum* proofs in which the mathematician sets up as true what she wants to prove false and manipulates it as true, in the hope of arriving at an internal contradiction that is taken as proving the original assertion's falsity. Counterfactuals include statements like "If this water had been heated at one hundred degrees Celsius, it would have boiled." The philosopher Nelson Goodman points out, with fiendish precision, the great importance of counterfactual thinking and the pitfalls involved in their use: "The analysis of counterfactual conditionals is no fussy little grammatical exercise. Indeed, if we lack the means for interpreting counterfactual conditionals, we can hardly claim to have any adequate philosophy of science." Following Goodman, philosophers, political scientists, linguists, and psychologists have devoted considerable efforts to the study of counterfactual reasoning. Their approach is to think of the counterfactual as setting up an alternative world whose differences from the actual world consist of only the difference specified by the linguistic expression (e.g., "If the water had been heated . . . ") and its direct consequences. Intuitively, this sounds easy, but the intractable aspect of the problem is to specify the consequences of change in one little part of our actual world. How do we compute the ripples of a single hypothetical change throughout a vast and intricate world where everything is interrelated? So the logic of counterfactuals as conceived by theorists is already a formidable problem. But examples like the Iron Lady actually reveal an even higher order of complexity. In the Iron Lady, we are not setting up a possible world in which Thatcher is a candidate in U.S. presidential primaries and elections, nor are we concerned with all the consequences of such an election (e.g., a British woman's being head of the United States). What we are doing is matching the U.S. and British systems in very partial ways and producing a blend that, far from being a complete world, is itself very partial, dedicated only to the purpose at hand. We have discovered, and will document in detail in Chapter 11, that counterfactuals are in general complex blends with emergent structure, and that the case on which theorists focus—minimal modification of one world—is only a special case.

So the answer to our question is that counterfactuals are widely considered to pose an exceptionally difficult problem, logically and semantically. Moreover, the Iron Lady type of counterfactual is of an even higher order of complexity than the ones usually studied and seems to put the lie to the usual methods of analyzing counterfactuals. This is because the usual methods have paid scant attention to the dynamic powers of the imagination. Accordingly, counterfactuals—however simple they may feel to us—turn out to have deep complexities we are only beginning to understand.

The problems counterfactuals pose are important not only to the philosopher and the linguist. As political scientists, economists, anthropologists, sociologists, and other social scientists have often noticed, frequently to their dismay, the most basic methods of social science seem to depend inescapably on counterfactual thinking. An assertion like "The shipping industry in Greece prospered after World War II because Greece had developed good infrastructure during the war" is unintelligible except as a claim that "If Greece had not developed such-and-such infrastructure, its shipping industry would not have prospered," which is of course an explicit counterfactual. Most of the analytic claims in the social sciences turn out to be implicit counterfactuals, as has been explicitly recognized in recent books and articles.

HUMAN-COMPUTER INTERFACES

The desktop interface for the computer is very common, and nobody seems to have much trouble learning to use it.

Questions:

- Are computer engineers really unaware of what we have said about the computer desktop?
- Doesn't the fact that anybody can learn this interface show that it can't be as complicated as we say?

Our answers:

Computer engineers and cognitive scientists are very well aware of issues of metaphor in the design of interfaces, and regard them as posing difficult questions for both the designer and the psychologist. But the role of unconscious blending in the design of interfaces, and in the construction of the metaphors they use and develop, has gone unnoticed.

The question "Can this interface really be so imaginative if everybody can do it?" can be asked of many examples we will present in this book. Our answer by now is predictable: The unconscious mental processes we take for granted deliver products and performances to our conscious minds that seem completely simple but whose invention is much too complicated for feeble consciousness to begin to apprehend. Just as talking, walking, seeing, grasping, and so on have come to be recognized as involving astonishingly complicated and dynamic unconscious processes, so the simplest feats we learn to perform, like using the computer desktop, are the hardest to analyze. That every human being can do it, but no member of any other species can do it, should tip us off immediately to the evolutionary development that was needed to make such feats possible. Only really big brains connected in special ways, and doing a lot of dynamic work as trained by their cultures, can even begin to handle these feats, and even those big brains cannot know consciously what it is that they are doing.

The form approaches, interestingly, did discover and explore the erroneous assumption behind the "everybody can do it" question insofar as form was concerned. The child's handling of syntax and phonology is universally known to be a marvel of the human mind. It was by studying syntax formally and methodically that Noam Chomsky convinced psychologists that learning language went far beyond simple associations. The fact that every child can master grammar now counts as a strong reason for studying it and for expecting it to pose much greater difficulties for the analyst than mere exceptional performances by a few special individuals. From the standpoint of cognitive science, the everyday capacities of the well-evolved human mind are the best candidates for complexity and promise the most interesting universal generalizations.

Our major claims in this book are radical but true: Nearly all important thinking takes place outside of consciousness and is not available on introspection; the mental feats we think of as the most impressive are trivial compared to everyday capacities; the imagination is always at work in ways that consciousness does not apprehend; consciousness can glimpse only a few vestiges of what the mind is doing; the scientist, the engineer, the mathematician, and the economist, impressive as their knowledge and techniques may be, are also unaware of how they are thinking and, even though they are experts, will not find out just by asking themselves. Evolution seems to have built us to be constrained from looking directly into the nature of our cognition, which puts cognitive science in the difficult position of trying to use mental abilities to reveal what those very abilities are built to hide.

THE PSYCHOLOGY OF ERRORS

As we saw in the example of the Skiing Waiter, people use their mental capacities in seemingly unusual ways to learn to perform physical actions and social routines correctly.

Questions:

- Is there anything to learn from these seemingly unusual events?
- Is the skiing example far-fetched? Is it representative?
- How does it relate to work by psychologists on human action?

Our answers:

The Skiing Waiter is an example where the blending yielded good results in human action, and so counts as a success. But blending, always at work unconsciously, can also recruit patterns and operations that end up having unfortunate consequences. In the psychological study of errors, one focal question is how agents who made mistakes could have made them, and that is usually a question of how they could have brought to bear meanings and operations that turned out to yield the wrong results. Consider a situation in which blending gave bad results: A driver, paying attention to the job of driving and having trouble hearing the people in the back seat with whom he was conversing, "absent-mindedly" reached for the volume button on the radio to turn it up. For a fleeting moment, the radio volume and the voice volume were blended as both controllable by the radio knob. Of course, this was not an absent-minded action at all: The driver's mind was fully at work doing quite ingenious and opportunistic blending that solved the problem conceptually but happened not to play out properly in the real car. The blending itself was not a success, but it might inspire a device that could be used to amplify (or mute) sounds coming from the back seat—a very useful invention for drivers and parents. The study of errors is for psychologists a precious source of evidence about actual mental operations and the kinds of invisible connections they carry with them.

In the case of the skier, we saw that the instructor was able to prompt the novice to perform the correct motion by using an inventive blend. The novice was performing the incorrect motion of looking in the direction of his feet, which in turn is the direction of the skis. This incorrect motion is itself a blend, though much less noticeable, of skiing and walking. The instructor asked the novice explicitly to abandon that blend and make a new, if unusual, blend of skiing and carrying a tray of champagne and croissants through a café. In doing so, the instructor was using a technique of coaching that is basic to any domain involving human bodily action, from learning to ride a bicycle to learning a martial art. In the last few decades, elaborate simulation technologies have been developed to train tank drivers, airplane pilots, air traffic controllers, and even automobile drivers to make the right blends instead of the wrong ones. People learning to fly a plane must learn very early that even though they are looking out of a windshield and holding a steering device, motor routines for driving a car can be lethal in this new situation. Part of learning a motor action is displacing old blends with new.

ANALOGY

All the examples in this chapter involve aligning elements in two or more inputs and forging analogies between them.

Question:

- Isn't blending just a kind of analogy, and don't we already know how analogy works?

Our answer:

Analogy theory is about analogical projection. In standard analogical reasoning, a base or source domain is mapped onto a target so that inferences easily available in the source are exported to the target. We can thus reason about the target. But in the skiing example, this is not what is happening. The instructor is not suggesting that the skier move "just like" a waiter. A skier who did this would remove his skis and start walking. It is only within the blend, when the novice tries to mentally carry the tray while physically skiing, that the appropriate emergent action pattern arises. That pattern does not inhere in either the skiing input or the waiting input, as it would if it were only analogically transferred.

The Image Club is an even more striking example. Here, it is entirely clear that the customer must not determine his behavior in the image club analogically through projection from a teacher's behavior in a classroom, or conversely. Analogy in either direction would be disastrous. In both the Skiing Waiter and Image Club examples, blending is not just manipulation or projection of inferences. Rather, it leads to genuine novel integrated action. Integration of this kind is not a feature of models for analogical reasoning, which typically rely on structure-mapping only. Sexual fantasy, whether or not enacted, is a vast and important area of systematic human cognition that is imaginative but not explained by metaphor or analogy.

In the case of the evolution of the notion of number, there was at one point in history a purely formal use of "imaginary" numbers (such as the square root of negative one). To the surprise of mathematicians, formal manipulation of imaginary numbers worked in equations, even though it violated fundamental conceptual properties of numbers. Then a structure-mapping analogy was developed between points in the Euclidean plane and numbers, including imaginary numbers. Again, the structure-mapping analogy worked, in the sense of yielding mathematically appropriate actions and computations. But the analogy was not enough: It took mathematicians another century to achieve the integrated blend in which complex numbers are a coherent category subsuming imaginary numbers. The achievement of this blend gave integrated conceptual structure to a now-extended category of "number," and so the objection to imaginary numbers disappeared.

THE OLD AND THE NEW

We have discussed many fields in which blending phenomena arise and pointed to research in these fields that has to do with aspects of blending.

Question:

- So, hasn't everybody known about blending forever?

Our answer:

In one sense, we all know everything about blending and are complete masters of it, just as we all have complete unconscious "knowledge" about vision but almost no conscious knowledge of our unconscious ability. The 30,000-year-old art of the Upper Paleolithic found on the cave walls of the Grotte Chauvet reflects elaborate creative blending in the mind of the artist.

Since the products of blending are ubiquitous, sometimes visibly spectacular, it is natural that students of rhetoric, literature, painting, and scientific invention should have noticed many specific examples of what we call blending and noticed, too, that something interesting was going on. The earliest such observation that we have found comes from Aristotle. It occurs in Book 3 of Aristotle's *Rhetoric:*

> The address of Gorgias to the swallow, when she had let her droppings fall on him as she flew overhead, is in the best tragic manner. He said, "Nay, shame, O Philomela." Considering her as a bird, you could not call her act shameful; considering her as a girl, you could; and so it was a good gibe to address her as what she was once and not as what she is.

The shameful act exists only in the blend: The act is impossible for the girl, and the shame is impossible for the swallow. It is not quite clear that Aristotle recognized the existence of this blend, or recognized the emergent meaning in "shameful act," or recognized that the emergent meaning exists only in the blend. He saw the performance of Gorgias moreover as an exotic achievement, not as an instance of a basic mental operation. This is presumably why he did not look into its theoretical consequences. Evidently, insight into blending as a general and routine mental operation of the imagination was simply unavailable to classical rhetoricians.

The traditions of art history, literary criticism, and rhetoric are replete with similar examples. Many writers quite insightfully notice that some creativity is going on but present it as idiosyncratic to the example at hand. Many Freudian analyses of identity and dreams can be viewed retrospectively as the study of blends that are central to the human condition. And more recently, special cases of what we are calling conceptual blending have been discussed insightfully by Erving Goffman, Len Talmy, H. Fong, David Moser and Douglas Hofstadter, and Z. Kunda, D. T. Miller, and T. Clare. All these authors, however, take blends to be somewhat exotic, marginal manifestations of meaning. They don't focus on the general blending capacity itself. In fact, they do not appear to recognize it as a general capacity. Rather, they point out the interesting aspects of the local product of blending, be it the painting, the poem, or whatever. Blending is taken for granted as an available resource instead of being properly viewed as presenting the challenge. How can Gorgias do what he does? This is not a question Aristotle even recognizes.

One writer who did succeed in going beyond the particular and local features of individual examples of striking creativity in the sciences and the arts is Arthur Koestler, in *The Act of Creation* (1964). Seeing a symptom shared by all these examples of remarkable creative invention, he gave it the name "bisociation of matrices." In doing so, he showed what the scientific challenge was—namely, to explain a notion that he left vague but which includes the idea that creativity involves bringing together elements from different domains. But Koestler did not make the next leap: to discover that the general mental operation involved in these striking cases is also ubiquitous in everyday thought and language. Nor did he offer, even for the striking cases, any specific characterizations of the structural and dynamic processes that produced them.

Our own research program has come up with decisive evidence from many fields that conceptual blending is a general, basic mental operation with highly elaborate dynamic principles and governing constraints. After launching this research program in 1993, we were heartened to discover that, coming from another angle and with very different kinds of data, several "creativity theorists" were insisting on the existence of a general mental operation—which Steven Mithen called "cognitive fluidity"—whose result is to bring together elements of different domains. As Mithen wrote in 1998:

> Margaret Boden has argued that creative thought can be explained "in terms of the mapping, exploration, and transformation of structured conceptual spaces." Her definition of conceptual spaces is vague; she describes them as a "style of thinking—in music, sculpture, choreography, chemistry, etc." In spite of this vagueness, the idea of transforming conceptual spaces is intuitively appealing. It has a close association with the earlier notions of Koestler, who has described creative thinking as arising from the sudden interlocking of two previously unrelated skills or matrices of thought, and the contemporary ideas of Perkins, who uses the terminology of "klondike spaces" and argues that these are often systematically explored in the process of creative thinking. In this regard, while creative thinking is clearly part of ordinary thinking, and not something restricted to "geniuses," we can nevertheless see the potential for how particularly creative thoughts may arise from quite unusual transformations of conceptual spaces undertaken by particular individuals in particular circumstances.

What has occurred, then, is a convergence toward the essential idea that there is a single mental operation involved in creativity in a number of different domains. Whereas Aristotle and others picked out some interesting features of a few strikingly creative examples, and Koestler proposed that there is a special operation that underlies all these striking cases, contemporary creativity theorists have argued that this operation is not reserved for geniuses or for extraordinary acts of creation. We will show that this operation is indeed fundamental to all activities of the human mind, and we will try to lay out a precise and explicit theoretical framework in which to study its nature.

We begin where Koestler left off, with the case of "The Buddhist Monk."

Three

THE ELEMENTS OF BLENDING

> A Buddhist Monk begins at dawn one day walking up a mountain, reaches the top at sunset, meditates at the top for several days until one dawn when he begins to walk back to the foot of the mountain, which he reaches at sunset. Make no assumptions about his starting or stopping or about his pace during the trips. Riddle: Is there a place on the path that the monk occupies at the same hour of the day on the two separate journeys?

THIS IS THE AMAZING RIDDLE that Arthur Koestler presents in *The Act of Creation*. What we have to say about this Buddhist Monk will be more effective if you close the book for a moment and try to solve the riddle without any hints.

Now that you have found your place again, try this: Rather than envisioning the Buddhist Monk strolling up one day and strolling down several days later, imagine that he is taking both walks on the same day. There must be a place where he meets himself, and that place is the one we are looking for. Its existence solves the riddle. We don't know where this place is, but we do know that, whatever its location, the monk must be there at the same time of day on his two separate journeys. For many people, this is a compelling solution to the riddle.

But solving this little riddle only poses a much bigger scientific one: How are we able to arrive at the solution, and why should we be persuaded that it is correct? It is impossible for the monk to travel both up and down. He cannot "meet himself." Yet this impossible imaginative creation gives us the truth we are looking for. We plainly don't care whether it is impossible or not—that's irrelevant to our reasoning. But the scenario of *two* people meeting each other is not only possible but also commonplace. Using that scenario is crucial to seeing the solution, even though it is nowhere in the original riddle, which describes just one person doing different things on different days.

The imaginative conception of the monk's meeting himself blends the journey to the summit and the journey back down, and it has the emergent structure of an "encounter," which is not an aspect of the separate journeys. This emergent structure makes the solution apparent.

THE NETWORK MODEL

The Buddhist Monk example reveals the central principles of the network model of conceptual integration. We will lay them out here.

Mental Spaces

Mental spaces are small conceptual packets constructed as we think and talk, for purposes of local understanding and action. In the Buddhist Monk network, we have a mental space for the ascent and another mental space for the descent. Mental spaces are connected to long-term schematic knowledge called "frames," such as the frame of *walking along a path*, and to long-term specific knowledge, such as a memory of the time you climbed Mount Rainier in 2001. The mental space that includes you, Mount Rainier, the year 2001, and your climbing the mountain can be activated in many different ways and for many different purposes. "You climbed Mount Rainier in 2001" sets up the mental space in order to report a past event. "If you had climbed Mount Rainier in 2001" sets up the same mental space in order to examine a counterfactual situation and its consequences. "Max believes that you climbed Mount Rainier in 2001" sets it up again, but now for the purpose of stating what Max believes. "Here is a picture of you climbing Mount Rainier in 2001" evokes the same mental space in order to talk about the content of the picture. "This novel has you climbing Mount Rainier in 2001" reports the author's inclusion of a possibly fictional scene in a novel. Mental spaces are very partial. They contain elements and are typically structured by frames. They are interconnected, and can be modified as thought and discourse unfold. Mental spaces can be used generally to model dynamic mappings in thought and language.

At various times along the way, we will use diagrams to talk about mental spaces and blends. In these diagrams, mental spaces are represented by circles; elements, by points (or icons) in the circles; and connections between elements in different spaces, by lines. In the neural interpretation of these cognitive processes, mental spaces are sets of activated neuronal assemblies, and the lines between elements correspond to coactivation-bindings of a certain kind. In addition, the frame structure recruited to the mental space is represented as either outside in a rectangle or iconically inside the circle.

Input Spaces. In the Buddhist Monk network, there are two input mental spaces. As shown in Figure 3.1, each is a partial structure corresponding to one

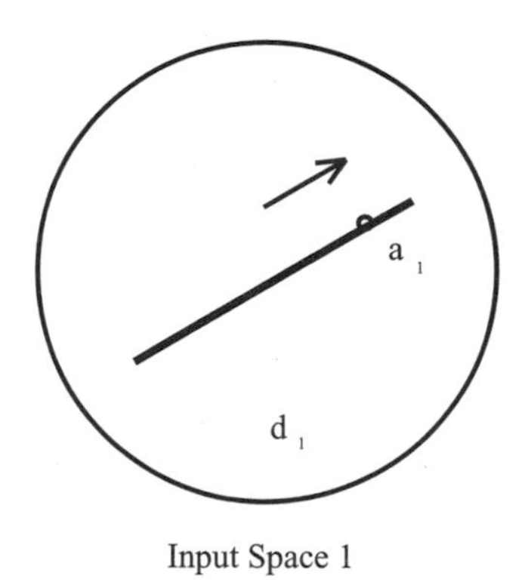

FIGURE 3.1 INPUT MENTAL SPACES

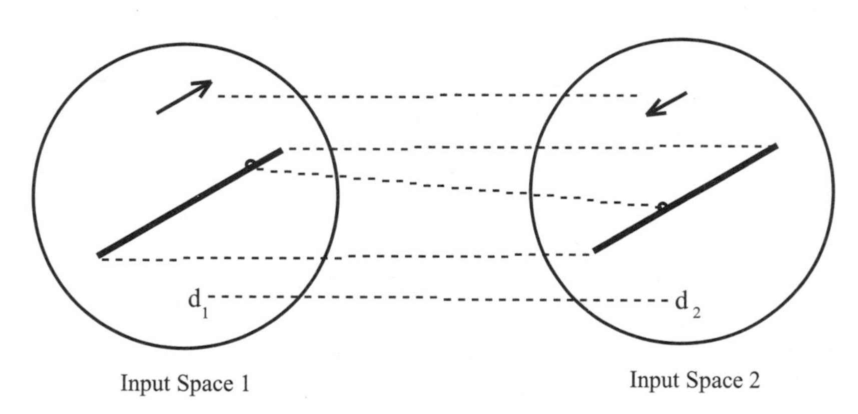

FIGURE 3.2 CROSS-SPACE MAPPING

of the two journeys. The day of the upward journey is d_1, the day of the downward journey is d_2, the monk going up is a_1, and the monk going down is a_2.

Cross-Space Mapping. A partial cross-space mapping connects counterparts in the input mental spaces (see Figure 3.2). It connects mountain, moving individual, day of travel, and motion in one mental space to mountain, moving individual, day, and motion in the other mental space.

Generic Space. A generic mental space maps onto each of the inputs and contains what the inputs have in common: a moving individual and his position, a path linking foot and summit of the mountain, a day of travel, and motion in an unspecified direction (represented in Figure 3.3 by a double-headed arrow).

Blend. There is a fourth mental space, the blended space, that we will often call "the blend" (see Figure 3.4). Each of the mountain slopes in the two input mental spaces is projected to the same single mountain slope in the blended

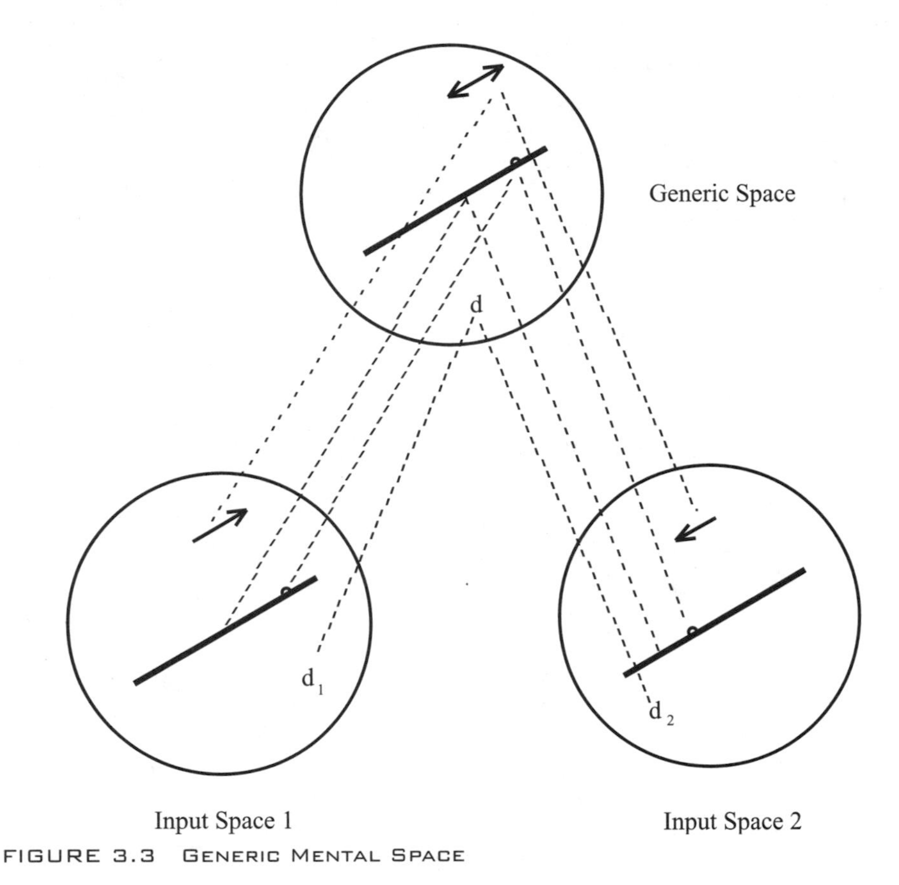

FIGURE 3.3 GENERIC MENTAL SPACE

space. The two days of travel, d_1 and d_2, are mapped onto a single day d' and are thus fused. But the moving individuals and their positions are mapped according to the time of day, with direction of motion preserved, and therefore cannot be fused. Input Space 1 represents dynamically the entire upward journey, while Input Space 2 represents the entire downward journey. The projection into the blended space preserves times and positions. The blended space, which has time t and day d', contains a counterpart of a_1 at the position occupied by a_1 at time t of day d_1 as well as a counterpart of a_2 at the position occupied by a_2 at time t of day d_2.

Emergent Structure

The blend develops emergent structure that is not in the inputs. First, *composition* of elements from the inputs makes relations available in the blend that do not exist in the separate inputs. In the blend but in neither of the inputs, there are two moving individuals instead of one. They are moving in opposite directions,

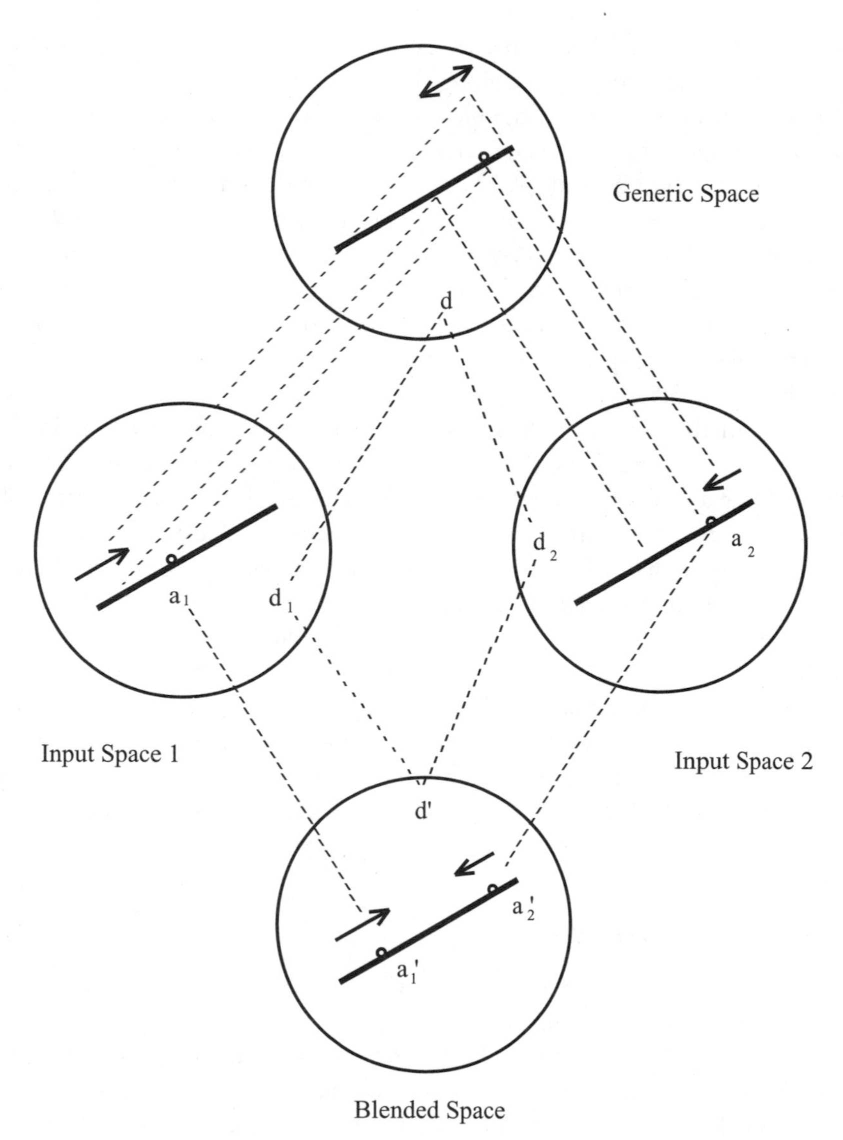

FIGURE 3.4 BLENDED SPACE

starting from opposite ends of the path, and their positions can be compared at any time of the trip, since they are traveling on the same day, d'. Second, *completion* brings additional structure to the blend. This structure of two people moving on the path can itself be viewed as a salient part of a familiar background frame: two people starting a journey at the same time from opposite ends of a path. Third, by means of *completion*, this familiar structure is recruited into the blended space. At this point, the blend is integrated: It is an instance of a particular familiar frame, the frame of two people walking on a path in opposite directions. By virtue of that frame, we can now run the scenario dynamically: In the blend, the two people move along the path. This "running of the blend" is called *elaboration*. Running of the blend modifies it imaginatively, delivering the actual encounter of the two people. This is new structure: There is no encounter in either of the input mental spaces, even if we run them dynamically. But those two people in the blend are projected back to the "same" monk in the two input mental spaces. The meeting place projects back to the "same" location on the path in each of the inputs, and, of course, the time of day when they meet in the blend is the same as the time of day in the input spaces when the monk is at that location. The mapping back to the input spaces yields the configuration suggested by Figure 3.5.

As we run the blend, the links to the inputs are constantly maintained, so that all these "sameness" connections across spaces seem to pop out automatically, yielding a flash of comprehension, Koestler's magical "act of creation." But for this flash to occur, counterpart links must be unconsciously maintained even as they change dynamically across four mental spaces. In particular, there are geometrical regularities across these spaces. Given the way we have built the blend, we know that any point on the path in the blend projects to counterparts in the input spaces. More generally, anything fused in the blend projects back to counterparts in the input spaces. But this "geometric" knowledge of correlations among time, position of the monk, and location on the path in the different spaces is completely unconscious. What comes into consciousness is the flash of comprehension. And it seems magical precisely because the elaborate imaginative work is all unconscious.

WHAT HAVE WE SEEN?

Blending in the riddle of the Buddhist Monk has features that turn out to be universal for conceptual integration.

Building an integration network involves setting up mental spaces, matching across spaces, projecting selectively to a blend, locating shared structures, projecting backward to inputs, recruiting new structure to the inputs or the blend, and running various operations in the blend itself. We will talk about these operations in sequence, but it is crucial to keep in mind that any of them can run at any time and that they can run simultaneously. The integration network is trying to achieve equilibrium. In a manner of speaking, there is a place where the network is "happy." Context will typically specify some conditions of the equilibrium, as when we are instructed to find a solution to the riddle of the Buddhist Monk. The network will achieve equilibrium if structure comes up in the blend that projects back automatically to the inputs to yield the existence of the special point on the path. More generally, what counts as an equilibrium for the

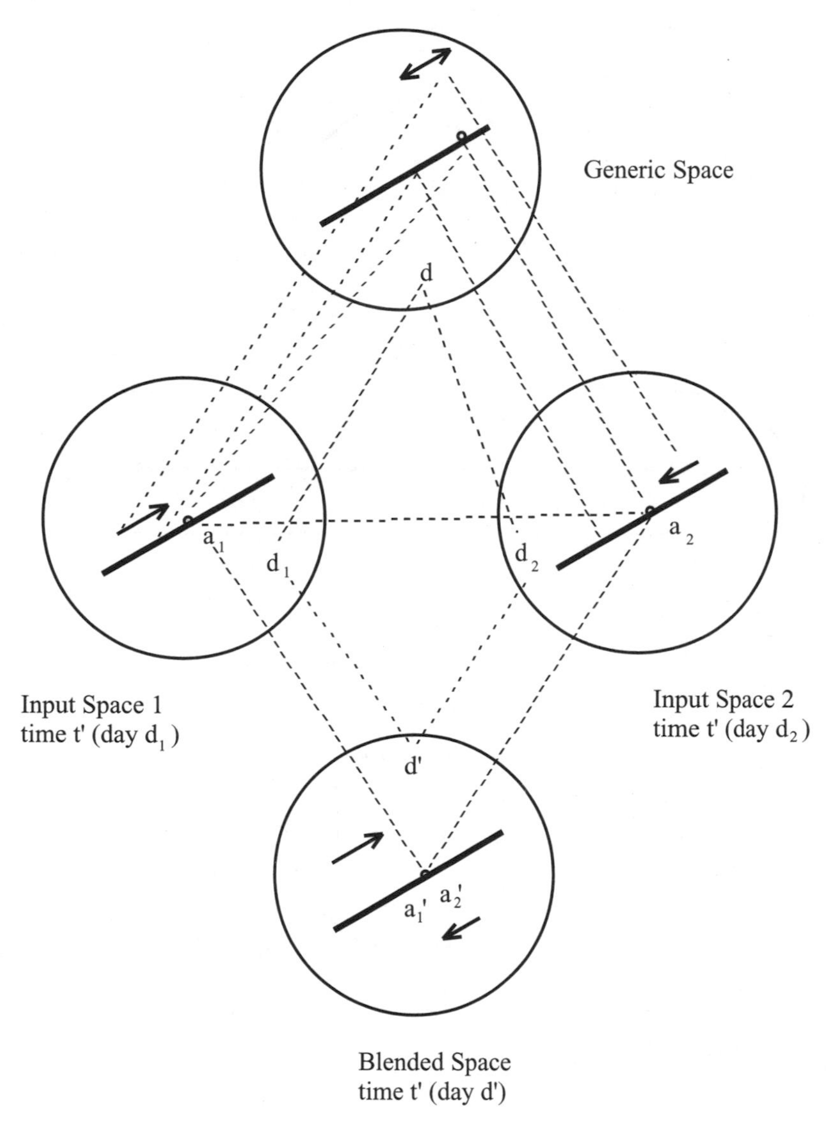

FIGURE 3.5 Mapping Back to Input Spaces

network will depend on its purpose, but also on various internal constraints on its dynamics.

The Basic Diagram in Figure 3.6 illustrates the central features of conceptual integration: The circles represent mental spaces, the solid lines indicate the matching and cross-space mapping between the inputs, the dotted lines indicate connections between inputs and either generic or blended spaces, and the solid square in the blended space represents emergent structure.

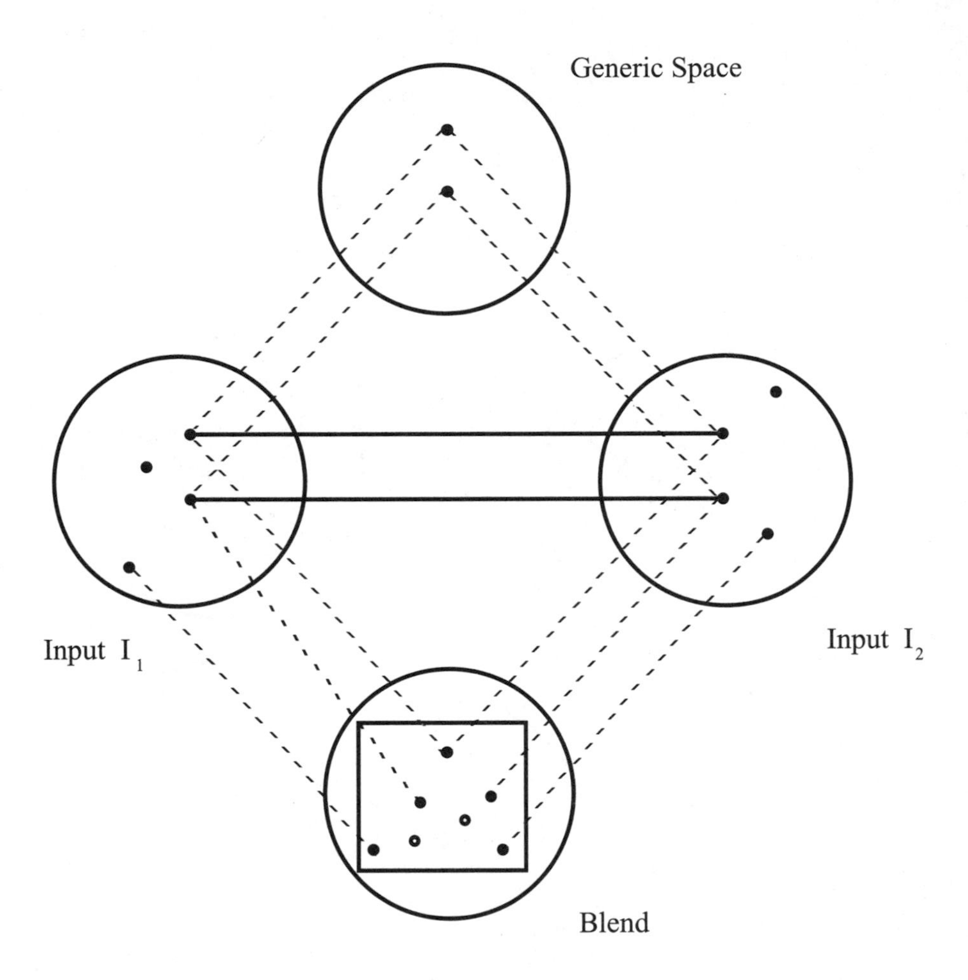

FIGURE 3.6 THE BASIC DIAGRAM

While this static way of illustrating aspects of conceptual integration is convenient for us, such a diagram is really just a snapshot of an imaginative and complicated process that can involve deactivating previous connections, reframing previous spaces, and other actions. We think of the lines in this diagram (lines that represent conceptual projections and mappings) as corresponding to neural coactivations and bindings. Here, then, are the essential aspects of blending, presented in a sequence not meant to reflect actual stages of the process:

- *Conceptual integration network.* Blends arise in networks of mental spaces. In the network illustrated in the Basic Diagram, there are four mental spaces: the two inputs, the generic space, and the blend. This is a minimal network. Conceptual integration networks can have several input spaces and even multiple blended spaces.
- *Matching and counterpart connections.* In conceptual integration, there is partial matching between input spaces. The solid lines in the Basic Diagram represent counterpart connections produced by matching. Such counterpart connections are of many kinds: connections between frames and roles in frames, connections of identity or transformation or representation, analogical connections, metaphoric connections, and, more generally, "vital relations" mappings (as explained in Chapter 6). In the Skiing Waiter case, for example, ski poles are counterparts of a tray. When matches are created between two spaces, we say that there is a cross-space mapping between them.
- *Generic space.* At any moment in the construction of the network, the structure that inputs seem to share is captured in a generic space, which, in turn, maps onto each of the inputs. A given element in the generic space maps onto paired counterparts in the two input spaces. In the Iron Lady case, the generic space is something like "Western democracy with labor unions and voters." *Labor unions* in the generic space maps onto *American labor unions* in one input and *British labor unions* in the other, which are accordingly counterparts. In the Skiing Waiter case, the generic has a moving individual carrying something in his hands. The carried object in the generic space maps onto the ski poles in one input and onto the tray in the other. They, too, are accordingly counterparts.
- *Blending.* In blending, structure from two input mental spaces is projected to a new space, the blend. Generic spaces and blended spaces are related: Blends contain generic structure captured in the generic space but also contain more specific structure, and they can contain structure that is impossible for the inputs, such as two monks who are the same monk.
- *Selective projection.* Not all elements and relations from the inputs are projected to the blend. The calendrical time of the journey in the Buddhist Monk case is not projected to the blend. In the Skiing Waiter case, neither walking nor the customer nor the price of the champagne is projected from the waiter input. Sometimes two counterparts are both projected (both paths, both monks), sometimes only one (in the Iron Lady example, only American voters are projected, not British voters), sometimes none (in the Buddhist Monk example, calendrical dates). Sometimes counterparts in the input spaces are fused in the blend (the two paths), but often not (the two monks). And, finally, sometimes an element in one input without a

counterpart in the other gets projected to the blend (skis in the Skiing Waiter case).

- *Emergent structure*. Emergent structure arises in the blend that is not copied there directly from any input. It is generated in three ways: through *composition* of projections from the inputs, through *completion* based on independently recruited frames and scenarios, and through *elaboration* ("running the blend").
- *Composition*. Blending can compose elements from the input spaces to provide relations that do not exist in the separate inputs. In the Buddhist Monk example, composition yields two travelers making two journeys at the same time on the same path, even though each input has only one traveler making one journey. Counterpart elements can be composed by being included separately in the blend, as when the monks from the inputs are brought into the blend separately, yielding two monks; or by being projected onto the same element in the blend, as when the two days in the two inputs are projected onto the same day in the blend. We refer to this kind of projection as "fusion."
- *Completion*. We rarely realize the extent of background knowledge and structure that we bring into a blend unconsciously. Blends recruit great ranges of such background meaning. Pattern completion is the most basic kind of recruitment: We see some parts of a familiar frame of meaning, and much more of the frame is recruited silently but effectively to the blend. Figure 3.7 demonstrates this well-known psychological phenomenon, where we see two line segments and a rectangle and, through pattern completion, infer that there is a straight line running "behind" the rectangle. A minimal composition in the blend is often automatically interpreted as being a richer pattern. In the Buddhist Monk case, the composition of two monks on the path is completed so automatically by the scenario of two people journeying toward each other that it takes some thinking to see that the "journeying toward each other" scenario is much richer than the "two monks" composition.
- *Elaboration*. We elaborate blends by treating them as simulations and running them imaginatively according to the principles that have been established for the blend. Some of these principles for running the blend will have been brought to the blend by completion. We run the Buddhist Monk blend to get the "encounter" in the blend that provides the solution to the riddle. We are able to run the blend because we know the dynamics of the scenario of two people making opposite journeys along a path, which was brought in by pattern completion. That scenario gives us principles having to do with the passage of time, the possibilities of self-locomotion, and so on. Part of the power of blending is that there are always many different possible lines of elaboration, and elaboration can go

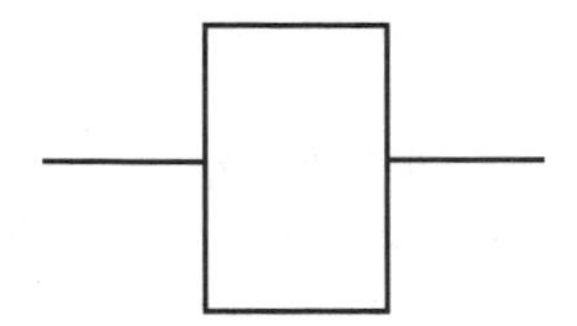

FIGURE 3.7 PATTERN COMPLETION

on indefinitely. We can run the blend as much and as long and in as many alternative directions as we choose. For example, the two monks might meet each other and have a philosophical discussion about the concept of identity. That particular elaboration would divert us from the purpose of solving the riddle, but it could also lead to something interesting and useful. The creative possibilities of blending stem from the open-ended nature of completion and elaboration. They recruit and develop new structure for the blend in ways that are principled but effectively unlimited. Blending operates over the entire richness of our physical and mental worlds.

Composition, completion, and elaboration lead to emergent structure in the blend; the blend contains structure that is not copied from the inputs. Note that in the Basic Diagram (Figure 3.6), the square inside the blend represents emergent structure.

- *Modification.* Any space can be modified at any moment in the construction of the integration network. For example, the inputs can be modified by reverse mapping from the blend, as in the Buddhist Monk case where we add to the inputs the existence of the location asked for in the riddle by backward projection of the spot of "encounter" from the blend.
- *Entrenchment.* Blends are often novel and generated on the fly, as in the Buddhist Monk case, but they recruit entrenched mappings and frames. Blends themselves can also become entrenched, as in the Complex Numbers blend, giving rise to conceptual and formal structures shared throughout the community.
- *Event integration.* Blends are a basic instrument for achieving event integration. In the Skiing Waiter and Image Club examples, the event integration is the purpose of the imaginative construction. But in the Buddhist Monk case, it is only a means for solving the riddle about the existence of a location with certain properties.
- *Wide application.* Though uniform in their dynamics, integration networks can serve many different goals. In the examples we have seen so far, these goals include transfer of emotions (Image Club) and inferences

(Buddhist Monk and Computer Desktop), counterfactual reasoning (Iron Lady), conceptual change and creativity in science (Complex Numbers), integrated action (Computer Desktop and Skiing Waiter), and construction of identity through compression (Graduation).

Are there different kinds of integration networks and blended spaces? What is the relationship between imaginative integration networks and formal systems of language? Are there constraints on blending? If so, what are they? Where do the inputs come from, in the first place? Just how is blending related to categorization, analogy, metaphor, metonymy, logic? How complicated can networks get? Can blends be reblended or serve as inputs to other blends? What exactly is a generic space? Does it participate in the dynamics of the network, and can it evolve?

Blending is child's play for us human beings, but we are children whose games run deep. We will now try to find out more about those games.

CHAPTER 3
ZOOM OUT

WILD COGNITION

The Buddhist Monk blend is a fantastic and impossible scenario.

Question:

- So is blending a special device for producing fantastic thoughts?

Our answer:

The Buddhist Monk example might make you think that. And although the Skiing Waiter and Image Club cases have some connection to reality, they still look quite exotic. On the other hand, there is nothing fantastic about the resulting blends in the "safe beach" and "red pencil" examples, which go completely unnoticed. Clearly, blends can be either fantastic or not, but when they are fantastic, they may stand out. As we will see repeatedly, unnoticed blends are much more common than noticeable or fantastic ones. In fact, it takes some work to find an example whose blending is immediately apparent. Why do we begin with so many exotic-looking examples? Because we have to make the phenomenon of blending visible before we can begin to analyze its operation, and wild examples like the Buddhist Monk show us blending in a way no one can deny. Consider, by way of analogy, the topic of electricity. Although electricity is a pervasive and usually completely unnoticed aspect of the universe, schoolbooks introduce the subject with exotic and memorable illustrations, such as Benjamin Franklin's drawing down lightning by flying a kite in a storm. Indeed, it is quite reasonable that we do not become aware of a general, pervasive process until we see a case that looks exceptional.

The blend in the Iron Lady counterfactual is completely fantastic but, in context, is not noticed as unusual. We will see this repeatedly: Expressions that seem completely normal and literal to people (e.g., "This beach is safe") turn out to involve intricate blending, which stands out once we have analyzed it. Here are some other completely normal examples that have fantastic blends similar to the Buddhist Monk but go unnoticed in context: "This chapter is writing itself," "You're getting ahead of yourself," "Normal Mailer loves to read himself," "I can't keep up with the schedule," and "My heart is disagreeing with my head."

The Buddhist Monk blend, rather amazingly, shows up in real life. Ed Hutchins studies the fascinating mental models set up by Micronesian navigators to sail across the Pacific. In such models, it is the islands that move, and virtual islands serve as reference points. Hutchins reports a conversation between Micronesian and Western navigators who have trouble understanding each other's conceptualizations. As described by David Lewis, the Micronesian navigator Beiong comes to understand a Western diagram of intersecting bearings in the following way:

> He eventually succeeded in achieving the mental tour de force of visualizing himself sailing simultaneously from Oroluk to Ponape and from Ponape to Oroluk and picturing the ETAK bearings to Ngatik at the start of both voyages. In this way he managed to comprehend the diagram and confirmed that it showed the island's position correctly. [The ETAK is the virtual island, and Ngatik is the island to be located.]

It is not just the blends that can be fantastic. Real life, from a certain point of view, can be fantastic, too. The Micronesian navigator's tour de force is especially fantastic for the anthropologist, Lewis, who knows the standard Western techniques of navigation, but what is fantastic for the Micronesian, and at first completely unintelligible to him, is the Western chart with the intersection of bearings. The spectacular blend invented by the Micronesian navigator Beiong exists only to make sense of the bizarre Western conceptualization. Ed Hutchins and Geoffrey Hinton, in their article "Why the Islands Move," explain how the seemingly mysterious Micronesian system actually works and reveal its great sophistication and power.

One might object that this case, even if real, is quite exceptional. But even so, it is no less representative of human cognition than more ordinary cases. Cognitive scientists take all such examples as revealing the hidden aspects of cognition: The navigator had to have blending readily available as standard equipment in order to perform his tour de force. Evolution did not give us advanced powers of blending just so we could understand weird concepts held in other cultures or solve a riddle like the Buddhist Monk. Our robust powers of blending are useful throughout our everyday lives. Unusual examples and everyday examples are equally important, for our purposes here. Unobjectionable idioms like "getting ahead of oneself" and "catching up with oneself" are built on the same principles of blending as the Buddhist Monk. Such blends are more common than one might think. Paul Rodriguez reports that when he was driving guests around the city, one of them said, "You're driving us around so much you're going to meet yourself coming down the street." The blended space in this quip, he observes, is almost exactly the same as in the Buddhist Monk riddle. He then notes,

> It is interesting that very little discourse context was needed to set up the blend. The phrase *driving around so much* implies more than one trip, just as the monk riddle explicitly states there are two separate journeys. Realizing this was a blend, I asked my guest to explain and she said, "When the kids were in school I would drive them around so much that I used to say "One of these days I'm going to meet myself coming down the street."

UNIFORMITY OF THE OPERATION

We have seen cases that appear intuitively simple and others that appear intuitively complex.

Questions:
- Is commonplace blending simple?
- Is fantastic blending complex and therefore more interesting?

Our answers:

As we just saw, the very same principles and complexity apply in Koestler's strikingly creative Buddhist Monk riddle, Beiong's tour de force, and the passenger's comment on driving around. Blending finds structure where it can. There is a variant of the Buddhist Monk riddle that recruits a quite different set of frames to create a quite different integration network, yet the same solution. It is contained in the Spanish short story "Páginas inglesas," in which a man must prove he was twice in the same spot at the same hour. He has just run down a hill in twenty minutes. The day before, he had climbed the hill in five hours. But the twenty minutes are contained in the period of the day spanned by the five hours. The solution comes in the form of imagining that yesterday a car had driven down the hill in the "same" twenty minutes. So first we have a nonfantastic blend of today's quick journey and an automobile: In the blend, it is yesterday, and the automobile makes the journey. Then we are told that inevitably, yesterday, the person would have been passed by the automobile. This is a second blend, again nonfantastic.

By comparing this variant solution with that given by Koestler, we can see three general points.

First, impossibility need not be crucial to the blend: Koestler's blend is impossible, but the more complicated double blending in the Spanish story gives us two nonfantastic blends.

Second, the schemas used for pattern completion and emergent structure in the blend are experiential. In Koestler's solution, the experiential frame recruited is two people meeting each other on a path. The Spanish version recruits a richer frame: an oncoming car passing a pedestrian, where the car is moving much faster than the pedestrian. The encounter schema looked almost trivial in Koestler's solution, but finding the right encounter schema looks like a much more imaginative achievement in the Spanish version. In fact, the story frames the person who finds this solution as exceptionally intelligent for being able to deliver a blend that is familiar and immediately intelligible. Because the Spanish version has the descent taking much less time than the ascent, finding the automobile schema gives us a way to exploit our everyday knowledge about differential speeds. Even though the Spanish riddle is "mathematically" more complex, given the differences in the time it takes to ascend and descend and the corresponding differences of average speed, some people find the solution easier to accept because the final blend is a straightforward instance of a familiar situation: The oncoming car zooms past the pedestrian. It is worth noting that in both the original and Spanish solutions, if "the same place" were interpreted literally, we would have to imagine two objects inhabiting the same space. In the Spanish solution, the car would run over the pedestrian. But everyone is willing in both cases to exploit, unconsciously, the convention that if two moving bodies pass each other closely, they are momentarily in "the same place."

Third, both solutions begin from the same inputs (ascent and descent of one person) and deliver the same topological solution, but the imaginative networks they put together are different. So the inputs do not determine the integration network.

THE SCIENCE OF BLENDING

We have seen that blending is not deterministic.

Questions:

- Does the fact that we can't predict the integration network from the inputs mean that just any projections from the inputs are OK?

- Is the process of blending fuzzy and unscientific? Is the science of blending fuzzy and unscientific?
- Can we hope to have a science of something so uncontrolled and "magical" as blending?

Our answers:

This is a very good set of questions. It is part of our research program to show that not just any projections are OK. Indeed, as we will see in Chapter 16, there are quite strong constraints on projections. We have made some headway on this very hard problem and, in that later chapter, will suggest lines for further research.

A view often implicit in form approaches holds that only generative algorithmic models, specifying unique outputs from given inputs, are scientific, so that the underdetermined nature of blending, as we analyze it, brands our theory as unscientific. This objection is simply wrong. Theories of probability, subatomic particles, chaos, complex adaptive systems, evolution, immunology, and many others could not get off the ground as sciences if they were required to offer models in which the specified inputs determined unique outputs.

Is "descent of species" a fuzzy and unscientific concept? You can never predict what new species will emerge or even how a given species will evolve; the outputs of evolution do not fall into clear-cut categories (although there are central examples for each category); and the outcome of natural selection cannot be specified algorithmically. But evolutionary biology is by now a pillar of science.

FIRST THINGS FIRST?

We have begun with striking examples so the blending can be easily seen.

Question:

- Wouldn't it be in better scientific taste to start by explaining the meaning of "The cat is on the mat" rather than trying to tackle the science fantasy of the Buddhist Monk who can be in two places at the same time? First things first!

Our answer:

The scientific study of meaning would be easier if there were simple theories for simple meanings, which could be supplemented by more complicated theories for the more complicated meanings. But the simple meanings, it turns out, are also complicated, and as we will show repeatedly, you must have *all* of the operation of blending in order to put them together. Blending is not special in this way. It takes all of our cognitive powers to have common sense.

More generally, explanation does not come from taking simplified accounts of simplified observations and extending them by adding some bells and whistles to give better explanations. Science moves in the other direction: A general scientific account explains the supposedly simple examples as special cases of the general account. For instance, it is an intuitive primitive for us that if we drop something, it falls to our feet. But that intuitive primitive is really only a very special case of the general theory of gravity, which also explains planetary motion and the parabolic trajectory of projectiles. We cannot explain scientifically how a dropped object falls to our feet without the general theory of gravity. There isn't a part of the theory of gravity that applies only to apples. Similarly, what we need for blending is a general theory that explains both the everyday and exotic examples.

FALSIFIABILITY

We have so far given analyses of blends, but we have not framed our analyses in terms of prediction and confirmation.

Questions:

- Doesn't science involve making falsifiable predictions?
- What falsifiable predictions come from the theory of blending?

Our answers:

Actually, sciences like evolutionary biology are not about making falsifiable predictions regarding future events. Given the nature of the mental operation of blending, it would be nonsense to predict that from two inputs a certain blend must result or that a specific blend must arise at such-and-such a place and time. Human beings do not think that way. Nonetheless, in the strong sense, we hope to make many falsifiable predictions, including predictions about types of blending, what counts as a good or bad blend, how the formation of a blend depends on the local purpose, how forms prompt for blending, what possibilities there are for composing mappings, what possibilities there are for creating successive blends, how other cognitive operations (such as metonymy) are exploited during blending, and how categories are extended.

In fact, we take it that we have already falsified existing accounts of counterfactuals by showing the centrality of counterfactuals like the Iron Lady, which such theories are on principle unable to handle. We will do more of the same as we go along, by showing that, to the extent that they do not have a place for blending, existing accounts of metaphor, analogy, or grammar are in certain respects "false." Sociologically and psychologically, theory tends to define data, so it is hard to use data to falsify a theory while staying conceptually inside that theory. A logician whose specialty is counterfactuals might honestly believe that an example like the Iron Lady falls outside the range of relevant data for logic and is only an anomaly of natural language, a "way of speaking," not a real example of counterfactual *thinking*. Philosophers like Donald Davidson who claim to study meaning have decreed in a similar vein that the study of metaphor is not part of the study of meaning.

VISIBILITY OF THE BLENDING

We have tried to make consciously visible some of what happens during blending.

Questions:

- How can meaning be consciously apprehended at all if it is constructed unconsciously?
- What parts of blending are visible to consciousness, and when do invisible aspects rise to the surface?
- The Buddhist Monk example gives us a "Eureka" or "aha!" effect in which the whole solution pops out in a way that we seem to understand all the way through. How does that happen, and how is it different from moving step by step through a long argument to reach an earned conclusion? Is there a difference between meanings that we apprehend all at once and those we build up step by step?

Our answers:

Consciousness—like identity, sameness, and difference—appears to the conscious mind to be a primitive. For millennia, the fact of consciousness and unconsciousness has been recognized and thought to be interesting, but theorists have only recently begun to be able to ask why people should have consciousness at all. Previously, this seemed a nonsensical question. Now it is recognized as a central and extremely difficult question for cognitive neuroscience, and several hypotheses have been proposed. We refer readers to the *Journal of Consciousness Studies* and *Brain and Behavioral Sciences* for the rich but short history of this controversy.

The question "How can meaning be consciously apprehended at all if it is constructed unconsciously?" makes the false assumption that what is apprehended consciously must be the output of a conscious process. But as we saw with the perception of a coffee cup, it is in the nature of consciousness that it gives us effects we can act on, and these effects are correlated with the unconscious processes. In the case of meaning, the apprehension of these effects typically induces us to reify meaning. Consciousness sees an effect and reifies the effect to provide a cause: I see a cup, and, in the folk theory, the reason I do is that there is a cup that causes me to see a cup. In the same way, I hear a sentence and

I "see" a meaning for it, and, in the folk theory, the reason I do is that there is an abstract thing, the meaning, that causes me to "see" it. We have already shown the falsity of this view. In the case of blending, the effects of the unconscious imaginative work are apprehended in consciousness, but not the operations that produce it. In the case of the Buddhist Monk, the ultimate meaning pops out when we realize that the inevitable encounter in the blend yields the solution to the initial problem. The dynamic web of links between blend and inputs remains unconscious. What is registered consciously is the encounter in the blend and the "consequent" alignment between the two inputs.

The moment of tangible, global understanding comes when a network has been elaborated in such a way that it contains a solution that is delivered to consciousness. That delivery in nonroutine cases can produce the "Eureka!" or "aha!" effect, and it is to be contrasted with the pattern in which we proceed step by step through a long analysis and are persuaded by each step, including the last, so that we accept the conclusion but do not feel that we have a global understanding of the whole. Instead, we know only that we did approve each step, whatever it was, and so the conclusion must be true even if we do not actually grasp why. This seems paradoxical: In cases of step-by-step analysis, we do part of each step consciously, yet are left without a feeling that we understand the truth deeply, while in cases of blending, most of the analysis is done unconsciously, yet we can end up with a deeper satisfaction. We suggest that in the case of blending, at the moment of solution, the entire integration network is still active in the brain, even if unconsciously, while in the case of step-by-step analysis, at the moment of solution we have already lost most of the structure of the preceding steps.

Four

ON THE WAY TO DEEPER MATTERS

The authors of these books take part in the great conversation.

—*Mortimer J. Adler*

CONCEPTUAL INTEGRATION UNDERLIES BOTH remarkably creative feats that look as if they need explanation and everyday mental actions that look completely simple. Performances that seem too simple to need explanation turn out to be devilishly hard to explain. Cognitive scientists have shown that many feats that we find easy—categorization, memory, framing, recursion, analogy, metaphor, even vision and hearing—are exceptionally resistant to scientific analysis. They turn out to be the things that are hardest to explain. Syntax before Harris and Chomsky, framing before Bateson and Goffman, and analogy before Gentner, Hofstadter, and Holyoak were surely recognized as pervasive, but the need for their systematic study was not perceived. In a sense it could not be perceived, because there was no framework or set of techniques within which to ask questions systematically.

Now that we have the beginnings of a framework for conceptual integration, we can begin to show how it works systematically in human cognition.

THE DEBATE WITH KANT

The Buddhist Monk example presents a very salient and intuitively apparent blend, precisely because it includes the strange event of someone meeting himself. But blending, with all its structural and dynamic imaginative properties, is rarely so visible. Let us look at a case where the invisible blending becomes visible as we analyze it. Imagine that a contemporary philosopher says, while leading a seminar,

> I claim that reason is a self-developing capacity. Kant disagrees with me on this point. He says it's innate, but I answer that that's begging the question, to which

> he counters, in *Critique of Pure Reason,* that only innate ideas have power. But I say to that, What about neuronal group selection? And he gives no answer.

As a straightforward report, this passage describes an actual historical event in which Kant is tongue-tied when confronted by the modern philosopher. That no one interprets the passage this way presents us with a bigger question: How can reporting an argument with a dead man count as a sane expression of one's philosophical position? Philosophers pride themselves on their logical thought and expression. The modern philosopher here is not experiencing a mental breakdown in which he has delusions of speaking to Kant and humiliating him in debate. On the contrary, this passage, or the first half of it anyway, is supposed to be a good instance of logical thought and expression. How can a passage about talking to someone who has been dead for centuries be logical? How can Kant lose the debate by not answering the modern philosopher, since presumably long-dead people do not answer anybody?

Our reader will not be surprised to hear that the reasoning of this passage is the reasoning that is available from integration networks, in which we can have a blended space that is useful but not about something real. The Buddhist Monk blend lets us see the truth of the situation in the inputs, but not by asking us to believe that people meet themselves on mountain paths. Just so, the Debate with Kant integration network tells us about the modern philosopher's relationship to Kant and his ideas, but it does not require us to believe that the philosopher and Kant talk to each other, or even, in this case, to notice that there is any blending going on.

The Debate with Kant blend has two input spaces. In one, we have the modern philosopher, making claims. In a separate but related input space, we have Kant, thinking and writing. In neither input space is there a debate. The blended space has both people. In addition, the frame of *debate* has been recruited to frame Kant and the modern philosopher as engaged in simultaneous debate, mutually aware, using a single language to treat a recognized topic.

The *debate* frame comes up easily in the blend through pattern completion, since so much of its structure is already in place in the composition of the two inputs. Composition provides two philosophers, both discussing the issue of reason but holding different views. This scenario gives us much of the structure of the *debate* frame, which is recruited to give us further structure, such as Kant's awareness of the modern philosopher, questions and answers, and a potential victor in the debate. Once the blend is established, we can "run the blend"—that is, operate cognitively within it, developing new structure and manipulating the various events as an integrated unit. The *debate* frame brings with it conventional expressions, available for our use. We use such expressions to pick out structure in the blend directly, but because the connections between the blend and the inputs remain active, applying imagination to the blend has consequences for the inputs. For example, even though the debate is fictional, we assume that the modern philosopher's opinions in the blend are also his opinions in the input. In the blend, the modern philosopher wins the debate, and his victory establishes that he has the better idea, which projects back to both inputs. It projects back to the modern philosopher input to establish there that his idea is very good, that he argues very well, and that he is a very good philosopher. It projects back to the Kant input to establish there that, however good Kant's ideas are, they are not at the top.

The Debate with Kant has all the expected properties of blending:

There is a cross-space mapping linking Kant and his writings to the philosophy professor and his lecture. The counterparts include Kant and the professor, their respective languages, topics, claims, times of activity, goals (such as the search for truth), and modes of expression (such as writing versus speaking).

There is selective projection to the blend: Kant, the professor, some of their ideas, and the search for truth are projected to the blend. Kant's time, language, mode of expression, the fact that he's dead, and the fact that he was never aware of the future existence of our professor are not projected.

There is emergent structure through composition: We have two people talking in the same place at the same time. There is emergent structure through completion: Two people talking in the same place at the same time evoke the cultural frame of a conversation, a debate, or an argument. When they disagree, we tend to pick the frame of *argument* or *debate*. In this case, we pick the *debate* frame to structure the blend. It is evoked by the professor's syntax and vocabulary ("disagrees," "answer," "counters," "What about. . . ?"). And there is emergent structure through elaboration: "Running the blend" in this case is a matter of running the *debate* frame by elaborating questions and answers, retorts and concessions, with corresponding emotions like defensiveness, aggression, and elation. Running the blend gives us a sequence of arguments by the modern philosopher, their superiority to Kant's arguments, and, therefore, their ultimate validity, since Kant is already at the top of the scale.

We also see integration of events: Kant's ideas and the professor's claims are integrated into a unified event, the debate. Like the Buddhist Monk and Skiing Waiter, the Debate with Kant integrates into a single scenario various events of uncertain relation spread out over time. Blends provide a space in which ranges of structure can be manipulated uniformly. But the other spaces do not disappear once the blended space has been formed. On the contrary, the blended space is valuable only because it is connected conceptually to the inputs. The Buddhist Monk blend and the Debate with Kant blend lead us to alter the inputs imaginatively. The Buddhist Monk looks exotic, but we do not even notice the blending in the Debate with Kant since the general blending template it deploys is conventional for engaging the ideas of a previous thinker.

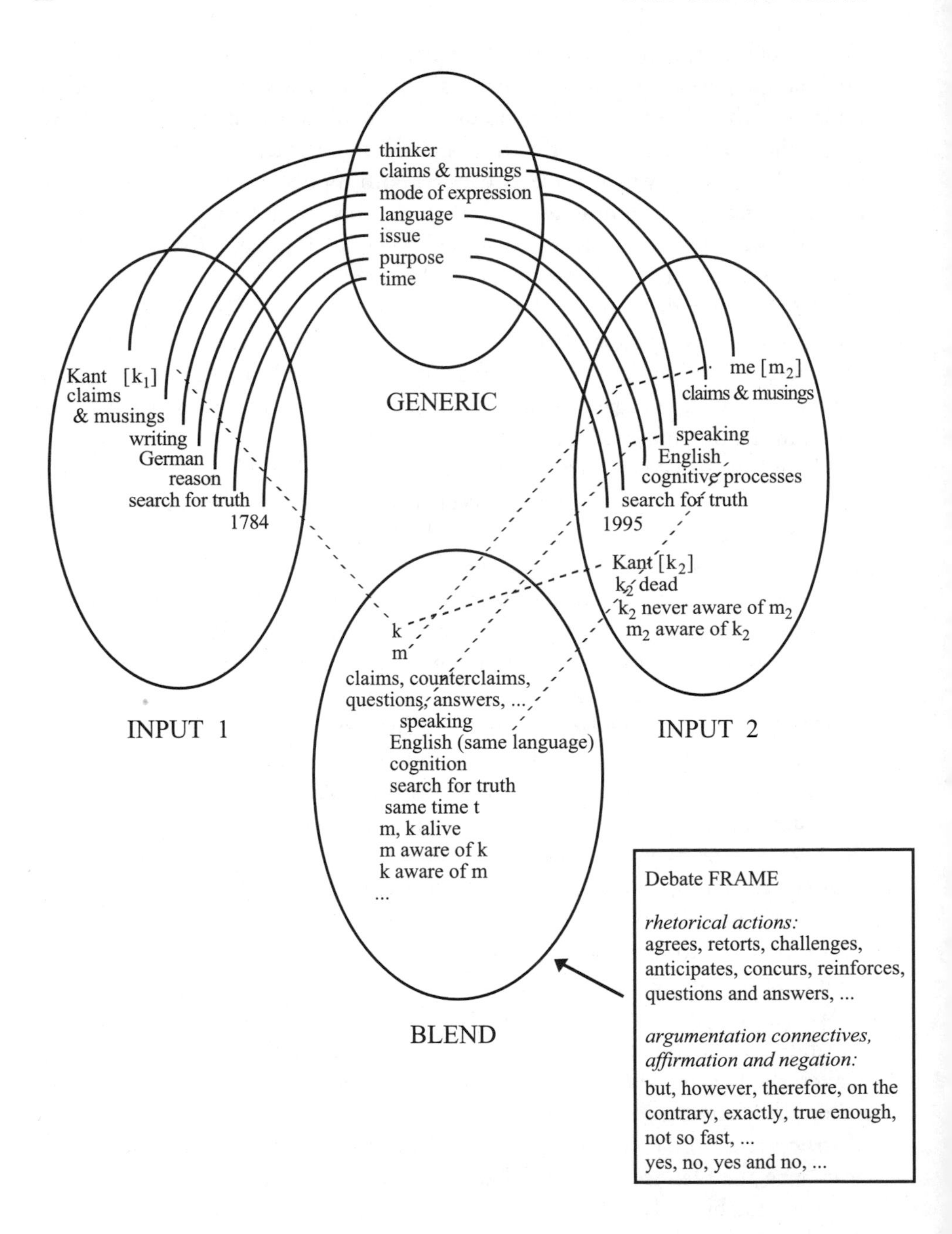

FIGURE 4.1 THE DEBATE WITH KANT NETWORK

REGATTA

The clipper ship *Northern Light* sailed in 1853 from San Francisco to Boston in 76 days, 8 hours. That time was still the fastest on record in 1993, when a modern catamaran, *Great American II,* set out on the same course. A few days before the catamaran reached Boston, observers were able to say:

> At this point, *Great American II* is 4.5 days ahead of *Northern Light.*

This expression frames the two boats as sailing on the same course during the same time period in 1993. It blends the event of 1853 and the event of 1993 into a single event. There is a cross-space mapping that links the two trajectories, the two boats, the two time periods, positions on the course, and so on. Selective projection to the blend brings in the two boats, the course, and their actual positions and times on the course, but not the 1853 date, the 1853 weather conditions, or the fact that the clipper ship was engaged in transporting cargo. The blend has rich emergent structure: Like the traveling monks, the boats are now in a position to be compared, so that one can be "ahead" of the other. The scenario of two boats moving toward the same goal on the same course and having departed from San Francisco on the same day fits into an obvious and familiar frame, that of a *race,* which is automatically added to the blend by pattern completion. That frame allows us to run the blend by imagining the two boats in competition. Just as in the Buddhist Monk case, elaboration of the blend is constrained by projection of locations and times from the inputs. The race frame in the blended space may be invoked more noticeably, as in:

> At this point, *Great American II* is barely maintaining a 4.5-day lead over *Northern Light.*

"Maintaining a lead" is an intentional part of a race. Although in reality the catamaran is sailing alone and the clipper's run took place 140 years before, the situation is described in terms of the blended space. No one is fooled: The clipper has not magically reappeared. The blend remains solidly linked to the inputs, and inferences from the blend can be projected back to the inputs: In particular, if we know that *Great American II* is 4.5 days ahead of *Northern Light* in the blend, then we know that the corresponding location of *Northern Light* in its input is less far along the course than the corresponding location of *Great American II* in its input, and we know that it takes 4.5 days of sailing (by one of the boats) to get from the first location to the second. Another noteworthy property of the *race* frame is its emotional content. Sailors in a race are driven by emotions linked to winning, leading, losing, gaining, and so forth. This emotional value can be projected to the *Great American II* input. The solitary run of *Great American II*, conceived as a race against the nineteenth-century clipper, can be experienced with corresponding emotions, which in turn can change the course of events. The crew of *Great American II* can draw courage and commitment from seeing themselves as engaged in a historic competition or, if daunted by *Northern Light's* performance, may be cowed into failure.

The exact words that prompted our interest in the "boat race," taken from a news report in *Latitude 38*, were these:

> As we went to press, Rich Wilson and Bill Biewenga were barely maintaining a 4.5-day lead over the ghost of the clipper *Northern Light*. . . .

The word "ghost" points explicitly to the blend. Its effect is to indicate how to build connections over three separate spaces: There are people in the temporally later input (in this case, the one with *Great American II*) who remember an element in the temporally earlier input (in this case, *Northern Light* in 1853), and that element is not in the temporally later input but it has a counterpart in the blend (here, the "ghost" ship). As we will see, this use of "ghost" to indicate how to build an integration network is quite conventional. It cannot be explained as just predicating a feature of a single element; it also tells us something important about the web of that element's connections. In Chapter 2, we saw a similar lexical phenomenon with the word "safe," which also could not be explained as just predicating a feature of a single element; instead, it told us something important about the web of connections across spaces in an integration network involving a counterfactual scenario of harm. In addition, the word "ghost" signals that events involving the "ghost" in the blend are constrained by the events involving its ancestor counterpart. In the Regatta example, then, the run of the ghost must be the same as that of its ancestor. It cannot go faster than it went in 1853, benefit from 1993 weather, collide with *Great American II*, and so on. So, again, "ghost" is telling us not about specific features of the events in the two spaces but only that those events have a particular cross-space relationship: This kind of ghost, at least, must copy its ancestor counterpart.

Here, too, nobody is fooled into confusing the blend with reality. There is no inference that the sailors actually saw a ghost ship or imagined one. The construction and operation of the blend are creative, but also conventional in the sense that readers know immediately and without conscious effort how to interpret the blend.

Because blending is neither deterministic nor compositional, there is more than one way to construct an acceptable blend, and this is confirmed in Regatta. The preferred reading seems to be that *4.5 days* is the difference between the time *N* it took *Great American II* to reach its current position (point A), and the time *N* + 4.5 it took *Northern Light* back in 1853 to reach point A. Under that interpretation, the boats' positions in the initial spaces (*1853, 1993*), as well as in the blend, are their positions (point A for *Great American II* and point B for *Northern Light*) after *N* days, where *N* days is the elapsed time in the *1993* space at the time of writing. In this reading, the 4.5 days are a time in the *1853* space—the time it took *Northern Light* to get from B to A. Another conceivable reading has this reversed, taking the elapsed time in the *1853* space and the 4.5 days in the current *1993* space. Under the latter interpretation, *Northern Light* got to point B' after *N* days, *Great American II* got to point A after *N* days, and it took *Great American II* 4.5 days to get from B' to A. In other words, *Northern Light* in the blend is just reaching the point that *Great American II* passed 4.5 days ago.

Still other readings are possible. Suppose *Great American II* is following a course different from its illustrious predecessor's, so that positions on the two journeys cannot be directly compared but experts can estimate, given current positions, how long it "should" take *Great American II* to reach Boston. Then, "4.5 days ahead" could mean that, given its current position, *Great American II* should arrive in Boston with an elapsed time of 76 days, 8 hours minus 4.5 days (i.e., in 71 days, 20 hours). This time, in the blended space of *1853* and the experts' hypothetical *1993* space, *Great American II* reaches Boston 4.5 days ahead of Northern Light.

The three readings of "4.5 days" involve integration networks that differ minimally in how the "4.5 days" is computed over the inputs and the blend. Each network yields a precise quantified evaluation of the truth conditions it imposes on the actual world. The blended space is different in each case, and its structure accounts for the corresponding difference of truth values in the interpretations. This is a nice point: Far from being fuzzy and fantastic, the blends allow an exact specification of truth values.

THE BYPASS

The Buddhist Monk, the Iron Lady, the Debate with Kant, and Regatta all invite us to blend input spaces that are separated in time. In the Buddhist Monk, the identical monks in each input are brought in as separate elements to the blend. Now consider a case in which identical elements in the inputs are brought into the blend and fused there, as illustrated in Figure 4.2. The advertisement in this figure is meant to persuade readers to help in the fight to raise standards in American schools. It shows three doctors in an operating room, who seem to be looking in the direction of the person reading the ad. The headline is a voice introducing the doctors to the reader, who is also the patient. It says, "Joey, Katie and Todd will be performing your bypass." The only odd thing about this scene is that Joey, Katie, and Todd are about seven years old. The body of the ad explains that doing anything sophisticated, like practicing medicine, requires sophisticated learning, but that America's kids are getting dumbed-down curricula. They won't understand chemistry or laser refraction or immunology, so they won't be good doctors, and the public, personified by the reader, will be at risk. Specifically, Joey, Katie, and Todd will operate on you, and you will probably die. Therefore, you should help get standards raised.

Joey, Katie, and Todd in one input space are children yet to be educated. In the other input space, they are doctors whose formal education lies behind them. The cross-space mapping connects child to adult. Both are projected to the blend, partially, and fused there. We also project to the blend the frame of surgery that comes from the space with the adults. The surgeons in the blend are seven-year-olds, a scenario that is naturally terrifying. We want our surgeons to be more competent than seven-year-olds, which leads us directly to the question of how to turn the children into competent adults. If we do nothing, the ad tells us, these children will grow up in a system that will not teach them what they need to learn to be doctors. But we have a choice: We can leap in now and provide the education that will make this integration network no longer terrifying, since it is the input with the *adult doctors* that finally counts. It is a question of how much distance, measured in terms of education, there will be between the blend and the space with the adult doctors. In the blend, the appearance of the doctors matches their competence: They have young bodies and they are incompetent. In the input with the adult doctors, they have adult bodies, and the question is, What kind of competence will they have? Doing nothing leaves the adult doctors with low competence. Improving education in our schools will give us a situation in which the adult doctors have high competence. As the ad says, "If we make changes now, we can prevent a lot of pain later on."

The ad is powerful because it uses blending brilliantly to bring together children as they are now with the frames they will inhabit much later on. The reader is also projected into the blend, as the patient. This makes a distant situation urgent by bringing it into the immediate present. In the inputs, the lethal consequences of the children's poor education emerge only much later, when you are old and need a coronary artery bypass. In the blend, you need a bypass now and the operation is just about to be performed. You might be apathetic about what will happen to you in twenty years, and you might be apathetic about the education of children who are not yours, but it is hard to be apathetic about the incompetence of doctors who are about to open up your chest. Interestingly, the emergent meaning in this blend includes fear and anxiety, which are not necessarily attached to the inputs.

FIGURE 4.2 THE BYPASS (SOURCE: EDUCATION EXCELLENCE PARTNERSHIP WEBSITE, 2001.

CHAPTER 4
ZOOM OUT

IS KANT CONSCIOUS?

The Debate with Kant is an elaborate blend that goes unnoticed until the blending is pointed out. Other blends, like the Buddhist Monk and perhaps the Image Club, are intuitively more obvious. Still others, such as "The beach is safe," do not feel intuitively like blends at all.

Questions:

- Why is there variation in the degree to which blends are intuitively visible to consciousness, and does this variation matter?
- Do blends become visible because more blending is going on?

Our answers:

Visibility and complexity are different issues. In the Buddhist Monk case, the blend is very visible because the reader is explicitly instructed not only to do the blending but also how to do it. In the Debate with Kant, the general debate blend is ready-made and available whenever we compare a present thinker to a past thinker. We don't need explicit direction to do the blending, and so we don't ordinarily notice it. But once pointed out, the *debate* frame is very salient and makes the blend intuitively accessible. The same holds for "*Great American II* is 4.5 days ahead of *Northern Light.*" This blend goes unnoticed until we use a word like "ghost" that is specialized to pick out counterparts in three separate spaces in a blending network. "Ghost" cues us both to do blending and to be aware of the *process*, whereas "safe," as in "The beach is safe," cues us only to do blending, not to be aware of the process.

GRICE'S ERROR

In the Debate with Kant, Regatta, the Buddhist Monk, and many other networks, the statement that prompts us to construct the blend, interpreted literally, is false, but it receives an interpretation that is considered to be true. For instance, it is considered to be true that the catamaran is 4.5 days ahead of the clipper ship and that Kant disagrees with the modern philosopher.

Question:

- Won't a simple application of Grice's Maxim of Quality take us from the false meaning of the expression to the true interpretation we give it?

Our answer:

This question is prompted by the well-known approach to irony, metaphor, and other kinds of "figurative" meaning offered by the philosopher Paul Grice. He proposed the "cooperative principle," according to which a speaker's behavior is assumed to be helpful, accurate, and relevant. We assume, for example, that a speaker obeys the Maxim of Quality: "Do not say what you believe to be false." Accordingly, when a speaker says "You are the cream in my coffee," we take it that the speaker knows the statement is literally false but intends for us to come up with a true interpretation, such as "I love you." Recognizing that the speaker knows (and knows that we know) that the statement is false prompts us to see that the expression is ironic, or metaphoric, or somehow figurative, and to look for an alternative meaning.

Yet Grice himself would have been the first to say that his proposal was not meant to explain how alternative meanings are actually reached. For us, on the contrary, this is the central issue. As we have shown, constructing the blended meaning is no simple task. It depends on complex available cognitive capacities. But putting the blend together does not require going through an intermediate stage of forming and rejecting a literal meaning. In fact, it is hardly plausible that, for example, we would form a literal interpretation in which a ghost of *Northern Light* appears on earth and runs a race with *Great American II*, judge that interpretation to be false, and then cast about for some other meaning. Grice's account intends to go only so far, but it is not right even as far as it goes.

Moreover, what gets called literal meaning is only a plausible default in minimally specified contexts. It is not clear that the notion "literal meaning" plays any privileged role in the on-line construction of meaning.

PROJECTION TO THE BLEND IS NONTRIVIAL

Projection to the blend often looks automatic and fairly trivial. A monk is a monk, a path is a path, an opinion is an opinion.

Question:

- What, if anything, is tricky about projection to the blend?

Our answer:

Let's first consider the Debate with Kant. Intuitively, it seems obvious that the ideas of the modern philosopher in the blend are the same as the ideas of the modern philosopher running the seminar. Intuitively, it seems equally obvious that the ideas of Kant in the blend are the same as the ideas of the historical Kant. But not so fast. The Kant of the blend understands the language and terminology of the modern philosopher, including such notions as "self-developing capacity" and "neuronal group selection," which may not have been concepts Kant ever thought of. He shares the modern philosopher's priorities and concerns and contexts. In consequence, the ideas of Kant in the blend cannot be indistinguishable from the ideas of the historical Kant, even when copied verbatim from one of his treatises. It is a general property of mental-space configurations that identity connections link elements across spaces without implying that they are identical or have the same features or properties. When someone says, "When I was six, I weighed fifty pounds," he prompts us to build an identity connector between him now and "him" when he was six, despite the manifest and pervasive differences. The Debate with Kant is an asymmetric integration network: It's the ideas of the modern philosopher that we care about, and they have priority for preservation. Since something must give if two people from different centuries are to debate, Kant's ideas are adjusted as we run the blend.

In the Bypass example, there is an identity connector between the doctors in the medical input and the doctors in the blend, but they could hardly be more unlike. The doctors in the blend have none of what we associate with doctors—adulthood, training, legal status—except for the garb and equipment and authority and duty to cut you up. In the Debate with Kant, we try to preserve the content of philosophical ideas in the obvious way, but in the Bypass, we make a different choice for projection, with spectacular results: The kid-doctors are having ideas that neither the real kids nor the real doctors would have. They are looking expectantly at the reader, who is also the patient; that is, they are looking at one person who is both their potential benefactor and their potential beneficiary (or victim).

AQUINAS, KANT, AND US

Imagination is often thought of as an optional ability, employed more for color than for rigor.

Questions:

- Isn't blending always parasitic on more basic ways of thinking?
- Isn't blending always optional?

Our answers:

The general debate blend is the basic way we have for conceiving of the relationship between two thinkers in different times. (Aquinas used it as the explicit rhetorical form of the *Summa Theologica*.) Developmentally, in fact, there has been no way to conceive of this relationship except through some sort of blend in which the two thinkers can interact. And such blends typically go unnoticed precisely because they are conventional. For example, in the Zoom Out sections of our chapters, we are using a blend in which readers—some of them previous thinkers with whom we agree or disagree—can ask us questions, and we can answer. Yet this form often goes unnoticed as a blend until we point it out. There are many creative variants of the general debate blend. Newspaper editorials often present a disagreement with an elected official as an implicit debate between the official and the writer. The electronic summary of the French national press *Le Petit Bouquet* routinely presents opposed editorials as a debate between the different writers who wrote their pieces during the night, each totally unaware of the other. For example, in summarizing the French diplomatic response to the taking of French hostages in Cambodia, *Le Petit Bouquet* said:

> *Or, rappelle* José Fort, "on ne traite jamais avec des tueurs, on les combat, on les isole, on les met hors d'état de nuire, mais c'est tout le contraire qui a été fait lors de ces accords". "*Non*, la France n'a rien à se reprocher, *lui rétorque* Alain Danjou dans le Courrier de l'Ouest. La France a encouragé le retour d'un régime plus attentif aux droits de l'homme et la négociation engagée rendait possible une issue heureuse."

The words in italics refer directly to a debate frame, but Fort and Danjou, in the original editorials, were not addressing each other or using words like "Non."

There are other cases where conventional expressions evoke the debate blend, as in "The bean burrito is California's *answer* to France's Croque Monsieur" and "Stag's Leap Chardonnay is California's *answer* to Corton-Charlemagne." Expressions like these bring in the idea of a competition between California and France, framed as a strange debate in which challenges and responses are food and wine.

SELECTIVE PROJECTION

We have used the phrase "selective projection" to talk about what comes into the blend from the inputs.

Question:

- "Selective" suggests that someone is carefully selecting something. Is that right?

Our answer:

When we see the final integration network with all of its connections in place, it may look like a *tour de force* showing the mastery of its creator in selecting just the right projections. In looking at the result, we miss much of what went into creating it. There is always extensive unconscious work in meaning construction, and blending is no different. We may make many parallel attempts to find suitable projections, with only the accepted ones appearing in the final network. As we project to a blend, we are also working on the entire network, and we may, for instance, recruit new structure to the inputs precisely to make it available for possible projection to the blend. After the fact, it looks as if that structure was in the inputs to begin with, and as if building a network is a sequence of discrete operations; but that appearance is misleading. Input formation, projection, completion, and elaboration all go on at the same time, and a lot of conceptual scaffolding goes up that we never see in the final result. Brains always do a lot of work that gets thrown away. In Chapter 16, we will discuss the governing principles by which we unconsciously decide what to keep and what to throw away.

But blending does not happen on-line from scratch. Cultures work hard to develop integration resources that can then be handed on with relative ease. As we just saw, the debate blend is a general template widely applicable to specific cases. With such templates, the general form of the projections and the completion are specified in advance and do not have to be invented anew. The creative part comes in running the blend for the specific case. In cultural practices, the culture may already have run a blend to a great level of specificity for specific inputs, so that the entire integration network is available, with all of its projections and elaborations. This is what we saw with complex numbers, but in such cases, it takes the culture centuries to achieve the desired blend. In hindsight, it looks as if the right projections were there all along and merely had to be "selected," but the discovery process was actually a painful, unpredictable, and messy series of trials and errors, with plenty of lucky and unlucky accidents thrown in. The optimality principles that govern individual brains operating on-line also apply to communities of brains working together in a distributed fashion to come up with suitable shared networks.

In addition to specific blends and blend templates, cultures offer us methods for setting up a blend. The cognitive anthropologist Edwin Hutchins points to the method of loci used by many cultures as a way of creating blends. He describes the method of loci as follows: "In order to remember a long sequence of ideas, one associates the ideas, in order, with a set of landmarks in the physical environment in which the items will have to be remembered. The method of loci sets up a simple trajectory of attention across a set of features . . . of the environment." The method of loci invites us to create a blended space in which items to be remembered are objects along a familiar path and, as such, take on the inevitable sequence that would be perceived by a traveler along it. This is a very old method in Western culture. Cicero discusses it at length for the case where the path goes through the speaker's own house.

Other blends, such as the Buddhist Monk, seem relatively unprepared for in the culture and thus highly imaginative, but they do not get picked up as useful for other important purposes and so remain curiosities rather than deep insights. But history is rife with curiosities that later became culturally important.

Although it can be hard to come up with good projections, once the culture has them they are easily learned, precisely because cultures have invented systems of form, such as language, whose purpose is to prompt for various kinds of imaginative work like selective projection. Finding a blend for which the culture has no previous recipe can involve considerable amounts of unconscious cognitive exploration, but using the formal prompts provided by culture to reconstruct such a blend once it has been found is much easier. The imaginative Achilles puts his formal armor to good use.

Five

CAUSE AND EFFECT

> Any mathematical argument, however complicated, must appear to me as a unique thing. I do not feel that I have understood it as long as I do not succeed in grasping it in one global idea.
>
> —*Jacques Hadamard*

THERE IS NOTHING MORE basic in human life than cause and effect. It has been a triumph of mathematics, science, and engineering to break up unified events into causal chains made up of much more elementary events, such that each is the effect of the previous one and the cause of the next. This kind of analysis gives us the feeling that we understand the complex event, having consciously reduced it to a set of basic events that are taken as self-evident. This type of explanation lends itself especially well to expression in form approaches because we can code the basic events and the causal relations symbolically, in such a way that manipulation of the formal objects correlates with the changes in the complex system they code.

For example, life and death are the biggest mysteries, and the moment of death is dramatic and punctual, but medical science breaks the event into complicated causal chains involving simple cellular and metabolic events based on heart rates, blood flow, oxygen delivery, neuronal activation, and so on. Similarly, a mathematical truth can seem striking and even mysterious, but the method of proof breaks the mystery down into a sequence of logical steps, each seemingly simple and obvious, and each leading to the subsequent step until the conclusion is reached. This was Aristotle's insight, which also drives modern computers.

GLOBAL INSIGHT

But as we saw at the end of Chapter 3, this step-by-step understanding is only one side of the coin. The breaking down of an event into a set of smaller events, each understood consciously and separately, can paradoxically give us a feeling of less understanding, because we feel we have not grasped the essential whole. It is a strength of human understanding to be able to do both, and our greatest assurance comes when we feel we understand the same event both ways. For a mathematician like Jacques Hadamard, understanding a complex mathematical result would demand both an intuitive apprehension of the whole and a detailed step-by-step proof. The former is commonly viewed as the essence of all creativity. The latter is also valued of course, but often viewed as a way to keep intuition in check and publicly share the results of a discovery.

Evolutionarily, our ancestors have typically been in situations where they needed to be able to recognize, in a flash, the potential integrated event: The falling tree has to be connected instantly to potential harm and to the right and wrong places to be standing when it falls; the roaring tiger needs to be connected—hypothetically and, one hopes, counterfactually—to its killing us. It is a commonplace in psychology that higher animals are evolutionarily equipped to recognize various facial expressions, postures, gestures, and voice tones as indicative of ensuing behavior. We even say that someone *looks* or *sounds* "violent" or "criminal" when what we mean is that their appearance makes us think they might cause violence or commit a criminal act. In short, it is evolutionarily advantageous to be able to unite cause and effect in our understanding. It's good to see potential effects in a cause, and it's good to see potential causes in an effect—it's probably a tiger that is behind that roar. We recoil from the scene of death, we like the scene of eating. When we recoil from the rattle of the rattlesnake, it is because we are integrating it with its potential effect. When we are attracted to luscious fruit, it is because we are integrating it with its potential effect. To be sure, it is not impossible that some small fraction of this ability was genetically assimilated, but higher animals can do such integration very broadly, even in novel cultural domains.

In fact, the canonical case of stimulus-response conditioning—Pavlov's dog, who, once he had been presented several times with a ringing bell followed by a piece of meat, salivated for the bell alone—shows how common it is to blend cause and effect. Evolution set up the part involving salivation for the meat, but not salivation for the bell, which is purely learned. The dog looks stupid but is in a way quite smart. It takes sophisticated cognitive operations to become conditioned in this way. Compressing cause and effect is, on its own, no sign of stupidity. It would be stupid only if the dog entirely lost the distinction between cause and effect and ate the bell.

It is not trivial to bring cause and effect together. They have to be brought together in one mental space, in the right way, while being kept distinct in other spaces. Appropriate blending of cause and effect is a matter of having your cake and eating it too.

Pavlov's dog, who has learned to blend the food with the ringing bell, and the fleeing antelope, who has a genetic disposition to blend death and the tiger's roar, have their analogues in the highest intellectual realms. Consider again the comment of the great mathematician Jacques Hadamard: "Any mathematical argument, however complicated, must appear to me as a unique thing. I do not feel that I have understood it as long as I do not succeed in grasping it in one global idea." Such a global understanding is a matter of blending cause and effect. Each step of a proof is the effect of the previous steps. All of these causes and effects can be followed as a sequence, but achieving a global understanding requires blending them all into one space. This is easily seen in an example known since antiquity: the visual apprehension of the Pythagorean theorem. In Figure 5.1, we see one square on the left that consists of a smaller square and four identical right triangles. The smaller square is obviously the square on the hypotenuse (h) of the triangle. The square on the right consists of the same four identical right triangles and now two other squares, which are obviously the squares on the legs (a and b) of the triangle. Obviously, the total square on the left and the total square on the right have the same area. When we subtract the four triangles from each of them, we are left with the square on the hypotenuse (h^2) on the left and the squares on the two legs ($a^2 + b^2$) on the right. So they must be equal; that is, $h^2 = a^2 + b^2$.

This could also be proved algebraically in steps, but most people, in keeping with Hadamard's intuition that a feeling of understanding requires a global grasp, find the visual representation more compelling. Cause and effect are brought together. The cause of this theorem, the reason why it is true, lies ultimately in geometric properties, some of which are accurately conveyed by the visual representation. The effect of these properties—the theorem itself—is not directly perceptible if all one sees is the right triangle. But in the presence of the creative visual representation, we see the theorem as self-evident, which is to say, we see the effect directly in the cause. Presenting the effect directly in the cause is a matter of finding the right representation, which in itself is highly creative. Once found, it can prompt and guide us to corollary creativity and global understanding. In the *Meno*, Socrates prompts a boy innocent of geometry to discover a theorem merely by asking him a few questions. The boy's discovery is supposed to be proof that, since Socrates did not actually assert anything to the boy, he could not have learned the knowledge from Socrates and so must already have known it in some way and have been merely reminded of it by Socrates' questioning. But this is not at all what it shows. Rather, it shows that someone like Socrates who already has the knowledge can use very efficient representations and expressions to prompt and guide someone else to develop it relatively quickly. In this case, the conclusion is the effect and the steps of the deduction are the cause. The boy sees the effect as carried in the cause and has global insight into their blend.

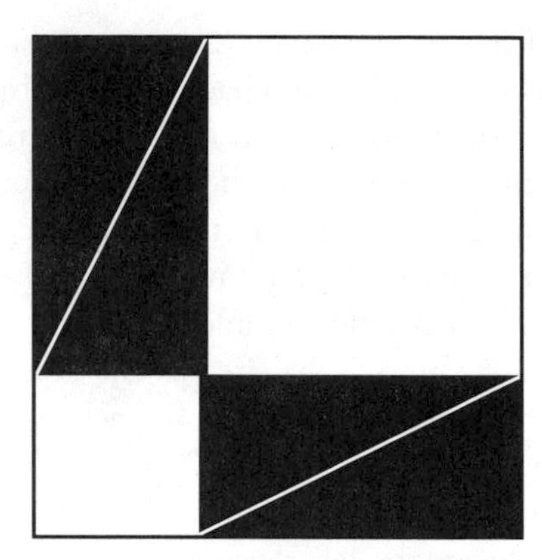

FIGURE 5.1 VISUAL APPREHENSION OF THE PYTHAGOREAN THEOREM

Of course, presentation of the effect in the cause is exactly what blending the Buddhist Monk's two journeys provides. The cause is the dynamics of the two input journeys; the effect is the existence of a location on the path they occupy at the same time of day. In the blend, the location and the encounter are presented directly as part of the causal dynamics of motion.

Finding proper representations is recognized in cognitive science as crucial for problem solving of all types. We are now beginning to see how imaginative blending that integrates cause and effect can provide illuminating representations. These representations typically take the form of dynamic scenarios that one can run.

PERCEPTION AND SENSATION

The integration of cause and effect is the central feature of perception. As we discussed in Chapter 2, the perception of a single entity, such as a cup, is an imaginative feat still very poorly understood by neurobiologists. The perception available to consciousness is the *effect* of complicated interactions between the brain and its environment. But we integrate that effect with its causes to create emergent meaning: the existence of a *cause*—namely, the cup—that directly presents its *effect*—namely, its unity, color, shape, weight, and so on. As a consequence, the effect is now in its cause: the color, shape, and weight are now intrinsically, primitively, and objectively in "the cup."

In perception, at the level of consciousness, we usually apprehend only the blend of cause and effect. We cannot fail to perform this blend, and we cannot see beyond it. Consequently, this blend seems to us to be the most bedrock reality. There are a few ways to make the distinction between cause and effect minimally, and sometimes upsettingly, perceptible. Brain damage, psychoactive drugs, and certain neurobiological syndromes can cause breakdown of these integrations and consequent bizarre perceptions. But for the most part, when we are functioning normally, consciousness cannot see the rest of the blending network. Accordingly, we are likely to think that the perception of a spot in the visual field is caused by the light coming to our eye from it, but that is false. The amount of light reflected to our eye from a black letter in a newspaper headline in sunlight is about twice the amount of light coming from the white paper in a dimly lit office, but we still see the letter as black and the paper as white. We are likely to think that we see the letter as black because it is invariantly "black," but that perception is an integration of cause and effect. A large spot of uniform illumination, such as a purple disk on a white wall, seems uniformly vivid, but ganglion cells are in fact active for only its border, not its interior, making the vividness of the interior, no matter how apparently real, a downstream cerebral computation. We think the purpleness of the center is directly causing our perception of it as purple, but our perception of it as purple is an integration of cause and effect. Similarly, when two lights flash in succession, provided they are the right distance apart and the flashes are timed correctly, we see, and cannot help seeing no matter how hard we try, a beam of light sweep from one to the other. If the first is one color and the second is another, we see the beam switch colors at the midpoint, before the beam has reached the second point, even though the light from the second does not exist until the second flash. Our perception of the beam, and our feeling that we are watching the beam in real time, are both effects; so here, too, we integrate effects and causes, creating an objective sweeping beam of light that changes color before the second flash. The effects seem inherent in the cause. In this case, unusually, by covering up one light or the other, we can get evidence that our perception is an integration.

Sensory projection, a universal feature of our perceptual life, also arises from integration. A sensation of pain in an ankle is constructed in our central nervous system, but we "feel" the pain, of course, as located exclusively in the ankle. We have conceptually integrated part of the cause with the mental effect, to create a "painful ankle," so that cause and effect are now located together in our mental conception of the ankle. The neurobiological effects that constitute the "pain" are distributed throughout the central nervous system, but the integrated cause and effect have only the single, undistributed location of "the ankle." Obviously, we do cause-effect integration for the sensation of pain because it makes sense: The anatomical ankle really does need attention and care. Phantom-limb phenomena famously show the same kinds of integration, except that the ankle is actually absent in such cases. An amputee can feel not only pain in the missing ankle but also that he has the ankle as a result. He may reach "absent-mindedly" down to rub the ankle that is in fact not there. What has happened is that neurons that used to connect to the ankle are still firing, causing the sensation of pain and inducing the usual integration of cause and effect. The pressure for the brain to achieve integration of cause and effect is so great that it will use conceptual blending to give an emergent structure not only a localized pain but also a phantom body part in which that pain takes place. The amputee cannot control this sensation even though it is outrageously opposed to everything he believes. Integration of sensory cause and effect is in this case badly misleading, but the integrating brain charges ahead unimpeded.

HUMAN RITUALS

The integration of cause and effect is often the central motivation of ritual, and here we can perceive consciously the difference between cause and effect even as we see their integration. When the bride at a wedding reception throws the bouquet into a crowd of single women, whoever catches the bride's bouquet is the next to be married. Part of the effect—carrying the bridal bouquet—becomes part of the cause—the event of catching. An effect distant in time—namely, the next marriage—becomes a feature of the person who has caught the bouquet: She is now the next in line. The ritual also creates the cause of the next marriage: There is now a woman who has caught the bouquet and is therefore next in line. A parallel ritual is the groom's throwing of the garter. And it is often stipulated as part of the combined ritual that the single man who caught the garter and the single woman who caught the bouquet might be considered, at least for symbolic purposes, a possible match. The ritual originally contains two spaces: The first is the present wedding, with roles for bride and groom filled by two specific people; the second is the unavoidable subsequent wedding, with unfilled roles for bride and groom. Catching the bouquet and the garter fills these unfilled roles. It creates a symbolic fiction representative of that inevitable future wedding. The single man and the single woman become counterparts of the future bride and groom (themselves, by identity) and, therefore, counterparts of the present bride and groom by analogy. In the integration network, these complex mappings launch a rich blended space in which the single woman and man take on aspects of the present bride and groom. In the blend, causes and effects are brought together: The single woman and man as causes of the future wedding are the imaginary bride and groom as effects, at least for the symbolic purposes of the ritual. The bouquet and garter, which cause them to be the future bride and groom, are also the symbolic equipment needed for the virtual courtship and wedding.

Eve Sweetser discusses a very rich public ritual, the Baby's Ascent, in which the newborn baby is carried up the stairs of the parents' house. This ritual, meant to promote the child's chances of rising in life, gets its meaning by virtue of an integration network. In one space, the baby is going to be carried up the stairs. In the other space, which is schematic, someone is going to live a life of some sort. This schematic space is already structured so that living a life is moving along a path in some manner, and good fortune is up and misfortune is down. The stair ritual has been chosen because it has many elements that can map naturally to the schematic motion in the space of life. In the cross-space mapping, the path up the stairs corresponds to the "course" of life, the baby is the person who will live the life, the manner of motion up the stairs corresponds to how the person "goes through" life, and so on. The main parts of the "life course" space are projected into the blend, so that an easy ascent of the stairs determines the child's easy rise in life. The goal is of course to climb the stairs smoothly all the way to the top. Running the blend is now imbued with deep symbolic meaning, because in the blend whatever happens is the baby's future life. This has interesting effects: While it would be insignificant in actually carrying a baby up the stairs if one stumbled slightly on the third step, this manner of motion takes on enormous significance in the blend, and is quite different from stumbling on, say, the last step.

The ritual of the baby and the stairs integrates in a single, very brief, concrete event the complicated and extended causal patterns of a human life. In the blend, reaching the top of the stairs is the desired effect, a successful life. But reaching the top of the stairs is also the ritual cause of the successful life because the ritual is performed to bring about success in life. The blend presents the effect directly as contained in its cause. This ritual is fairly representative of rituals in general, suggesting that such fundamental and elaborate human activities, unparalleled in the animal world, make use of the operation of conceptual blending as their basic instrument of imaginative invention.

PERSUASION AND REVELATION

Like complex numbers in mathematics, the ritual of the baby's ascent is an entrenched integration network developed by a culture and taught to its young. But these cause-effect blends can also be invented and communicated on the fly. The state of California, for instance, recently undertook a billboard campaign to persuade people to stop smoking. One such display presents the stark, virile profile of a cowboy above the words "WARNING: SMOKING CAUSES IMPOTENCE." It turns out that his cigarette is drooping. Understanding this ad requires an integration network with several input spaces. A Virile Cowboy space has the well-known smoking cowboy as presented in cigarette ads. He enjoys smoking, and smoking is part of his social independence and sexual authority. A Smoking Man space has a schematic man with a continued history of smoking. In both of these input spaces, smoking a cigarette symbolizes "being a smoker." And an Impotent Man space has a man with an incomplete erection. First, under instruction by the ad, we blend the Smoking Man space and the Impotent Man space to create a blend in which being a smoker causes impotence. In this Impotent Smoker blend, the cigarette is the cause of the unfortunate posture of the bodily organ. The Impotent Smoker now becomes an input to a second blend, whose other input is the original Virile Cowboy space. In this second network, the Virile Cowboy's cigarette, the Impotent Smoker's cigarette, and the posture of the Impotent Smoker's penis all project to the same element, yielding a cigarette that droops. The Virile Cowboy and the Impotent Smoker are cross-space counterparts who are fused in the blend, giving an Impotent Smoking Cowboy. In the blend, cause and effect have been merged and present each other directly. To recognize one is to recognize the other. The blend is where the truth resides. It conflicts with the implications of the original Virile Cowboy input space, and that is the point: We are to discount the Virile Cowboy space as false and instead see things as the blend presents them.

Ads like this are not thought of as serious in the way that human rituals, global mathematical insights, and human perception are. Yet all these human activities are using the same basic mental operation, conceptual blending, to achieve global insight through effective integration of cause and effect in imaginative ways. And in any case, the ad itself is more serious than we might think. It is the culmination of inventive work by many people involved in a multimillion-dollar campaign directed against rich and powerful industries. Its quite serious purpose is to bring about a major shift in people's social habits and conceptions and thereby eliminate a range of major health problems. The ad is a matter of life and death.

These cause-effect blends are also a matter of spiritual life and death. No work of literature is more serious than Dante's *Divine Comedy*, which offers comprehensive instruction in the most fundamental moral issues of human existence. In the first section, the *Inferno*, we are given a tour of hell, its geography, its principles, and its characters so that we may understand the varieties and nature of sin. One of the most famous of Dante's sinners is Bertran de Born, who, when we meet him, is carrying his own head in his hand like a lantern. Dante, en route through hell, has a conversation with the head. While living, Bertran de Born had created strife between the English king and his son, so his place in hell is with the "stirrers-up of strife":

> *Perch'io parti' così giunte persone,*
> *partito porto il meo cerebro, lasso!*
> *dal suo principio ch'è in questo troncone.*
> *Così s'osserva in me lo contrapasso.*
>
> *Because I parted people so joined,*
> *I carry my brain, alas, separated*
> *from its root, which is in this trunk.*
> *Thus is to be seen in me the retribution.*

One input space has Bertran de Born as he was in life, the cause of the division between the king and his son. The other input space is the schematic frame of someone who divides a physical object into two parts. According to a standard metaphoric connection in which social "division" corresponds metaphorically to physical "division," Bertran is the counterpart of the divider, king and son are as a unit the counterpart of the single unified physical object, Bertran's sinful separating of king and son is the counterpart of dividing the single unified physical object, and the alienated king and son are the counterparts of the two halves of the divided object. In the blend, Bertran is blended with the divided physical object, thus fusing the cause (Bertran) from one space with the counterpart effect (divided physical object) from the other. The punishment fits the crime. Bertran's body, a physical object, fits naturally in the blend into the schema of physical division as a man who gets decapitated. The imaginative connections achieved in the network work together to give us in the blend an image that integrates cause and effect. This is in general Dante's goal: to lead us to see the consequences of a sin in the sin itself. On earth, we may have trouble seeing correctly into the nature of our soul and its relation to God. In the *Inferno*, the proper inferences are presented directly.

CHAPTER 5
ZOOM OUT

LIVING IN THE BLEND

As long as our perceptual and sensory systems are working properly, it is almost impossible for consciousness to see outside the blend of cause and effect. In other cases, such as rituals and ads, it is easier to separate cause and effect consciously: Catching a bouquet is different from getting married, going up stairs is different from leading a successful life, smoking is different from impotence. In activities involving high-level abstract thinking, such as mathematics, we need both a step-by-step separation of cause and effect and a global understanding in which they are integrated. How thoroughly our conscious apprehension is limited to the blend depends on the kind of activity that blending serves. In the case of sensation and perception, our conscious experience comes entirely from the blend—we "live in the blend," so to speak. In other activities, conscious apprehension has more leeway to go back and forth, to "live in the full integration network."

Question:

- Why would it ever be a good thing to be condemned to live only in the blend?

Our answer:

As we discussed at the opening of this chapter, we live in the blend for activities that are crucial to survival—perception, sensation, arousal, immediate reaction to basic environmental threats. In the face of such threats, global and immediate insight is the priority, and there is little survival value in checking step by step how that global insight is achieved. Thus we evolved to be conscious of only the blend. For obvious reasons, we also live in the blend for basic mathematical and physical reasoning: We see globally and instantly that three apples are more than one and that the limb falling from the tree is going to hit us unless we jump out of the way. These kinds of global insights are most impressively displayed in sports events—for instance, when the centerfielder runs to catch a fly ball. In contrast, the development of sciences leads consciousness to live in the entire network: Global and creative insights require the blended space. Proofs, analysis, and verification and communication of theories require explicit unpacking of cause and effect.

It is often desirable to decrease the extent to which consciousness lives in the full network. The acquisition of expertise is in many respects the achievement of successful integration networks in which living in the blend gives you the desired effects with no conscious attention to the other spaces. Recall that the Skiing Waiter begins with two quite separate spaces and an instruction to try to blend them. Expert performance consists in having acquired the blended pattern in such a way that it is felt consciously as primitive. The child learning letters maintains a forceful distinction between seeing a shape and seeing a letter: The first causes the second. But very soon, the child cannot distinguish the cause from the effect: She can no longer see the shape without seeing the letter.

HAVING YOUR CAKE AND EATING IT TOO

We have now seen a wide range of cases in which there is a need to be able to maintain simultaneously what look like contradictory representations. In the ritual of the baby's ascent, a full life in the blend passes in the space of time needed to climb the stairs, even as we remain aware of the huge difference between a lifespan and an ascent of stairs, as they are represented in the input spaces. And in Bertran de Born's case, we grasp the full force of Dante's instruction in the blend, where Bertran is decapitated, even though in the inputs he never loses his head.

Question:

- Isn't holding contradictory representations irrational?

Our answer:

No, obviously, since we have now seen many reasons why the multiple representations and the blends are efficient at many levels of cognition. Usually, each space in a network is internally consistent, even though the spaces themselves may contradict each other, as in counterfactuals. In some cases, blending may create blends that are *internally* contradictory, because that internal contradiction is significant for the rest of the network. In *reductio ad absurdum* blends, for example, the self-contradiction of the blended space shows that the new assumption, which is in one input, does not have a legitimate place in the mathematical system, which is comprehensively available for activation in the other input.

RITUAL PROJECTIONS

Rituals are performed carefully and accurately, in an atmosphere of seriousness and even solemnity. It is not only the core events of a ritual that count—such as getting the baby up the stairs. "Minor" aspects of the performance may be crucial.

Questions:

- Why are people such sticklers about rituals?
- Why do rituals have to be performed just so?
- If rituals are complex blends, what is the role of the "minor" aspects of the performance in such blends?

Our answers:

Rituals integrate cause and effect, so that any aspect of the performance can be experienced as simultaneously a cause and its effect in both the blend and the future life. Running the blend therefore assumes deep significance. Scripting it is one thing, running it is quite another, and the performance may contain events that are not in the script. This is an example of what we have called emergent structure in blends. An unexpected slight stumbling on the third step, obviously inconsequential in the input space of climbing the stairs, can nevertheless cause observers of the ritual to catch their collective breath in apprehension. An abnormally quick ascent might be experienced as a successful but too-brief life and, therefore, as a bad performance of the ritual. Since lives are rich and so are manners of performing ritual acts, the possibilities for projection from the ritual to the life are open-ended, and the conservative script of the ritual is meant to shut down the possibilities of unintended projections that could mean harm for the infant. The ritual is meant to give the best possible life, so any departure from the norm can count as a detriment. Having the integration network does not require belief in its efficacy, but having it *is* sufficient to activate emotions in the blend. Independent of any belief, one does not want the ritual to go wrong, because for both believer and nonbeliever its going wrong calls up real emotions that are part of an important social situation.

Although many participants may lack belief in the efficacy of a ritual, they have a shared interest in achieving optimal correspondence between the performance of the ritual and the reality it is meant to capture. The performance can label the participants, and the labels can have social effects over time, making the performance ultimately self-fulfilling. The blend, for social reasons, can create its own efficacy. For instance, someone who has undergone the ritual of being made a knight, including praying on his knees in full armor all night and then rising to leap onto his horse, might be expected to behave dutifully, piously, and courageously, and this expectation can call forth from him duty, piety, and courage. The members of the community will also be biased to judge even his ambiguous actions as dutiful, pious, and courageous. The label is a global insight integrating vast ranges of cause and effect. The knight's good qualities cause him to become a knight, but his being a knight causes him to have those qualities.

An extreme case of this integration is voodoo death, where belief in the efficacy of the network is basic for both the community and the victim; indeed, the effect of these beliefs is that the social body and the physiological body conspire to make the death happen. In the blend, the victim is dead, and so the body does what it can to match that blend.

Some rituals are trials: Elements of the blend are left undecided in the script, to be determined by acting out the ritual. Trial by water is a ritual in which guilt or innocence is left unscripted. But the ritual of the baby and the stairs is not meant to be a trial. The script for the blend includes the desired outcome, and the only question is whether the performance will match the script. The script is intentionally made easy to perform, so that the performance will match the script closely.

Clearly, then, a ritual blend needs additional specifications in order to count as a trial, a prediction of the future, a virtual enactment of what we want to happen or not happen, a glimpse of another world, or a technique of discovering history. Baptizing a baby by immersion is meant to make it eligible for spiritual salvation, not to protect it against drowning or to make it good at having deep thoughts. But even though rituals are constrained by social purposes, new meaning can arise inadvertently, as in the case of the stumbling on the stairs. This is a remarkable property of human cognition. An artificial intelligence program runs strictly according to its script, but human rituals, which seem highly scripted, depend on blends, and emergent meaning is always part of the blending process. The ritual takes place embedded in the full richness of human life, and the principles of emergent meaning can always recruit from that richness.

IMPLICIT COUNTERFACTUAL SPACES

Throughout this chapter, there have been alternative notions hanging in the background: The baby will have a good life or a bad life, the cowboy is virile or impotent, someone sins and is punished or does not sin and is rewarded.

Question:

- Where are these alternative notions in the theory of blending?

Our answer:

This is a subtle and deep question. It leads to a large theoretical principle of conceptualization—namely, that conceptualization always has counterfactuality available and typically uses it as a basic resource. As we discussed earlier, to understand even so simple a sentence as "There is no milk in the refrigerator" requires constructing a network with counterfactual spaces: The desired space has the milk in the refrigerator, the other input space corresponds to the present situation, and the blend has a counterpart for the milk but the disanalogy between the inputs corresponds to the property *absent* in the blend. The blend is counterfactual with respect to the desired input. Language offers many expressions by which we prompt listeners to construct counterfactuality:

> I do have a car; *otherwise*, I would ride by bike.
> I will *not* listen.
> You'll be gone tomorrow? *Too bad!* You *could have* come to my party.
> He *missed* the ball.

Far from being impeded by this pervasive feature of incompatibility in conceptualization, blending draws some of its power from being able to operate over incompatible spaces. In the Smoking Cowboy network, the Impotent Smoker space brings with it implicitly a contrasting mental space in which a nonsmoker is a stud. To the extent that the Virile Nonsmoker space is activated, it will clash with the space of the Virile Smoking Cowboy. And in fact it is strongly activated, as the approved and desired scenario.

The Impotent Smoking Cowboy blend has two inputs: the Virile Smoking Cowboy and the Impotent Smoker. The first is something to emulate, the second something to avoid. Thus they clash on a central element. Blending works effectively over this clash to produce the blended space with the drooping cigarette. In that space, the Cowboy retains all of his superficially attractive facial and sartorial features, but his true essence is revealed via the drooping cigarette. This blended Cowboy prompts the projection of impotence back to the original Virile Cowboy input, thus leading to crucial revision of its original implications.

Six

VITAL RELATIONS AND THEIR COMPRESSIONS

> To see a World in a Grain of Sand
> And Heaven in a Wild Flower,
> Hold Infinity in the palm of your hand
> And Eternity in an hour.
>
> —*William Blake*

THIS BOOK STARTED WITH the ambitious claim that we are now entering an age in which the key intellectual goal is not to celebrate the imagination but to make a science of it. Imagination is at work, sometimes invisibly, in even the most mundane construction of meaning, and its fundamental cognitive operations are the same across radically different phenomena, from the apparently most creative to the most commonplace. These operations are characteristic of the human species. Though taken for granted by human beings, they are extraordinary by any other standard.

Conceptual integration is at the heart of imagination. It connects input spaces, projects selectively to a blended space, and develops emergent structure through composition, completion, and elaboration in the blend. This fundamental cognitive operation has not previously been studied. What would it mean to study this operation? Is it enough to recognize the phenomenon and describe it broadly? Should this book end here? What is left to do?

THE ANALOGY WITH CHEMISTRY

That atoms combine to form molecules by sharing electrons is surely the fundamental principle of chemistry. But when we consider the world of molecules familiar to us—water, salt, citric acid, serotonin, glucose, granite—this world is not explained by that principle. Knowing the principle will not provide a categorization of molecules, or tell you which molecules are possible and which are not. It will not predict which forms are more likely than others, characterize the stability of an individual molecule, or describe how to go about making new molecules for medical purposes. The principle is true and basic, but it does not solve any particular problem of chemistry. Rather, it opens a way to formulate certain problems and think about possible solutions. Over many years, chemistry and its later incarnations of organic chemistry and biochemistry developed a very careful and elaborate taxonomy of atoms. It studied the principles of their distinctions, which atoms would combine with others and in what proportions, and the complicated nature of the chemical bond.

A science of blending needs to follow the same track. We turn now to the study of recognizable types of integration networks, principles and pressures that guide the formation of integration networks, constraints and biases involved in making connections, and, further, subprinciples covering such topics as emergent meaning in the blend and emergent meaning in the network.

The marvelous systematic products of chemistry—acids and bases, colors from titration, metabolism—are not foretold in the principle that atoms combine to make molecules. And the marvelous systematic products of blending—"living in the blend" during perception, reasoning through counterfactual spaces, the emergence of mathematical concepts like complex numbers, the design of computer interfaces, the evolution of language forms that prompt for specific patterns of blending—are not foretold in the principle that input spaces blend to make new spaces with emergent meaning. Just as explaining the marvelous systematic products of chemistry requires pursuing the details of what happens when atoms combine, so explaining the marvelous systematic products of blending requires pursuing the details of what happens when spaces blend. We hope to show that the study of blending, like chemistry, has the potential to change our view of the world, subsuming many disparate phenomena for which we had partial descriptions, connecting them, and branching out to discover new phenomena we had not seen. Many phenomena for which we had partial descriptions—categorization, mathematical invention, metaphor, analogy, grammar, counterfactual thinking, event integration, various kinds of learning and artistic creation, global insight integrating vital relations like cause and effect—are products of the same well-defined imaginative operation. The unity of chemical processes is the scientific basis for explaining the diversity of their products, the different chemicals, because the diversity flows from the many ways in which the processes play out. This diversity is not explained by listing surface symptoms: Water is clear, vinegar tastes "acidic," arsenic is toxic, iron rusts but gold does not. Although these surface symptoms are crucial to our lives and much of the motivation for true chemistry lies in trying to explain them, the science of chemistry cannot take them as explanations themselves or even as primitives of scientific explanation. Similarly, the products of blending are different in crucial ways. Just as in the case of chemistry, the unity of the operation of blending is the scientific basis for explaining the astounding diversity of its products, because the diversity flows from the many ways in which the basic principles of blending play out.

THE ANALOGY WITH EVOLUTION

The simple and general principle of biological evolution is that a system whose elements vary and descend with inheritance under selection pressures produces great biological variety. Richard Dawkins writes, "Never were so many facts explained by so few assumptions." But Dawkins is overstating the theory's simplicity. The general principle by itself explains nothing. Explaining any of the interesting facts of biological evolution requires a large, detailed, and systematic theory about how that general principle plays out in different niches, along paths of development, under different boundary conditions. The fascinating parts of the inquiry lie in connecting the real world around us to this mysterious general process of mutation, descent, and selection. We want to know where birds came from, how the panda got its thumb, why large fierce animals are rare, why the American pronghorn still runs from the ghosts of its predators. We want to know how the evolutionary river flowed out of Eden. This fascinating pursuit of knowledge is a matter of discovering what counts as an element, a variance, a descent, an inheritance, or a selection pressure, and developing theories of how these parts of the evolutionary process interact.

The theory of conceptual integration faces the same hard challenge. The simple and general network model of conceptual integration presented in Chapter 3 covers a great deal of ground and, on its face, seems to fit great ranges of human thought and language. It is a powerful scheme that covers many conceptual facts. But by itself it explains nothing.

Explaining the interesting facts of conceptual evolution requires a large, detailed, and systematic investigation of the ways in which the network model plays out in different conceptual niches, along paths of conceptual development, with different purposes, in different contexts, with different affordances, between different people with different hopes, beliefs, and desires. As with evolutionary theory, the fascinating parts of the inquiry lie in connecting the real mental world around us to this mysterious general network model. We want to know where scientific discoveries come from, how we think about what might have been, how we can make sense of the American pronghorn's flight from the ghosts of predators past. We want to know why we, alone among all biological species, have rich, emotion-laden rituals. We want to know the bends and turns in the conceptual rivers of our mental lives. We want to know the paths imagination took and explore the most challenging enigma in the universe: the way we think!

COMPRESSION OF VITAL RELATIONS

We do not establish mental spaces, connections between them, and blended spaces for no reason. We do this because it gives us global insight, human-scale understanding, and new meaning. It makes us both efficient and creative. One of the most important aspects of our efficiency, insight, and creativity is the *compression* achieved through blending. In the previous chapter, we saw spectacular compressions of cause and effect in the Impotent Smoking Cowboy, the Baby's Ascent, and Bertran de Born.

Certain conceptual relations, such as cause-effect, show up again and again in compression under blending. We call these all-important conceptual relations "vital relations."

Let us take another look at the Bypass. We have talked generally about cross-space links between the inputs. The finer-grained structure of these links is extremely interesting. It includes links from cause to effect, links through time and through space, links through change, and links through identity. The input with the children in school and the quality of their education is causal for the input with doctors of a certain level of competence. This is a Cause-Effect link between the inputs. There is an interval of at least a couple of decades between the children and the doctors. This is a Time link between the inputs. There is a displacement between the physical space of the schoolroom in one input and the physical space of the operating room in the other. This is a Space link, where in this instance we mean physical space. There is a counterpart link between the children at one stage of life and the doctors later. This is an Identity link. And, finally, there is a transformation of the children into the doctors. This is a Change link.

Blending plays marvelous and imaginative tricks with these links. Look once more at the blended space in the Bypass. Every one of these "outer-space" links between the inputs to the conceptual integration network has a compressed counterpart in the blend! There is still cause-effect in the blend, but now the children must learn all at one shot. There is still a time interval between now and the surgery, but it has been compressed from over twenty years into the few minutes between the time on the clock and the time of the surgery. In the blend, the schoolroom is the operating room. This is space compression. In the blend, the children are the doctors. The "outer-space" link of personal Identity running over thirty years between people whose appearance, experience, and belief are very different is compressed into what we call "uniqueness" in the blended space. The "outer-space" protracted change of the youngsters into employed adults is also compressed in the blend into uniqueness.

The Bypass example shows very vividly how links between the input mental spaces—what we call "outer-space" links—can be compressed into relations inside the blend—what we call "inner-space" relations. Cause-Effect and Time are scaled down to tighter Cause-Effect and briefer Time in the blend. Incompatible physical spaces are compressed into the same physical space. Identity and Change are compressed into Uniqueness. This example gives us in miniature a demonstration of the compression of "outer-space" links into "inner-space" relations under blending. And it gives us the beginning of a list of "vital relations" that show up repeatedly in compression under blending. Cause-Effect, Time, Space, Identity, Change, and Uniqueness are Vital Relations that we see in the Bypass.

TYPES AND SUBTYPES OF VITAL RELATIONS

Change

Change is a very general Vital Relation, connecting one element to another and suites of elements to other suites. In the Bypass, the children change into adults. Conceptually, a sapling and the tree it grows into set up two mental spaces connected by Change. Age changes a person, translation changes a text, "Americanizing" something foreign changes it into something suitable for Americans. Mental spaces are dynamic, so change can be located within an individual mental space, as when we have a mental space for a flash, or pushing an object, or becoming hungry or cold, or for a match's being blown out.

An outer-space Change link is often bundled with an outer-space Identity link. In the Bypass, for example, the child who changes into a doctor remains the identical person. Change, with or without Identity, can be compressed into Uniqueness in the blend. If the devil "changes" Hitler into the Virgin Mary, there was no Identity connector between them, but in the blend they are a Unique being. There are less obvious but more noticeable cases where Uniqueness arises in the blend even though there is no Identity relation between particular elements in the outer-space links. Consider, for example, the presentation of the evolution of dinosaurs into birds in Figure 6.2. Here, we see a dinosaur chasing a dragonfly, unable to catch it. We also see the dinosaur at a series of locations along a path. At each location, the dinosaur looks increasingly bird-like. The dragonfly is always the same. At the last step, the dinosaur is now a bird and the dragonfly is in its beak.

In the outer-space relations between the stages of the dinosaur evolution, there are no Identity relations between particular dinosaurs. Instead, there are relations of Cause-Effect (genetic evolution) and Analogy (one dinosaur is analogous to another) and Disanalogy (generation-to-generation differences in the dinosaur phenotype). These outer-space relations are related: The Disanalogy is

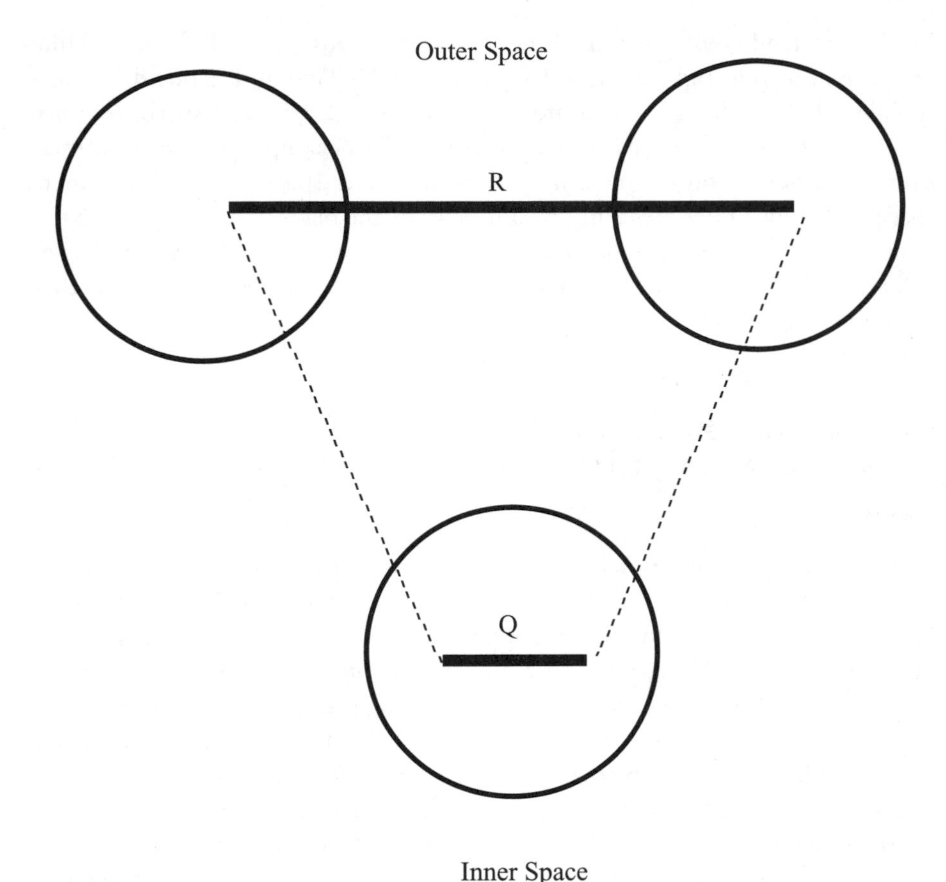

FIGURE 6.1 GENERAL SCHEME FOR THE COMPRESSION OF AN OUTER-SPACE RELATION TO AN INNER-SPACE RELATION

accounted for by the Cause-Effect relation of biological evolution. But in the blend, we have a unique dinosaur that has undergone change during its lifetime. We are not fooled by this presentation: We know there is no unique dinosaur who became a bird. But the blend gives us an efficient compression of change into uniqueness at human scale, which allows us to manage the network of evolutionary change from dinosaurs to birds. The connections between the blend and the inputs never disappear. We work not just with the blend but with the entire integration network.

Everyday language includes expressions for automatically compressing Disanalogy into Change. "My tax bill gets bigger every year" presents a compressed blend at human scale in which there is a single thing whose size changes. We automatically decompress Change in the blend into outer-space Disanalogy between quite distinct tax bills.

FIGURE 6.2 THE EVOLUTION OF DINOSAURS INTO BIRDS (SOURCE: WEXO, 1992.)

Identity

Identity may be the most basic Vital Relation. As we explained in Chapter 1, identity is taken for granted as primitive, but it is a feat of the imagination, something the imagination must build or disassemble. We connect the mental spaces that have the baby, the child, the adolescent, and the adult with relations of personal identity, despite the manifest differences, and we relate these identity connections to other vital relations, of Change, Time, and Cause-Effect. As we go along, we will see how blending is a powerful and supple instrument for creating and disintegrating identity.

In the Buddhist Monk, the monk in one input is "identical" to the monk in the other. This kind of connection seems absolutely straightforward. But identity connectors always involve interesting differences: For example, one monk is a few days older than the other, just as "Mary" as a baby and "Mary" as a fifty-year-old CEO and grandmother are linked by an identity connector across temporally separate spaces. That connection has nothing to do with objective resemblance, or shared visible features. Rather, identity across spaces is a stipulated connection. The complexity of identity connections across spaces leads to phenomena like referential opacity, which have long vexed philosophers of language. Identity connectors do not have to be one-to-one across inputs. "If he were twins, he would hate himself" prompts for a network in which one person in one input space has *two* identity counterparts in the other space. In the blend, the twins are counterparts by identity of the single person in the first input space. The "Mary" case is typical of identity connection, and a proper name like "Mary" is a way of keeping track of corresponding elements across spaces. But identity can also link less specific elements—in particular, roles. In understanding "In France, the president is elected for a term of seven years, while in the United States *he* is elected for a term of four years," we connect as identical the role *president* in the French space and the role *president* in the American space, and this is why we can use "he" in this way. Just as with "Mary," the identity connection is manufactured: The two presidential roles are remarkably different in the two countries. Identity connection can occur between highly unspecific elements, between a "leader" in one space and a "leader" in another, between a cause in one space and a cause in another, and so on. We can say, incorrectly of course, "The cause of the fall of the Roman Empire was the invasion of the Goths, but for the British Empire, *it* was failure to modernize."

Time

Time is a Vital Relation related to memory, change, continuity, simultaneity, and nonsimultaneity, as well as to our understanding of causation. The Buddhist Monk, Regatta, and the Bypass all have inputs separated in time, but the blends bring those times together. In the Baby's Ascent, a lifetime is compressed into less than a minute in the blend. Conceptually, New Year's Day in 2000 and New Year's Day in 2001 are two mental spaces connected by Time.

Space

Space is a Vital Relation much like Time. The Debate with Kant and many other blend networks, such as the Mythic Race, studied in Chapter 7, have inputs separated in physical space but a blend with a single physical space. Blends very frequently compress over Space.

Cause-Effect

Cause-Effect is a Vital Relation we have already discussed. A fire in a fireplace is connected by Cause-Effect to the cold ashes. It is not enough to see one thing as caused by the other; rather, we need the two proper mental spaces, the one with the logs burning and the one with the ashes. These are connected by Vital Relations of Time (one space is later than the other), Space (they are in the same place), Change (the logs become the ashes through the transformation of burning), and Cause-Effect (the fire causes the change and the existence of the ashes). There are subtypes of Cause-Effect. Producer-Produced is one.

Part-Whole

Useful integrations of Part-Whole Vital Relations are more common than we might think. We point to a picture of a face and say "That's Jane Doe," not "That's the face of Jane Doe." We have constructed a network mapping the individual to the picture of what seems to us her most salient part, her face. In the blend, the face is projected from one input and the whole person is projected from the other. In the blend, face and person are fused: The face is the personal identity. The part-whole connection between the face and person in the input spaces becomes uniqueness in the blend. The same could be done with fingerprints, X rays, and, in specialized circumstances, other body parts, as when we see an advertisement for a ring and say, seeing only the hands, "That's Jane Doe." The Vital Relation between the input spaces is Part-Whole, and it gets compressed by blending into Uniqueness. A more unusual example, for most of our readers, is Voodoo Death. One input space has parts of a person (hair, fingernails, photographs of the face); the other has the person. There are part-whole connections between the spaces. In the blend, the parts and the whole are fused, so that causation for the parts is causation for the whole. For example, burning the hair is, in the blend, killing the person. Burning the hair (the cause) and dying (the effect) happen simultaneously in the blend.

Representation

One input can have a representation of the other—as in a sketch of a person or a picture of a baby. When we think of an input as a representation, we build a conceptual integration network. One input corresponds to the thing represented; the other, to the element that represents it—as with paint of different colors on a canvas. In the blend, the representation link between the thing represented and the thing representing it is typically compressed into uniqueness. We look at the painted canvas and say "Here is Queen Elizabeth. She is dressed as Empress of India." We see the actor on stage and say "Richard II is in prison." We look at the marble statue and say "They are just about to kiss." We enter the "world of representations" by constructing blended spaces in integration networks. We do not lose sight of the inputs. We keep active the mental spaces in which the paint is just paint and not the Queen, in which the actor is an American employee in Hollywood and not Richard II, and in which the marble is only marble.

When we look at a picture of the face of Jane and say "This is Jane," we are actually using a complicated network with multiple blending and compression. Here we have a very basic and automatic blending network in which a person is linked to her body, and they are compressed in the blend into a unique element. There is also a systematic one-to-one mapping between bodies and faces. The neurobiological fact that faces are, through face recognition, the most salient part of the body for the purpose of maintaining identity favors another blend in which the Part-Whole and Identity connectors between the person and the face are compressed in the blend into uniqueness. If you have "seen Jane," "met Jane," or "caught a glimpse of Jane," the implication is that you have seen her face. It is this blend of person and face that underlies conceptions of "saving face," "faceless bureaucrats," and "defacing" something in the sense of altering its identity. In addition, we can have a representation link between Jane's face and a picture of Jane's face. But since we already have the blend in which Jane's face is the same as Jane, the picture of Jane's face is automatically a representation of Jane herself, and in the new blend, it is Jane. All of this is involved when we say, of the picture, "This is Jane."

Role

Role is a ubiquitous Vital Relation. Lincoln was *president*, Elizabeth is *queen*, and the president is the *head of state:* Roles have values. Lincoln, in 1863, is a value for president; Elizabeth, today, is a value for queen; and president, in the United States, is a value for head of state. Within mental spaces, and across mental spaces, an element can be linked, as a role, to another element that counts as its value. Elements are roles or values not in some absolute sense but only relative to other elements. *President* is a role for the value *Lincoln*, and a value for the role *head of state*. The canonical opportunities for compressing roles are fascinating. In the simplest-looking case, we have a role like *Pope* in one input and a value like *Karol Wojtyla* in the other, and a unique element, Pope John Paul II, in the blend. Once we have a serial string of such unique elements, Pope John XXIII, Pope Paul VI, Pope John Paul, Pope John Paul II, and all their precursors, we can make another blend in which there is one unique Pope who goes through many cycles and we can say "He was Italian for centuries but in 1978 he was Polish for the first time." When Pope John Paul, born Albino Luciani, died a month after Pope Paul VI, born Giovanni Montini, a newspaper ran the headline "Pope Dies Again!"

Analogy

Analogy depends upon Role-Value compression. Suppose we have one network with *Pope* in one input and *Giovanni Montini* in the other and *Pope Paul VI* in the blend, and we have another network with the same role, *Pope*, in one of its inputs and *Albino Luciani* in the other and *Pope John Paul* in the blend. In this case there is an identity connector between the Role input in one network and the Role input in the other. We say that *Pope John Paul* and *Pope John VI* are analogous, and what we mean, exactly, is that we have two Role-Value compression networks that have the same Role input. In other words, when, through blending, two different blended spaces have acquired frame structure in common, they are linked by the Vital Relation of Analogy.

Here's another example: Stanford is a West Coast "analogue" to Harvard. Both blending networks have the same American university frame with the role *prestigious private American research university*. One network has the value Stanford in the other input; the other network has a different value, Harvard, in the other input. The blends in the two networks are linked by Analogy because of the identity of those input roles.

As we saw in the case of the Pope who dies again, and as we will see again in detail in Chapter 7 in our discussion of the Pronghorn, Analogy is routinely and conventionally susceptible to compression into Uniqueness and Change. In the case of the dinosaur trying to become a bird so it can catch the dragonfly, the Analogy and Disanalogy of multiple dinosaurs through evolution is first compressed into Identity and Change between a much smaller number of mental spaces, corresponding to the small number of dinosaurs we see on the path chasing the dragonfly; the compression of Analogy into Identity makes all these dinosaurs identical. With one more step of compression, we now "see" a unique dinosaur changing as it chases a unique dragonfly, even though the cartoon merely shows us several animals that look quite different.

Disanalogy

Disanalogy is grounded on Analogy. We are not disposed to think of a brick and the Atlantic Ocean as disanalogous, but we are disposed to think of the Atlantic Ocean and the Pacific Ocean as disanalogous. Disanalogy is coupled to Analogy. Psychological experiments show that people are stymied when asked to say what is different between two things that are extremely different, but answer immediately when the two things are already tightly analogous. Disanalogy is often compressed into Change. So the disanalogy between the various dinosaurs at different stages of evolution corresponds, in the blend, to change in one unique dinosaur. Alternatively, if the blend has multiple slots with the same role, the disanalogous values can be brought in as separate values for those multiple slots, as in the case where the disanalogous boats are brought in as contesting boats in the Regatta blend, or the disanalogous Kant and the modern philosopher are brought in as debating philosophers in the Debate with Kant.

Property

Property is a Vital Relation in an obvious way: A blue cup has the property *blue*. A saint has the property *holy*. A murderer has the property *guilty*. The most obvious status of a property is as an inner-space vital relation: In the space of the blue cup, the cup is intrinsically blue. Blending often compresses an outer-space vital relation of some sort into an inner-space relation of Property in the blend. For example, outer-space Cause-Effect links can be compressed into Property relations in the blend. A warm coat is something that causes you to be warm; it is not something that is warm itself, but in the blend, it is has the Property *warm*.

Similarity

Similarity is an inner-space Vital Relation linking elements with shared properties. Human beings have perceptual mechanisms for perceiving similarity directly (extremely complex ones, from a neurobiological point of view), as when we place two pieces of fabric next to each other and see the similarity in their color. This direct perception of similarity is a human-scale scene. More diffuse outer-space links in mental space networks can be compressed into Similarity in the human-scale blend. For example, outer-space Analogy can be compressed into direct Similarity in the blend.

Category

Category is a Vital Relation like Property. The most obvious status of a category is as an inner-space vital relation: In the space of *Northern Light*'s run in 1853, *Northern Light* is a clipper ship. Blending can compress outer-space Vital Relations such as Analogy into Category in the blend. For example, what starts out as an outer-space analogy between a biological virus and an unwanted destructive computer program that invisibly comes to reside on your computer is compressed into a Category relation in the blend: In the blend, the computer program is a *virus*.

Intentionality

Intentionality covers a group of vital relations having to do with hope, desire, want, fear, belief, memory, and other mental attitudes and dispositions directed at content: We fear it will rain, hope we will get home, believe we are in California, remember that we were in France. We interpret each other on the basis of the view that people's actions and reactions are intentional in this technical sense. Intentionality is crucial because everything we do and think and feel is based on the relations it covers. It makes a difference to us whether the glass breaks accidentally or because we meant to break it. When something happens, our two major choices for framing the event, as Erving Goffman points out, are as a natural and unintentional "happening" or as a scripted "happening" that involves intentionality. We can say that "he died of cancer" or that "cancer took him." The second option adds some intentional framing. Religious thought, as part of its systematic global insight, frequently adds intentional framing to the natural world: Death is not a purely natural event but now something involving divine or infernal intentionality.

Intentionality is often heightened under blending. For example, in Regatta, the crew of *Great American II* is of course aware of the historical space in which *Northern Light* made its run. In the blend, the catamaran crew is competing directly against *Northern Light*. The *race* frame brings with it a tight relation of intentionality, involving not only knowledge of facts but also desire, fear, competition, and effort. The case is very similar in Debate with Kant. In other cases, compression creates Intentionality in the blend. In the Dinosaur Takes Flight example, the dinosaurs in the input spaces are not trying to evolve into birds in order to eat a dragonfly, but in the blend, the unique dinosaur is.

Uniqueness

Uniqueness obtains automatically for elements in the blend, and we take it for granted. The importance of Uniqueness in a technical sense is that many vital relations compress into Uniqueness in the blend.

The vital relations we will encounter repeatedly are these:

Change
Identity
Time
Space
Cause-Effect
Part-Whole
Representation
Role
Analogy
Disanalogy
Property
Similarity
Category
Intentionality
Uniqueness

There are canonical patterns of compression over these vital relations that we will encounter again and again. Compression can scale Time, Space, Cause-Effect, and Intentionality. Analogy can be compressed into Identity or Uniqueness. Cause-Effect can be compressed into Part-Whole. Identity itself is routinely compressed into Uniqueness. It is also a fundamental power of the way we think to compress Representation, Part-Whole, Cause-Effect, Category, and Role into Uniqueness. Other compressions will be studied as we progress.

There are also canonical patterns of proliferation of Vital Relations. Cause-Effect can be added to Analogy. Intentionality can be added to Cause-Effect. Representation can be added to Cause-Effect. Change usually comes with Uniqueness or Identity.

Vital Relations are what we live by, but they are much less static and unitary than we imagine. Conceptual integration is continually compressing and decompressing them, developing emergent meaning as it goes. Certain basic elements of cognitive architecture make blending and compression possible; we consider these next.

BASIC STRUCTURES AND CONNECTIONS

We now turn to some fundamental properties of mental spaces and the connections between them.

What Is a Space?

As we have seen, mental spaces are small conceptual packets constructed as we think and talk, for purposes of local understanding and action. They are very partial assemblies containing elements, structured by frames and cognitive models. As explained in Chapter 2, our hypothesis is that, in terms of processing, elements in mental spaces correspond to activated neuronal assemblies and linking between elements corresponds to some kind of neurobiological binding, such as co-activation. On this view, mental spaces operate in working memory but are built up partly by activating structures available from long-term memory. Mental spaces are interconnected in working memory, can be modified dynamically as thought and discourse unfold, and can be used generally to model dynamic mappings in thought and language.

Spaces have elements and, often, relations between them. When these elements and relations are organized as a package that we already know about, we say that the mental space is *framed* and we call that organization a "frame." So, for example, a mental space in which Julie purchases coffee at Peet's coffee shop has individual elements that are framed by *commercial transaction*, as well as by the subframe—highly important for Julie—of *buying coffee at Peet's*.

Spaces are built up from many sources. One of these is the set of conceptual domains we already know about (e.g., eating and drinking, buying and selling, social conversation in public places). A single mental space can be built up out of knowledge from many separate domains. The space of Julie at Peet's, for example, draws on all of the conceptual domains just mentioned. It can be structured by frames other than *commercial transaction*, such as *taking a break from work*, *going to a public place for entertainment*, or *adherence to a daily routine*. Another source for building mental spaces is immediate experience: You see the person Julie purchasing coffee at Peet's and so build a mental space of Julie at Peet's. Yet another source for building mental spaces is what people say to us. "Julie went to Peet's for coffee for the first time this morning" invites us to build a new mental space, no doubt one that will be elaborated as the conversation continues. In the unfolding of a full discourse, a rich array of mental spaces is typically set up with mutual connections and shifts of viewpoint and focus from one space to another.

Mental spaces are built up dynamically in working memory, but they can also become entrenched in long-term memory. For example, frames are entrenched mental spaces that we can activate all at once. Other kinds of entrenched mental spaces are Jesus on the Cross, Horatio at the bridge, and the rings of Saturn. An entrenched mental space typically has other mental spaces attached to it, in an entrenched way, and they quickly come along with the activation. Jesus on the Cross evokes the frame of Roman crucifixion, of Jesus the baby, of Jesus the son of God, of Mary and the Holy women at the foot of the Cross, of styles of painting the crucifixion, of moments of the liturgy that refer to it, and many more.

We will see that entrenchment is a general possibility not just for individual mental spaces but for networks of spaces. In particular, integration networks built up dynamically can become entrenched and available to be activated all at once. Indeed, much of our thinking consists of activating entrenched integration networks for dealing with present subjects.

For our present purpose—namely, to characterize sources of variation in networks—the most pertinent features of mental spaces are the degree of specificity of the elements, the degree to which they are framed, our familiarity with the space, the degree to which it is entrenched, and the degree to which it is tied to our experiences.

Specificity and Familiarity of an Input Space

Input spaces can be of any variety of specificity. A space that contains only an abstract notion of cause, an abstract notion of effect, and their abstract relation can be an input space in an integration network. So can a space that is extremely specific: the wedding of Allison and Chip in April 2000 in Annapolis. *Allison* is more specific than *daughter*, which is more specific than *woman*, which is more specific than *human being*, which is more specific than *physical object*, and an input space can be built around any of these levels of specificity.

Input spaces can differ in the degree to which the space is framed. A space with minimal framing might have only two abstract elements and no relationship between them. A space might have two unrelated but named people (*Paul* and *Elizabeth*). It might have minimal framing and elements that lack specificity (*this causes that*). It might have more developed framing, with elements that correspond to roles but have no specific values (the frame of kinship roles attached to a person: father, mother, uncle, etc.). It might have two people (*Paul* and *Elizabeth*) framed by a subframe (father-daughter) to give a space with *Paul* and *Elizabeth* and a father-daughter kinship relationship between them. A frame can itself be more or less specific. *Action* is a frame, but a very unspecific one. *Buying coffee at Peet's* is also a frame, an instance of *action*, but a much more specific one.

An organizing frame for a mental space is one that specifies the nature of the relevant activity, events, and participants. An abstract frame like *competition* is not an organizing frame, because it does not specify a cognitively representable type of activity and event structure. *Boxing* is an organizing frame that specifies an activity, its events and sequences, and its participants. An input space does not need to have an organizing frame.

An input space can be more or less familiar to us, more or less entrenched, more or less connected to our episodic experiences. Sometimes we can activate a frame we know well and use it to organize the entire space. In other cases, we may have to work to develop and modify a frame. To learn a mental space is in some cases to learn the frame that organizes it. A mental space can be more or less easy to activate. It can also have an interesting conceptual descent of its own. Many mental spaces are the products of blending.

These parameters of difference in the construction of a single mental space have consequences for the types of integration networks that can arise.

Topology of a Mental Space

A mental space consists of elements and relations activated simultaneously as a single integrated unit. Often, a mental space will be organized by what we have called a conceptual frame. Consider a particular boxing match. It is organized by the conceptual frame *boxing match*. Such a frame often includes *scales*—for example, how hard someone hits, how quickly the match ends, how much money the boxers make, how large the audience is, how corrupt the judges are. Such a frame often includes *force-dynamic* structure: An arm stops a jab, a fist strikes a jaw abruptly, a coach restrains a boxer in his corner, a man falls slowly to his knees, the floor stops him from dropping farther. Such a frame often includes *image-schemas:* The boxing "ring" is really a square and it is a *container*, the jab fits a particular dynamic image, the two boxers are *opposed*. There is massive interplay among scales, force-dynamic patterns, image-schemas, and vital relations, all of which are ubiquitously available in human conceptual structure and cognition. Human beings are set up to locate and attribute these structures: When we see or think of a scene, we cannot do so without premium attention to vital relations, scales, force-dynamics, and image-schemas.

To pursue our example, consider that the duration of the boxing match depends on the vital relation of Time but also, of course, on a scale of duration. We understand the event—that is, we structure the corresponding mental space—in terms of Roles such as *boxer* and *umpire*, Identity such as the identities of the boxers, Cause-Effect such as the relationship between an *uppercut* and a *knockout*, Change such as the deterioration of a boxer, counterfactual Disanalogy such as "misses," Intentionality such as the goal of hitting, physical Space such as the boxing arena, Time such as the interval between rounds marked by the bell, and Uniqueness.

A mental space may be organized by a specific frame such as *boxing* and a more generic frame such as *fighting* and a still more generic frame such as *competition*. Each of these may have its scales, image-schemas, force-dynamic patterns, and vital relations. One can also use a finer topology in a mental space, below the level of the organizing frame. The organizing frame *boxing match* does not tell us the shoe sizes of the boxers or how much the boxing gloves weigh or whether the boxers are wearing protective head gear, but a finer topology can include such details.

Mapping Spaces

The mapping of spaces is a crucial component of the imaginative construction of a network. In retrospect, the mappings often look like obvious matches, as if they were given immediately by the spaces themselves. This is an Eliza illusion, as Douglas Hofstadter noticed for the case of analogy; it is similar to thinking that the perception of a cup is directly caused by the objective existence of the cup, without any imaginative construal. The achieved full integration network is the imaginative product, which we are disposed, through the Eliza effect, to see as directly caused by preexisting "objective givens." But constructing both the input spaces and the connections between them is often a highly creative act.

The different topological properties of individual spaces we just saw give rise naturally to different possibilities for matches between them. On the one hand, we can look at what is inside spaces and build correspondences between one space and another based on similar topologies. Thus, for instance, we might map a linear scale in one space onto a linear scale in another space, or a source-path-goal image-schema in one space onto a source-path-goal image-schema in another space, or the force-dynamic pattern of Caused Motion in one space onto a force-dynamic pattern of Caused Motion in another space. In all such cases, the vital relations of Identity or Analogy apply across spaces to the topology of scales, image-schemas, and force-dynamic patterns inside mental spaces.

On the other hand, we can match mental spaces at the level of their internal vital relations. Change of one sort may be mapped onto Change of another sort, just as Time (or Part-Whole or Cause-Effect) in one mental space can be mapped onto Time (or Part-Whole or Cause-Effect) in another mental space.

The vital relations of Identity and Analogy can also provide connections between organizing frames. Consider a situation where the organizing frame of one space can be connected by Identity to the organizing frame of another space. In that case, the two spaces have many identical roles because they have the identical frame containing those roles. The identity of the frames lines up all the roles at one shot. For example, there is an Identity mapping between the frame of ocean voyage in the space with *Northern Light* and the frame of ocean voyage in the space with *Great American II*.

The organizing frames in two spaces are analogous when there is a more abstract frame that applies to both of them. In that case, Analogy connects the organizing frames of the two spaces. For example, there is an Analogy mapping between the frame of a boxing match and the frame of a cockfight.

An organizing frame in one space can be connected globally to values for its roles in another space through a bundle of Role vital relations, as when *father* and *daughter* in one space are connected to *Paul* and *Elizabeth* in the other.

TOWARD A TYPOLOGY OF NETWORKS

In the next two chapters, we will investigate types of integration networks that arise repeatedly. In particular, we will begin to distinguish them in the way chemistry distinguishes molecules and evolutionary biology distinguishes species, having laid out an array of the possible topologies an individual space can have, along with the possible ways to connect two spaces. These possibilities of spaces and connections give us in principle a large number of conceivable ways to go about building networks that have input spaces, a generic space, and a blended space, and of course more complex networks that have multiple inputs or multiple blends. Blueprints for different, theoretically possible kinds of integration networks can be drawn up, and in fact we will see that many of these theoretical possibilities already exist in the world of human meaning and expression. The result will be a fairly detailed typology of everyday conceptual integration networks, the workhorses of the way we think. We will then see that the main prototypes in this typology fall along various continua, and that these continua anchor our intuitive everyday notions about meaning to a unified understanding of the unconscious processes at work. Varieties of meaning that on their faces seemed unequal—such as categorizations, analogies, counterfactuals, metaphors, rituals, scientific notions, mathematical proofs, and grammatical constructions—turn out to be avatars of the spirit of blending.

CHAPTER 6
ZOOM OUT

TRUTHS, ERRORS, AND WARNINGS

Blends systematically scale down relations, compress relations into others, and even create new relations.

Question:

- Doesn't all this compression distort our apprehension of reality?

Our answer:

In its beginning stages, any science faces crucial obstacles that come from folk theories. Folk theories are elaborate and indispensable systems of thought in everyday life. They are also typically a starting point for science. Science must use the folk theories to get off the ground but must also overcome some of their most fundamental principles. In the development of chemistry, a folk theory holds that earth, air, fire, and water are primitive components of the world. In the development of physics, another folk theory holds that, plainly, you have to keep pushing something to keep it moving. But in chemistry as a science, neither earth nor air nor fire nor water is an element, and in physics as a science, the idea that a body in motion tends to stay in motion was a fundamental discovery, one of Newton's laws.

Compression of vital relations is often essential for a folk theory, but it needs to be undone by a science. The popular understanding of evolution has to surmount exactly such an obstacle. The outcome to be explained—the wonderful variety of complex biological species—depends crucially on an interval of billions of years—that is, on a span of time long enough for the mechanisms of evolution to be able to bring about this panorama of effects. Human beings have no direct experience of any comparable interval of time. But blending allows us, through compression of the vital relation of time, to think of such intervals. The problem for the folk theory is that in the common blend, evolutionary time is blended with "very large" human time. But very large human time isn't nearly long enough for variation and selection to bring about the biological world we see, and human beings know this because it is part of our experience that we do not see the birth of species. So the time compression, which is done in the service of global insight, produces a blend that is not compelling and leads many people to reject the theory of evolution. Time compression often gives us a useful global insight, but in this case it gives us an insight with the wrong inferences—namely, that the theory of evolution can't be right because there is not enough time for it to have worked the way it is supposed to work. The evolution blend compares badly with a different one in which the complicated biological systems we see around us are the products of an exceptionally masterful designer and craftsman. Popularizers of evolution such as Richard Dawkins and Stephen Jay Gould, in an effort to discredit the God-as-designer Blend, spend much of their time trying to guide people to undo or hold in abeyance the time-compressed blend that makes evolution look unlikely. It is striking that one of the strongest tools in their arsenal of persuasion is a blend in which the vast, diverse, distributed events over biological time are compressed and unified into a single force: Evolution with a capital E. In this blend, Evolution is described implicitly as a designer, but one with many strange and alien techniques, which it uses repeatedly over billions of years. The differences between the God-as-designer blend and the Evolution-as-designer blend are all-important, including differences in processes and differences in time scales, but the main conceptual shapes of the blends are very similar.

THE PERNICIOUSNESS OF ELIZA

Compression over vital relations is one of the central engines of human insight and understanding. But as we just saw, compression over vital relations creates Eliza effects. We compress cause and effect in perception and so think perception is easy: There is a "cup" with various features and it causes us to perceive a "cup" with just those features. What else could there be to explain? We saw in Chapter 1 that the Eliza effect leads us to compress forms with meanings, and the products of the imagination with the processes that produce them, leading us to think that meaning and imagination are just a matter of the combinatorics of the forms that we can apprehend in consciousness.

Question:

- How can we avoid Eliza's traps?

Our answer:

It is in the nature of the way we think that Eliza effects will recur repeatedly, no matter how vigilant we are. The main obstacle to the launching of the scientific study of blending is the stultification of the Eliza effect, which persistently hides from view the important imaginative operations to be explained. Just as we must see past our Time-compression blend if we are to respect the operations of evolution and understand how they could achieve our world of biological diversity, so we must see past all the many vital-relation compressions that produce Eliza effects.

This is not easy, given Eliza's seductiveness. One thinker who is exceptionally insightful about how the Eliza effect fools us as we look at computer forms is

Richard Dawkins, who carefully leads us past the many Eliza errors we make when we watch computer models of evolutionary processes. Dawkins pioneered the approach of modeling computer biomorphs. In this approach, computers simulate heredity and random mutation, so we see evolution happening before our eyes on the computer screen. The program, called "The Blind Watchmaker," compresses the time of a generation radically, so that many generations can live and die within a half an hour or so. What we see on the screen, the "biomorphs," are complicated shapes vaguely reminiscent of beetles, spiders, snow crystals, and tiny oak trees. Dawkins is careful to point out that we see much more in these shapes than is actually there. For instance, what we inescapably see as "legs" are not even remotely analogous to biological legs: They are not even functional. They play no unified role as "legs" in the evolution of the "biomorph"; that is, they do not make the biomorph more adaptive by being more stable, stronger, more agile, or any of the other things legs might do. As Dawkins points out, the computer model is not a model of natural selection, because it contains nothing like selection pressures.

The biomorph computer program is useful as a way of getting us past the misleading Time-compression blend. We see the biomorphs, formed and evolved with extremely simple mechanisms, take on an amazing range of fascinating and sophisticated shapes. But it leads us, as Dawkins shows, into another Eliza trap, Cause-Effect compression: We see natural variety in the output of the computer program, and so imagine that it comes from natural selection in the program rather than from our own imaginative construal of what we see.

Dawkins wants to save us from more than this particular computer program. He wants to save us from the Eliza error involved in thinking that a computer model can capture "Evolution." Dawkins ponders what such a computer model might look like:

> What we ideally should do is simulate a complete physics and a complete ecology, with simulated predators, simulated prey, simulated plants and simulated parasites. All these model creatures must themselves be capable of evolving. The easiest way to avoid having to make artificial decisions might be to burst out of the computer altogether and build our artificial creatures as three-dimensional robots, chasing each other around a three-dimensional real world. But then it might end up cheaper to scrap the computer altogether and look at real animals in the real world, thereby coming back to our starting point!

What Dawkins is saying here is that natural selection cannot be scaled down, and cannot be compressed. A genuine model of evolution cannot compress the number of generations or the richness of the world. Evolution depends for its very operation on extremely large numbers of generations and all the richness of the world that is actually in place during every moment of its unfolding.

Similarly, we think blending depends for its very operation on extremely large numbers of mappings and all the richness of the physical and conceptual worlds in place during blending.

If evolution and blending cannot be modeled computationally, are they therefore not amenable to scientific study? With respect to evolution, Dawkins draws the opposite conclusion. The abstract idea of evolution is implemented, by the natural world. If that idea existed in a Platonic heaven, what better way to implement it than to create our universe and watch what happens? Watching what happens in the natural world is what Darwin did, and what evolutionary biologists still do to this day. It is in Nature's giant laboratory that we can accurately investigate the properties, consequences, and regularities of evolution. In principle, a compressed model will not allow us to do this, although it might be useful for exploring some of the formal features of the processes involved. Evolutionary biologists can now perform laboratory experiments, where the laboratory is a part of Nature that the biologists have colonized and pushed in one direction or another. These experiments are still working directly with the stuff of evolutionary biology itself, so they are not like computer models, which are not instances of the biological world itself.

The science of chemistry advances according to the same view and the same method: Nature is a giant laboratory for chemistry, and we can investigate the properties, consequences, and regularities of chemistry by investigating what actually happens in the world. The chemistry laboratory in an office building is a chosen subset of Nature, in which the chemist looks directly at what Nature does, contrived along this or that line. The chemistry experiments, too, belong to the real world of chemistry itself, and they are accordingly not like computer models of chemical events, which are not instances of chemistry.

Scientific study of the imagination must follow the same lines. Nature is already a giant laboratory in which imagination is fully at work. Because blending depends crucially, not just incidentally, on the richness of the conceptual world, we can investigate its principles only by investigating the meanings that people actually do construct in real situations. Nature's giant laboratory of blending produces ads in magazines, hyperbolic geometry, grammatical constructions, counterfactual arguments, cause-effect compressions, literary allegories, computer interface designs, and many other inventions. Just as Darwin found the Galapagos Archipelago a useful real-world laboratory, precisely because it was isolated and strange, so we often go to something that looks exotic but is no less fully a part of the human world, in order to investigate the principles and parameters of blending. And just as the evolutionary biologist or the chemist can contrive an experiment *within nature,* so we can do the same, by asking human beings to do something, understand something, solve something, and so on, and watching what they actually do under those circumstances.

MODULARITY, DIFFERENCE, AND UNITY

We have seen many different kinds of blends—the cultural invention of mathematical theories over centuries, the instant-action blend of the skiing waiter, counterfactual arguments that arise in discourse, and so on. These blends not only look different but also are clearly not identical in the processing they involve.

Question:

- How could anyone be foolish enough and simplistic enough to want to lump all of this diversity together?

Our answer:

Perhaps the greatest strength of chemistry and evolutionary biology is their discovery that behind striking diversity—water and nylon, beetles and elephants—are general principles. These strikingly different products can arise through strikingly different processes. On the one hand, burning is a routine chemical phenomenon, while on the other hand the creation of nylon in the laboratory is a strange event requiring massive cultural effort. On the one hand, asexual reproduction happens all the time as a routine evolutionary process, while on the other hand cloning is a strange event requiring massive cultural effort. Some chemical and evolutionary processes happen relatively quickly; others, like evolutionary convergence or the deformation of glass under gravity, happen relatively slowly. But chemistry and evolutionary biology have shown that behind the most diverse products and the most diverse processes there are common chemical and evolutionary principles.

In this book, we examine strikingly different kinds of conceptual products that arise often through strikingly different processes. Jokes and mathematical inventions look quite different, and we understand one instantly while the other may take centuries of conscious effort to achieve, but both depend on the principles of conceptual blending.

This common, basic operation underlies all our examples. It is well recognized in cognitive neuroscience that the human brain is not an undifferentiated mass: No one has ever imagined that we can hear with our retina or see with our cerebellum. But it is also undisputed that important physiological regularities occur throughout the entire central nervous system. Even the retina and the cerebellum share those physiological regularities. In the same way, conceptual integration is pervasively at work behind the many innovative and imaginative capacities of cognitively modern human beings.

Seven

COMPRESSIONS AND CLASHES

I see my life go drifting like a river
From change to change; I have been many things—
A green drop in the surge, a gleam of light
Upon a sword, a fir-tree on a hill,
An old slave grinding at a heavy quern,
A king sitting upon a chair of gold—
And all these things were wonderful and great;
But now I have grown nothing, knowing all.
Ah! Druid, Druid, how great webs of sorrow
Lay hidden in the small slate-coloured thing!

—*William Butler Yeats*

LIFE, VARIOUS AND DIFFUSE, courses over large expanses of time and space. The human mind constructs intelligible meanings by continually compressing over vital relations.

When we see a Persian rug in a store and imagine how it would look in our house, we are compressing over two different physical spaces. We leave out conceptually all of the actual physical space that separates the real rug from our real house. When we imagine what answer we would give now to a criticism directed at us several years ago, we are compressing over times. In some cases, a larger history includes the two spaces over which we are compressing. In the case of shopping for the rug, the history includes our desiring a rug, looking at the spot where it might go, making notes, going to the store, and looking at several other rugs. We choose two mental spaces in that history for compression. When we imagine a present answer to an old criticism, there is a history in which we have been thinking about an idea and dealing with criticisms. We choose two temporally separated mental spaces in that history and compress over them.

Blending is a compression tool *par excellence*. Selective projection from different related spaces and integration in the blend provides an exceptionally strong process of compression. We have seen blending perform temporal compression in the Buddhist Monk example (where two different journeys occurring at different times in history are brought together in a blend, thus in effect suppressing the time interval that separates them in history) and in Regatta (where two ocean voyages distant in time are brought together and the century that separates them is obliterated). In the Bypass, the entire span of time separating the schoolchildren from the surgical act of incision is scaled down to a few minutes. In the Baby's Ascent, the entire lifetime of the baby is scaled down into the time it takes to climb the stairs.

In Chapter 5, we saw many compressions of cause and effect through blending. In the Impotent Smoking Cowboy, the cigarette as cause is fused with the drooping effect. In the Bypass, the long causal chain between the early education of children and the competence of adult doctors is compressed down to the briefest span.

We saw blending perform compression over geographical spaces in the Debate with Kant, which brings Kant and a modern philosopher into the same room. The Debate also achieves sophisticated time compressions, most obviously in putting people separated by centuries into the same conversational moment, but also in compressing disparate assertions, scattered throughout Kant's life, into the few minutes of the debate.

Some vital relations bring with them an interval, expanse, or chain that we call a "string." Those vital relations are Time, Space, Cause-Effect, Change, Part-Whole, and Intentionality. In the Dinosaur Takes Flight, for example, the Time string running across all the dinosaurs is scaled down to a single lifetime. Furthermore, only certain moments in that lifetime are activated. This partial activation of points on a string we call "syncopation." Similarly, in the Bypass, the thirty years or so between elementary school and performing surgery are scaled down to a few minutes. The Causal string is also syncopated: Only the school learning and final professional competence are activated. The Spatial string between your living room and the rug store is obliterated when you imagine what the rug would look like if it were in your living room. In Graduation, the long string of Change that a college student undergoes over four years is compressed by syncopation: Only certain college events are activated, such as attending class, hearing a lecture, and seeing one's fellow classmates. There is also a radical scaling of the change the student undergoes during four years, down to the simple and quick change of moving the tassel from one side of the mortar board to the other. A Part-Whole string can be compressed under blending: Consider the case where an international airline company represents itself by having several ticket agents, obviously of different nationalities, say "Welcome" in their different languages one after another. We know that there are many levels of part-whole containment for the company. The international company has national parts; the national parts have many different regional offices; each of those regional offices has many different kinds of employees engaged in different activities, including ticket sales as one part of the operation; the group of ticket agents in a regional office has many members. In the blend, the entire sequence of part-whole steps that lead from the international company down to the individual ticket agent becomes one step: There is one whole, the international company, and it consists of these parts. A long string of Intentionality can also be compressed. In Regatta, the crew of *Great American II* can know about *Northern Light* only because there has been a long string of reporting between 1853 and 1993, but in the blend, the crew can observe the clipper ship directly. Diffuse transmission of knowledge has been compressed into direct observation.

IDENTITY, COMPRESSION, AND DECOMPRESSION

As we said in the opening chapter, identity is one of the mind's three *I*'s. It is not only a vital relation but perhaps the primary vital relation, without which the others are meaningless. Human mental life is unthinkable without continual compression and decompression involving identity. A linguistic system, to be useful at all, has to have a wide and powerful array of resources for prompting such compressions and decompressions. Identity seems to be a primitive, an unanalyzable notion, but instead it is an achievement of the imagination. To get a feel for the role of imagination and the importance of blending and compression in building up identity, consider the following story, which appeared on the front page of the science section of the *New York Times* on Tuesday, December 24, 1996. The story, titled "Ghosts of Predators Past," was illustrated by a large photograph of a small American pronghorn antelope chased by pen-and-ink prehistoric cheetahs and long-legged dogs. The American pronghorn is excessively faster than any of its modern predators. Why would evolution select for this costly excessive speed when it brings no additional reproductive benefit? The scientists propose that

> the pronghorn runs as fast as it does because it is being chased by ghosts—the ghosts of predators past. . . . As researchers begin to look, such ghosts appear to be ever more in evidence, with studies of other species showing that even when predators have been gone for hundreds of thousands of years, their prey may not have forgotten them" (p. C1).

The conceptual integration network for this passage is exceptionally complex. We will focus on some of the compression and decompression involving identity.

In the prehistoric story, the American pronghorn barely outruns nasty predators such as cheetahs and long-legged dogs. In the modern story, it easily outruns all its modern predators. In the blend, the pronghorn is being chased by nasty ancient predators, marked as "ghosts" to signal that they have no reference in the modern world. We are not confused by this felicitous blend. We do not expect to see ghosts chasing a real pronghorn; we do not think any living pronghorn remembers the prehistoric predators. Instead, we know how to connect the blend to the story of the pronghorn: Great speed was adaptive for the animal's ancestors, who faced nasty predators, and although those predators are now extinct, the physiological capacity for speed survives.

In the blended space of this integration network, there is a single individual pronghorn and that pronghorn remembers the nasty predators that once chased it. It was conditioned by those chases, and so now, when any predator tries to chase it, it runs with its old speed.

But wait a minute. Who is this pronghorn? Clearly not any individual animal, but also clearly not just a typical representative of the pronghorns in the world today. And it is not a representative of the modern American species, because no member of that species has seen any of these nasty predators, and so none could "remember" them. What gives us a global insight into an evolutionary truth is a massive compression of identity over species, individuals, and time.

Clearly, to achieve this blend, we must have as inputs a space in which there is a modern pronghorn that runs fast, but not from fast predators, and a space in which there is a prehistoric pronghorn that runs fast from fast and nasty predators. Notice that the one pronghorn in the blend corresponds to two quite different pronghorns in those inputs. But where do these input spaces come from? Where do we get "the prehistoric pronghorn" being chased by a cheetah and "the modern pronghorn" with no cheetahs around? Neither "the prehistoric pronghorn" nor "the modern pronghorn" is an actual individual. Each one is itself a compression into uniqueness over an epoch in the history of the species. Each one is chosen from an idealized set of "prototypical pronghorns," a set that is itself imaginatively constructed by attributing to each "prototypical pronghorn" the traits that are characteristic of the group of actual pronghorns during the relevant epoch.

Even more compression and decompression are needed to get to the prototype, but let us consider only the complex of compressions we have seen so far. At the outer limit, we have a decompression into all the individual pronghorns existing during either of these two epochs. First, they are compressed into two separate groups, each constituting a subcategory—ancient pronghorns versus modern pronghorns. Each group is compressed over time and space (and evolutionary cause and effect, for that matter, since each group consists of all the pronghorns during the epoch as either ancestors or descendants). We compress each of these groups by pretending that there is a homogeneous nature, experience, and behavior for all the members of the group. So now all the members are the same in each of the two groups. Then we compress each group through the nifty device of picking one member (any one, since they are all the same). This compression turns the two groups into two unique animals—the Ancient Pronghorn versus the Modern Pronghorn. Now, at last, after several compressions, we have what we need to do the actual noticeable blend, of the prototypical ancient pronghorn and the prototypical modern pronghorn, to get one pronghorn in the blend who can have today's flight behavior (the same as the ancient flight behavior) and memory of the cheetahs who used to chase it. This memory is an emergent structure in the blend, since neither the idealized Modern Pronghorn nor any actual living pronghorn can possibly have such a memory.

The input spaces are connected by vital relations of time and causality. The reason the Modern Pronghorn runs so quickly is that it has inherited that capacity through the generations that connect it back to the Ancient Pronghorn. This cross-spatial relation of evolutionary inheritance *between* the input spaces is compressed into structure *inside* the blended space—namely, the memory of the pronghorn. This is the general strategy available in building integration networks that we saw in Chapter 6: A vital relation between spaces is compressed into structure inside the blend. This very example of the pronghorn illustrates a second use of the general strategy. It is easy to think of the Ancient Pronghorn as having been conditioned by the nasty predators to run fast, as if it had *learned* that it had better run fast. But of course no learning was involved. The reason the Ancient Pronghorn runs so fast is that the pronghorns in the line of descent above it were the ones fast enough to escape the predators. Their slow cousins got eaten. This is a story of *adaptation* that connects many different generations of pronghorns, and this connection is a cross-space vital relation of *change*. Change through *adaptation* across spaces is compressed into change through *learning* for the Ancient Pronghorn.

When we get to the last compressions we need to do, we have one input with the Ancient Pronghorn and another input with the Modern Pronghorn, related by inheritance. First, the Category connection between the Ancient Pronghorn and the Modern Pronghorn is compressed into an outer-space Identity relation. The Ancient Pronghorn and the Modern Pronghorn become identical individuals, the Young Pronghorn and the Mature Pronghorn, at separate stages of one life. Evolutionary time between the Ancient and Modern Pronghorn has been compressed into the period of a lifetime. But now, Intentionality links can be added between the two spaces because it is the same individual in the two spaces. Therefore, the Modern Pronghorn can "remember" what the Ancient Pronghorn once "learned."

In the final blended space, the Identity and Intentionality links between these two pronghorns are compressed into Uniqueness and inner-space Intentionality:

There is a unique Mindful Pronghorn who remembers what it learned. Why can this mature pronghorn run so fast when there is no reason for it? Because it learned how when it was young and lived in a bad neighborhood. In other words, running fast, which is a property of the Ancient Pronghorn, is projected to the blend as a capacity of "the pronghorn," learned when young. The persistence of this capacity in the species is projected to the persistence through life of something one has learned when young. Just as we do not forget the taskmasters at the head of our classes or the bullies in the schoolyard who taught us to fight back and be self-reliant, so the mature pronghorn has not forgotten the cheetahs who taught it to run fast in its youth.

The compressions that give us the Mindful Pronghorn can be teased apart under analysis. They seem useful and fairly natural, if also perhaps a little supernatural. We have already seen compressions into uniqueness that seem even more natural. We compress over a person's lifetime to yield a unique person and a label that we call a proper name. We see the compression, but it seems somehow true to the essence of human life. An even less noticeable compression is the kind that gives us the perception of a single "blue cup." In this case, we need a neurobiologist to explain even the fact that identity compression—binding—is going on. But in the other direction from the Mindful Pronghorn, there are compressions into uniqueness that seem more supernatural. Metempsychosis is a case in point. You might be stunned at the compression involved in the discovery that you, Cleopatra, Saint Barbara, Queen Elizabeth the First, your great grandmother (the diva), and Sarah Bernhardt are the same person, but this explains why you dream in Egyptian, were once hit by lightning without harm, like to ride horses, think you can sing the Queen of the Night's aria flawlessly, and have a penchant (your friends say) for being histrionic. Unlike others, this compression into uniqueness is not primed for by a culture that gives us proper names for individual people, patronymics for family groups, and common nouns like "pronghorn" and "cup" for referring to the corresponding compressions, but no expression or construction for picking out the unique element that compresses the series of identical individuals extending from Cleopatra to you. Because it is not expressed conventionally within the language, it stands out in ways that the more conventional compressions of personal identity over time or of groups-to-single-individual do not. But that does not mean it is not present in our culture. We routinely use the conventional construction "You must have been So-and-So in a former life" to highlight that "you" are doing something that calls for *some* explanation, which would be supplied by your having been So-and-So, just as the pronghorn's speed calls for *some* explanation, which would be supplied by the pronghorn's memory of fast, hostile predators. This identity compression across history is a deep principle of life in any culture that supports it—for example, one with a belief system that includes reincarnation.

Compressions into uniqueness over history are meant to give us global insight into diffuse and various histories—of a species in the case of the pronghorn, of an organism in the case of biological identity, and of a human being in the case of personal identity. In the epigraph to this chapter, William Butler Yeats describes the metempsychosis that Fergus the Red Branch King is able to recognize once he has opened the "small slate-coloured thing" that the Druid has given him. In this case, the compression into identity and thence into uniqueness extends over the history of a people, their environment, and their culture.

HOW NETWORKS DO COMPRESSION AND DECOMPRESSION

Integration and compression are one side of the coin; disintegration and decompression are the other. The pronghorn blend does not provide the appropriate understanding on its own; it must be connected to the rest of the network, in which things that are compressed in the blend are decompressed and held separate. The same is true of the Buddhist Monk: The blend does not solve the riddle without its connection to the disintegrated input spaces. By itself, discovering that two monks traveling in opposite directions on the same path meet each other provides no insight. It is only when that encounter has projections back to the input spaces of the ascent held separate from the descent that we gain global insight into the riddle. When we are working with all the decompressions and compressions of the full network, the encounter in the blended space automatically connects into an alignment across the input spaces, and it is specifically this alignment of the two journeys that is the solution to the riddle. The understanding, therefore, is crucially a matter of activating and connecting compressions and decompressions simultaneously in the entire network.

In principle, a conceptual integration network contains its compressions and decompressions. Typically, in use and processing, only parts of the network are available and the rest must be constructed dynamically. In some cases, decompression will be the main avenue of construction, and in other cases, compression will. But in most cases, in processing or recognition or scientific discovery or artistic creation, there will be some of each.

The multiple possibilities for compression and decompression, for the topology of mental spaces, the kinds of connections among them, the kinds of projection and emergence, and the richness of the world produce a vast array of possible kinds of integration network. Amid this diversity, four kinds of integration network stand out: simplex, mirror, single-scope, and double-scope. The network model predicts their existence from theoretical principles, and, indeed, when we look at the laboratory of Nature, we find very strong evidence that they really exist.

Simplex Networks

An especially simple kind of integration network is one in which human cultural and biological history has provided an effective frame that applies to certain kinds of elements as values, and that frame is in one input space and some of those kinds of elements are in the other input space. A readily available frame of human kinship is *the family*, which includes roles for father, mother, child, and so on. This frame prototypically applies to human beings. Suppose an integration network has one space containing only this frame, and another space containing only two human beings, Paul and Sally. When we conceive of Paul as the father of Sally, we have created a blend in which some of the structure of the *family* frame is integrated with the elements Paul and Sally. In the blended space, Paul is the father of Sally. This is a simplex network. The cross-space mapping between the input spaces is a Frame-to-values connection—that is, an organized bundle of role connectors. In this case, the role *father* connects to the value *Paul* and the role *daughter* connects to the value *Sally*.

In a simplex network, the relevant part of the frame in one input is projected with its roles, and the elements are projected from the other input as values of those roles within the blend. The blend integrates the frame and the values in the simplest way. The frame in one input is compatible with the elements in the other: There is no clash between the inputs, such as competing frames or incompatible counterpart elements. As a result, a simplex network does not look intuitively like a blend at all. But it is a perfectly regular integration network, predictable in kind from the theoretical principles of blending. A sentence in English that will prompt the construction of this blend is "Paul is the father of Sally." We will see later that this grammatical construction, "X is the Y of Z," is a general prompt for constructing integration networks of any type.

But "Paul is the father of Sally" is taken to be the prototype of semantic composition, easily expressible in first-order Fregean logic as F(a,b), where F is father, a is Paul, and b is Sally. And this is correct! What we have just discovered is that Fregean logical forms correspond to the cross-space mappings in simplex networks. The blended space in such a network is compositional in the sense that the entirety of the relevant information from both inputs is brought into the blend. This composition is truth-conditional in the following sense: The sentence counts as "true" in a world if the blend fits the current state of that "world" (i.e., if Paul is indeed the father of Sally).

What a surprise to find that this type of complex integration network, the simplex network, is nothing but our old friend "framing" as studied in artificial intelligence and as captured formally in predicate calculus notation! If it is a compositional form, goes the thinking, then it can't be a blend. But on the contrary, it is wonderful to discover that first-order logic and blending are not

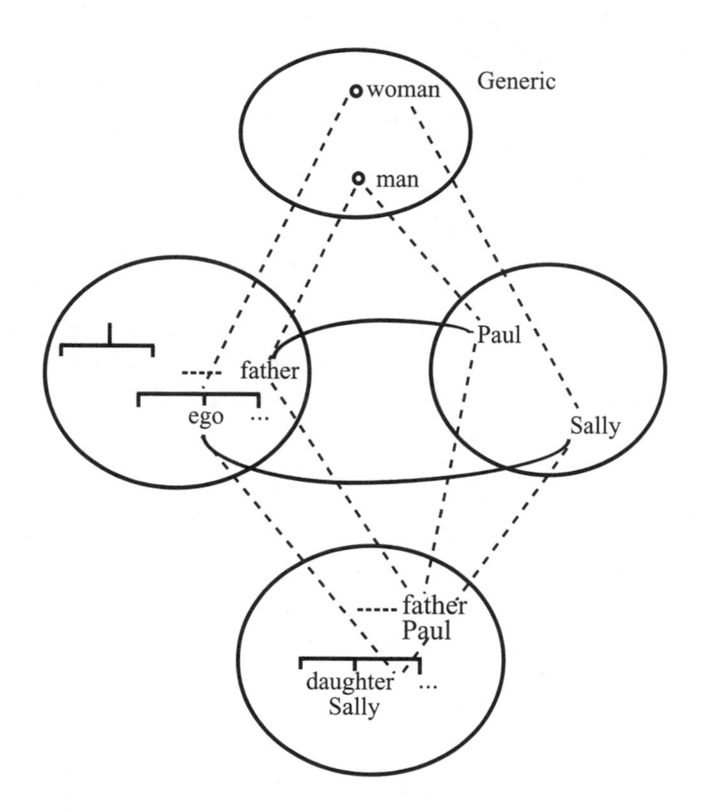

Here, one input contains a frame with roles but no values, the other input contains unframed elements, and the inputs are matched by a Frame-to-values connection.

FIGURE 7.1 A SIMPLEX NETWORK

antagonistic; one is a simple case of the other. It is similarly wonderful to discover that a point, a line, and a triangle are all special cases of conic section, as are a hyperbola, a parabola, a circle, and an ellipse. This brings parsimony to our description; but, more important, it captures a deep generalization about the phenomena being studied. There is a great variety of conceptual integration networks, and this variety accounts for the variety and creativity in the way we think. One such network, the simplex network, has special properties: It is for the most part compositional and truth-functional. We suggest that it has ideal properties for external computation and sequential symbolic manipulation, which is why it has been seized upon by form approaches as both tractable and basic. These properties explain the success of logic for computational purposes, while at the same time helping us to see why, despite decades of focused effort, it has proven to be impossible to reduce semantics and other important forms of thought to symbolic logic. Because the construction of meaning requires many kinds of integration networks in addition to simplex networks, a great deal of semantics falls outside the realm of symbolic logic.

On inspection, we find that even in simplex networks, more is going on than the logical approaches recognized. It turns out that one of the conceptual strengths of a simplex network is its ability to compress roles and establish the compression as a single new role in the blend. For example, in the inputs for conceiving that Paul and Sally are related as father and daughter, we have an abstract role *father* in one input and a concrete person *Sally* in the other. The simplex blend inherits these elements but also creates a new role: *father of Sally*. Once that new role is established, it can be referred to directly, as in "Paul is the father of Sally. As *such*, Paul is responsible for Sally's parking tickets." As we saw in Chapter 6, Role-Value is an essential vital relation. In simplex networks, it connects the two input spaces through Frame-to-values organization. Like other vital relations, Role-Value can be compressed through blending. This is what happens in simplex networks. The connection between *father* and *Paul* is an *outer-space* Role-Value connection between the inputs, as is the connection between *ego* and *Sally*. But under blending, the *inner-space father-ego* relationship inside one input is compressed with the *outer-space* Role-Value connection between *ego* and *Sally* to create a single compressed new *inner-space* role: *father of Sally*. In the overall blend, this role is additionally compressed with its unique value, Paul.

Of course, any value may itself have internal complex semantic structure. Take the simplex blend that has, as its first input, the frame of *sports*, which has a role for *fan* and a role for *sporting event*, and as its other input a specific sporting event, *bicycle racing*, and a specific person, *Sally*. In the blend, we can have Sally as a fan of bicycle racing. One role connector, from sporting event to bicycle racing, and another, from fan to Sally, are compressed in the blend, along with the frame relation of the fan to the sporting event. In this particular case, the value *bicycle racing* itself has frame structure as well as roles such as contestant, umpire, and mechanic.

In simplex networks there are no clashes between the organizing frames of the inputs, because the input with the values (*Paul* and *Sally*) has no organizing frame that competes with the organizing frame provided by the other input (*father-ego*). Accordingly, these networks perform indispensable role compressions.

Mirror Networks

We have seen several mirror networks already—the Buddhist Monk, the Debate with Kant, and Regatta. A mirror network is an integration network in which all spaces—inputs, generic, and blend—share an organizing frame. As we saw in

Chapter 6, an organizing frame for a mental space is a frame that specifies the nature of the relevant activity, events, and participants. An abstract frame like *competition* is not an organizing frame, because it does not specify a cognitively representable type of activity and event structure.

The input spaces mirror each other in the sense that they have the same organizing frame. So does the generic space. The blended space also has that frame, but often, in the blend, the common organizing frame of the network inheres in a yet richer frame that only the blend has. For example, in Regatta, the shared organizing frame *boat sailing along an ocean course* inheres in the more elaborate frame in the blend of *sailboats racing along an ocean course*. In the Debate with Kant, the shared organizing frame *philosopher musing on a problem* inheres in the more elaborate frame in the blend of *philosophers debating about a problem*. And in the Buddhist Monk, the shared organizing frame *man walking along a mountain path* inheres in the more elaborate frame in the blend of *two men meeting on a mountain path*.

An organizing frame provides a topology for the space it organizes; that is, it provides a set of organizing relations among the elements in the space. When two spaces share the same organizing frame, they share the corresponding topology and so can easily be put into correspondence. Establishing a cross-space mapping between inputs becomes straightforward.

While spaces in a mirror network share topology at the level of an organizing frame, they may differ at a more specific level. For example, in the boat race network, there are two elements that fit the role *boat* in the organizing frame and so have identical topology at the level of the frame. More specific relations, however, define finer topologies that often differ. For example, in Regatta, one of the elements fits the more specific frame *nineteenth-century clipper on a freight run* and the other fits the more specific frame *late-twentieth-century exotic catamaran on a speed run*. The two specific frames are different, so the topologies at that specific level are different.

A mirror network can integrate many different spaces, provided they share the same organizing frame. On July 8, 1999, the *New York Times* reported that Hicham el-Guerrouj had broken the world record for the mile, with a time of 3:43.13. An illustration accompanying the article shows a quarter-mile racetrack with six figures running on it, representing el-Guerrouj in a race against the fastest milers from each decade since Roger Bannister broke the 4-minute barrier in 1954. El-Guerrouj is crossing the finish line as Bannister, trailing everyone, is still 120 yards back. This illustration prompts us to construct a conceptual packet that blends structure from six separate input mental spaces, each with a 1-mile race in which the world record is broken by a runner. The blend places all six runners on a single racetrack, with a single starting time.

This blend has all the familiar features of conceptual integration networks. There is a cross-space mapping connecting counterparts in each of the six spaces: winners, racetracks, finish lines, the 1-mile distance, and so on. There is a generic space containing the structure and elements taken to apply to all these spaces, which constitute the fairly rich organizing frame of *running the mile and breaking the record.* There is selective projection to the blend: From each of the six input spaces, we project to the blend the entire frame of *running the mile,* but not, for example, a specific location for the race, or any of the runners except the winner. Some counterparts projected to the blend are fused, such as the racetracks; others, such as the record breakers, are not. And, finally, there is emergent dynamic structure in the blend—namely, structure that cannot be found in any of the inputs: The blend is a simulation of a mythic race between giants of the sport, most of whom never in fact raced against each other. In this mythic race, el-Guerrouj "defeats" Bannister by 120 yards.

This blend is immediately intelligible and persuasive, but its construction is remarkably complicated. Projecting to the blend el-Guerrouj, his location at the finish line, and his winning time as he crosses the finish line does not tell us how to locate the other runners behind him. The historical records do not indicate where they were at time 3:43.13. Their location on the track at that time must be calculated separately. In this case, the calculation is made by assuming that each runner ran his race at a uniform speed even though this never happens. We therefore see that the input mental spaces to the blend, however useful, are fictions that do not quite correspond to the real situations that inspired them. With these fictions in place, it is easy to compute the distance each runner has traveled at time 3:43.13 as the product of 1 mile and the ratio of el-Guerrouj's winning time to the runner's winning time. Subtracting the distance traveled from 1 mile yields the distance by which the runner trails el-Guerrouj. Specifically, Bannister, whose historic time was 3:59:4, trails el-Guerrouj by [1,760 yards] – [(3:43.13/3:59.4)(1,760 yards)] = 120 yards, rounding to the nearest yard.

To see further that there is nothing automatic or inevitable about this mirror network, we can compare it to the blend for the history of breaking the distance record for a fixed time in bicycling. In the standard one-hour competition in bicycling, the time of the performance is invariant but the distance varies. One breaks the record by going farther in one hour than anyone else has ever gone in one hour. For this blend, we can project both the time and the distance for each previous record-holder without having to perform any calculation. We simply place all the record-holders on the same track, each at the distance he achieved after one hour. The blend for the bicyclists looks just like the blend for the runners, but in the latter case some aggressive manipulations were required to achieve it. In the bicycle competition, the contestants in the inputs and in the blend do in fact stop after an identical time has elapsed—namely, one hour. In the runners' race, the contestants in the blend effectively stop competing to win the moment the winner crosses the finish line, even though their counterparts in the input spaces continue to compete, finish the mile, and break the world record that holds in each input space.

Clashes. In a mirror network, there are no clashes between the inputs at the level of organizing frame, because the frames are the same. But there will be clashes at more specific levels below the frame level. In Regatta, the centuries and the kinds of boats in the two spaces clash. In the Buddhist Monk, the directions and times of travel in the two spaces clash. In the Debate with Kant, the input spaces clash with respect to the languages used by the philosophers, the centuries in which they live, the mode of expression, and so on. These specific-level clashes can be resolved in two ways. In the first way, only one of the clashing elements is projected to the blend. For example, in the Debate with Kant, the frame-element *language* is the same in both spaces, and is projected from both spaces to be fused in the blend, but at the more specific level, there is a clash between its values in the two inputs—German and English, respectively—and only English is projected to the blend. In the second way of resolving specific-level clashes, the clashing elements are brought into the blend as separate entities. For example, in Regatta, the frame-element *boat* is the same in both spaces, and is projected from both spaces, but they are not fused in the blend. The more specific clashing elements *clipper ship* and *catamaran* are both brought into the blend, to yield two boats of different types. At the frame level in the blend, this projection of *two boats* satisfies the frame *ocean race* emergent in the blended space. But at the specific level in the blend, the result is an odd kind of race, because clippers don't usually race catamarans. The oddity has no effect on the purpose of constructing the mirror network—namely, to determine the relative speeds and positions of the boats—but it makes plain that this race is a conceptual blend with two input spaces.

Compressions. Mirror networks perform compressions over the vital relations of Time, Space, Identity, Role, Cause-Effect, Change, Intentionality, and Representation. Indeed, they make it exceptionally easy and straightforward to find the right candidates for compression and to perform acceptable compressions, because there are no clashes between the frames in the various spaces. Any footrace occurs on a single track located in a single location, so making a blend of footraces that has only a single track is easy.

The Buddhist Monk, Regatta, the Mythic Race, the Pronghorn, and the Debate with Kant all employ time compressions in which two spaces separated by the vital relation Time are made simultaneous in the blend. This simultaneity fits the frame used for the blend: Encounters require two people approaching each other at the same time, boat and foot races require competitors to perform simultaneously, predation requires the predators to be chasing as the prey is fleeing, and debating requires interaction. However, simultaneity is not the only possibility. Other patterns of selective projection of times to the blend can produce other results, such as time overlap, immediate succession, or separation by a short interval. As we have already seen, chunks of time can be syncopated: In the Buddhist Monk, fusing the times of his ascent and descent results in syncopation of the time between his arriving at the summit and descending from it.

The Buddhist Monk, Regatta, and the Mythic Race do not involve time scaling, since they preserve time magnitudes. Yet mirror networks can use time scaling, which preserves the topology of a temporal interval but changes its length. In the Pronghorn, which is a mirror network since all spaces have the frame in which a pronghorn is running from prey, the very long *outer-space* temporal expanse of evolutionary time is scaled down in the blend to the lifetime of the pronghorn.

Compression of Space is also easy in mirror networks. For example, spatial distances between the various inputs to the Mythic Race are compressed to zero in the blend, where all the runners are on a single track in a single location.

All the other vital relations—Identity, Change, Cause-Effect, and so on—are equally compressible into mirror networks. The Pronghorn is a marvel of vital-relation compressions. The most obvious is compression into Identity and Uniqueness, which turns, by stages, every pronghorn that ever existed into a single pronghorn in the blend, bent on survival and sprinting for its life. The long outer-space chain of many small evolutionary Changes of adaptation and inheritance is compressed into the blend as Intentional Change in the mind of a lone pronghorn, whose life has been a struggle to avoid death through feats of learning and memory. And the long outer-space chain of causes and effects across the entire phylogenetic descent of the pronghorns is compressed in the blend into being chased and learning very quickly to run fast and escape.

Conceptual integration in mirror networks routinely performs compressions of vital relations, both inner-space and outer-space, keyed by the shared frame of the network. The shared frame automatically provides linked roles.

Single-Scope Networks

A single-scope network has two input spaces with different organizing frames, one of which is projected to organize the blend. Its defining property is that the organizing frame of the blend is an extension of the organizing frame of one of the inputs but not the other.

The scenario of two men boxing gives us a vibrant, compact frame to use in compressing our understanding of two CEOs in business competition. We say that one CEO landed a blow but the other one recovered, one of them tripped and the other took advantage, one of them knocked the other out cold. This construal of the situation builds up a conceptual integration network. There is a cross-space mapping between the boxing input and the business input that maps, for example, each boxer to a CEO, a punch to an effort by one of the CEOs, a blow to an effective action, and staying in the fight to continuing the business competition (see Figure 7.2). In short, the projection to the blend in a simple single-scope network is highly asymmetric: One of the inputs but not the other supplies the organizing frame (*boxing*) and, therefore, frame-topology. (For example, there are just two agents, in close spatial and temporal proximity, engaged in physical adversarial action.)

Single-scope networks are the prototype of highly conventional source-target metaphors. The input that provides the organizing frame to the blend, the framing input, is often called the "source." The input that is the focus of understanding, the focus input, is often called the "target."

In one type of single-scope network, the inputs are not contained within a larger history. For example, in the Boxing CEOs example, we do not think that the boxers and the CEOs belong to one unified narrative. There are no historical connections between the input of boxing and the input of business. The CEOs are not former sparring partners, their businesses are not subsidiaries of the boxing business, and what they sell is not some transformation of boxing gloves. More generally, there are no vital relations of Time, Space, Change, Cause-Effect, and Intentionality that connect the input spaces directly, and no outer-space Identity connections between organizing frame roles or elements below that level of topology in the two different inputs. So, the individual boxer of the framing space is not directly identical to the individual CEO of the focus input space, and the role *boxer* in the framing space is not directly identical to the role *CEO* in the focus input space.

In a second type of single-scope network, however, the inputs are contained within a larger history. Outer-space vital relations can connect their organizing frames and elements below that level. The relevance of one input space to the other cannot be denied. Suppose a man tells his older sister about his present troubles, and she responds,

> "Do you remember how when you were little you were so intent upon hiding your treasures that you hid them so well even you could not find them again? Do you remember that you hid your new penny when you were four and we never found it? That's just what you have done with Angela. You've been talking for two hours about all your troubles, but what they boil down to is that you have hidden away your love for her so deeply that you can't see it. Once again, you've hidden your penny, even from yourself."

This is an example of a single-scope network. The frame that is exploited in the blend for purposes of understanding is the frame of one input (hiding the cherished penny too well), and the point of the blend is to cast light on the other

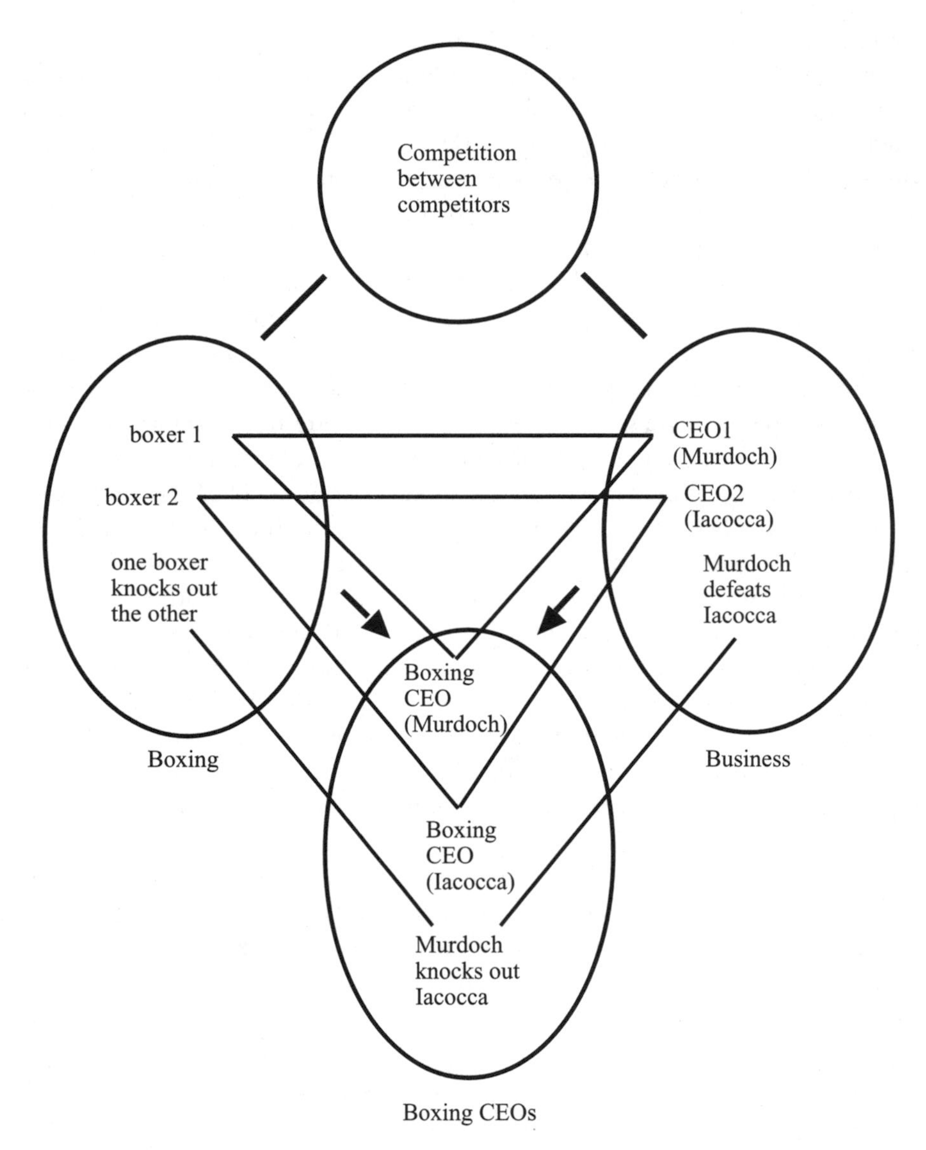

FIGURE 7.2 SINGLE-SPACE MAPPING

input (the adult brother's troubled life). Hiding the Penny is the framing input and the Troubled Life is the focus input. When there is such a vital relation between the two inputs, the effect of the network goes far beyond analogy. It adds to the temporal connection in the overall history a comprehensive pattern of causality. What has happened once will happen again in another guise. "Hiding the Penny" reveals a deeper principle of psychological character and action that is responsible for the later input as well. This single-scope network reveals something of the brother's deeper psychological essence. Indeed, such Vital-Relation single-scopes have many purposes. Suppose the sister says, "Watch out! Do you remember when you hid your penny . . . ? Well, you are on the verge of doing just that with Angela. Don't do it to yourself again! This time it's more than a penny you are going to lose!" In this case, she would be alerting her brother to the network and its potential, in the hope that he will take action now to stop that network from "coming true." Alternatively, it might happen that the brother himself, coming home to find Angela gone, has a sudden revelation, a sense of déjà vu, in which the image of himself as a four-year-old tearfully tearing up the house looking for the penny comes suddenly to his mind. The folk psychology behind this kind of network is either It Is Written ("A leopard can't change its spots") or Early Habits Persist ("As the twig is bent, so grows the tree"). In the It Is Written version, the person possesses an essence or abiding character that causes "the same thing" to happen repeatedly, so that the first such event becomes a warning signal for the later cases; in the Early Habits Persist version, the person early in life establishes a pattern that persists. Accordingly, in the It Is Written version, the causality in the network is flowing from the abiding personal character in the generic space to the two inputs; in the Early Habits version, the causality is flowing from the earlier framing input.

Single-scope networks offer a highly visible type of conceptual clash, since the inputs have different frames. They are cases where the clash is dealt with by giving the overall organizational power of the network to only one of the input spaces, the framing input. In the typical case, the framing space has a prebuilt superb compression that is exploited to induce a compression for the focus input. Naturally, then, single-scope networks give us the feeling that "one thing" is giving us insight into "another thing," with a strong asymmetry between them. This feeling of insight has three causes: The blend brings to bear inferences that are available from the framing input; it brings to bear useful compressions that already exist in the framing input; and it evokes emotions, seemingly anchored in the trustworthy framing input, that feel to us as if they are all-clarifying. As we have seen for blends in general, strong emotions emergent in the blend can induce the feeling of global insight, because the highly compressed blend remains actively connected to the entire network.

In prototypical single-scope networks, the blend does not disrupt the frame of the framing input. One feels that what is experienced in the blend was there all along and, therefore, that the insight captured is indeed some reliable discovery about the focus input.

The most obvious kind of compression in single-scope networks is the use of preexisting compressions from the framing input. A principal job of such networks is to project diffuse structure from the focus input into the already-compressed inner-space relations that have been projected to the blend from the framing input (see Figure 7.3).

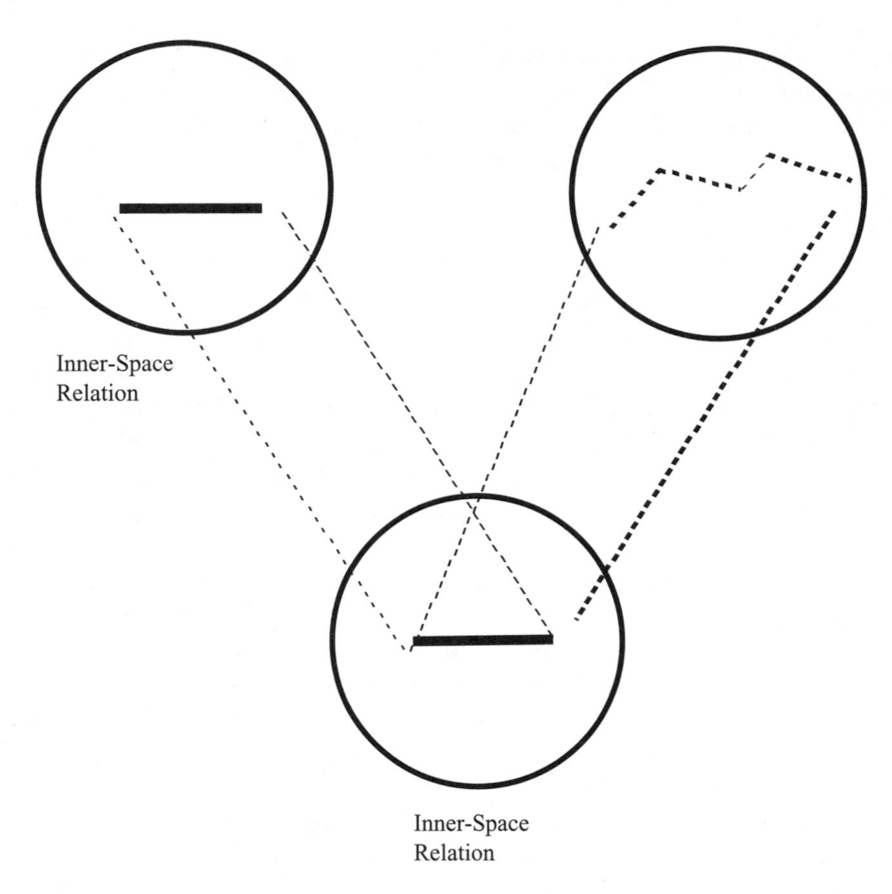

FIGURE 7.3 PROJECTION OF INNER-SPACE RELATION

In the boxing input, the identities, events, time, space, role-value relations, and causality individually have a tight compression, and the package of relations that frames them is also compressed. Two individual people who are boxers throw punches in a boxing ring for a half-hour or so, and the blows cause one of them to fall down. In the focus input, the identities, events, time, space, role-value relations, and causality are diffuse. The CEOs are part of a large organization of agents, and the businesses they head have complicated relations to other institutions and to external financial events. The relevant actions may occur over long periods of time, involve many agents, and take place in many physical locations. Despite these manifest clashes between the framing input and the focus input, the topologies of the two spaces are preserved in the cross-space mapping: Cause-effect relations, agent-action relations, and temporal ordering are aligned in the two spaces. For this crucial reason, projecting the business structure as values into the boxing frame (e.g., each CEO is a boxer) leaves that frame undamaged. This kind of projection is an imaginative achievement. Frequently, considerable work is required to find a way to project the focus input that leaves the frame undistorted in the blend.

"He digested the book" elicits a single-scope network that achieves a superb integration of events. On the one hand, the network blends conceptual counterparts in the two spaces—eating and reading. On the other, it helps us to integrate some distinct event sequences from the space of reading. In the framing input, digesting already constitutes an integration of several different events. But its counterpart in the focus input, if considered independent of the network, is a series of discrete events—taking up the book, reading it, parsing its individual sentences, finishing it, thinking about it, and understanding it as a whole. The integrity in the framing input is projected to the blend so that this array of events in the focus input acquires, in the blend, a conceptual integration of its events into a unit.

Double-Scope Networks

A double-scope network has inputs with different (and often clashing) organizing frames as well as an organizing frame for the blend that includes parts of each of those frames and has emergent structure of its own. In such networks, both organizing frames make central contributions to the blend, and their sharp differences offer the possibility of rich clashes. Far from blocking the construction of the network, such clashes offer challenges to the imagination; indeed, the resulting blends can be highly creative.

The Computer Desktop interface is a double-scope network. The two principal inputs have different organizing frames: the frame of office work with folders, files, and trashcans, on the one hand, and the frame of traditional computer commands, on the other. The frame in the blend draws from the frame of office work—throwing trash away, opening files—as well as from the frame of traditional computer commands—"find," "replace," "save," "print." Part of the imaginative achievement here is finding frames that, however different, can both contribute to the blended activity in ways that are compatible. "Throwing things in the trash" and "printing" do not clash, although they do not belong to the same frame.

Double-scope networks can also operate on strong clashes between the inputs. Consider the familiar idiomatic metaphor "You are digging your own grave." It typically serves as a warning that (1) you are doing bad things that will cause you to have a very bad experience, and (2) you are unaware of this causal relation. A conservative parent who keeps his money in his mattress may express disapproval of an adult child's investing in the stock market by saying "You are digging your own grave."

At first glance, this conventional expression looks like a straightforward single-scope network, where the organizing frame of graves, corpses, and burial is projected to organize the blend—a blend in which someone unwittingly does the wrong things, and ultimately fails. Failing is being dead and buried; bad moves that precede and cause failure are digging one's own grave. It is foolish to bring about one's own burial or one's own failure. And it is foolish not to be aware of one's own actions, especially those that may lead to one's very extinction.

A closer look, however, reveals that this cannot be a single-scope network, because in a single-scope network the cross-input mapping aligns the topologies of the inputs and that topology appears in the blend. But in Digging Your Own Grave, the topologies of the inputs clash on causality, intentionality, participant roles, temporal sequence, identity, and internal event structure. In all these cases, the blend takes its topology from the "unwitting failure" input, not from the "digging the grave" input! The causal structure in the blend comes from the "unwitting failure" input, not the "digging the grave" input. Foolish actions cause failure, but grave digging does not cause death. It is typically someone's dying that prompts others to dig a grave. And if the grave is atypically prepared in advance, to secure a plot, to keep workers busy, or because the person is expected to die, there is still not the slightest causal connection between the digging and the dying. Even in the exceptional scenario where a prisoner is coerced into digging his own grave, it is not the digging that causes the death. The prisoner will be killed anyway.

The intentional structure comes from the "unwitting failure" input, not the "digging the grave" input. Sextons do not dig graves in their sleep, unaware of what they are doing. In contrast, figurative digging of one's own grave is conceived as unintentional misconstrual of action.

The frame structure of agents, patients, and sequence of events likewise comes from the "unwitting failure" input. Our background knowledge is that a "patient" dies, and then the "agent" digs the grave and buries the "patient." But in the blend, those actors are fused and the ordering of events is reversed. The "patient" does the digging and, if the grave is deep enough, has no other option than to die and occupy it. Even in the unusual real-life case where one might dig one's own grave in advance, there would be no necessary temporal connection between finishing the digging and perishing.

The internal event structure comes from the "unwitting failure" input. In that input, it is certainly true that the more trouble you are in, the more you risk failure. Amount of trouble is mapped onto depth of grave. But again, in the "digging the grave" input, there is no correlation between the depth of someone's grave and that person's chances of dying.

The blend in *digging one's own grave* inherits the concrete structure of graves, digging, and burial, from the "digging the grave" input. But it inherits causal, intentional, and internal event structure from the "unwitting failure" input. The two inputs are not simply juxtaposed. Rather, emergent structure specific to the blend is created, with all the curious properties noted above. The existence of a satisfactory grave causes death and is a necessary precondition for it. It follows straightforwardly that the deeper the grave, the closer it is to completion, and the greater the chance for the grave's intended occupant to die. In the blend (as opposed to the "digging the grave" input), digging one's grave is a serious mistake that makes dying more likely. In the blend, it becomes possible to be unaware of one's concrete actions—a situation that is projected from the "unwitting failure" input, where it is indeed fully possible, and common, to be unaware of the nature or the significance of one's actions. But in the blend, it remains highly foolish to be unaware of such concrete actions—a judgment that is projected from the "digging the grave" input and will project back to the "unwitting failure" input to produce suitable inferences (i.e., to highlight foolishness and misperception of individual's behavior).

We wish to emphasize that in the construction of the blend, a single shift in causal structure—*The existence of a grave* causes *death*, instead of *Death* causes *the existence of a grave*—is enough to produce emergent structure, specific to the blend: undesirability of digging one's grave, exceptional foolishness in being unaware of such undesirability, correlation of depth of grave with probability of death. The causal inversion is guided by the "unwitting failure" input, but the emergent structure is deducible within the blend from the new causal structure and familiar common-sense background knowledge. This point is essential, because the emergent structure, though fantastic from a literal point of view, is supremely efficient for transferring the intended inferences back to the "unwitting failure" input and, thereby, for making real-world inferences. This emergent structure is not in the inputs—it is part of the cognitive construction in the blend. But, also, it is not stated explicitly as part of the blend. It just follows, fairly automatically, from the unstated understanding that the causal structure has been projected from the "unwitting failure" input, not the "digging the grave" input.

The integration of events in the blend remains linked to events in both input spaces. We know how to translate structure in the blend back to structure in the inputs. The blend is an integrated platform for organizing and developing those other spaces. Consider a slightly fuller expression: "With each investment you make, you are digging your grave a little deeper." In the "financial failure" input, there are no graves, but there are investments; in the "digging the grave" input, the graves are not financial, but one digs; in the blend, foolish investments are shovels, and what one digs is one's *financial grave*. A single action is simultaneously investing and digging; a single condition is simultaneously having finished the digging and having lost one's money. Digging your own grave does not kill you, but digging your own financial grave causes your death/bankruptcy.

Digging Your Own Grave is a double-scope network. Death and graves come from the "dying" input, but crucial framing is projected from the "unwitting failure" input of discretionary action and mistakes that lead to failure.

Complex numbers are also a double-scope network. The inputs are, respectively, two-dimensional space and real/imaginary numbers. Frame structure is projected from each of the inputs: angles, rotations, and coordinates from two-dimensional space; multiplication, addition, and square roots from the space of numbers. In the blend, there is an emergent structure of numbers with angles and of multiplication as involving rotation. We will return to this blend in Chapter 13.

We also see a double-scope network in *same-sex marriage:* The inputs are the traditional scenario of marriage, on the one hand, and an alternative domestic scenario involving two people of the same sex, on the other. The cross-space mapping may link prototypical elements such as partners, common dwellings, commitment, and love. Selective projection takes frame structure from each input. It takes social recognition, wedding ceremonies, and mode of taxation from the first input of "traditional marriage," and same sex, absence of biologically common children, and the culturally defined roles of the partners from the second input. Emergent properties will characterize this new social structure reflected by the blend.

Double-Scope Networks with High Asymmetry. In some double-scope networks, although the blend receives projections of organizing frame topology from both inputs, the organizing frame of the blend is nonetheless an extension of the organizing frame of only one input.

Suppose a person, observing that the Vatican seems to be flat-footed in the metaphorical boxing match over abortion, says, "I suppose it's hard to bob and weave when you have a mitre on your head." The Pope's competition with an adversary is portrayed as a boxing match where the Pope is impeded as a boxer by the mitre he is obliged as Pope to wear on ritual occasions. We interpret this as meaning (with respect to the input space with the Pope) that his obligation as Pope to remain dignified impedes him in his competition. In the input space with the Pope, there is a relationship at the level of the organizing frame between the Pope and dignified behavior and also between the Pope and his mitre. The cross-space mapping between inputs does not give counterparts in input 1 for the *required dignity* or *required headgear* elements in input 2. The Pope's obligation and his headgear in input 2 both project to the headgear of the boxing Pope in the blend.

In the organizing frame of the input of boxing, the boxers are not impeded by headgear that is an impediment. In the blend, the organizing frame is slightly different: It contains the role *heavy headgear that makes fighting difficult.* This organizing frame is an extension of the frame of boxing, not of the frame of Pope and Roman Catholicism. Specifically, the frame of the blend has all the roles of *boxing.* But, the headgear—the mitre—is projected from input 2. In that input

2 frame, there is a crucial relation *R:* The dignity of the Pope makes it harder for him to compete because he must always be honest and decorous. In input 2, the role *mitre* is directly linked (as a symbol) to the role *dignity and obligation* of *Pope*. The crucial relation *R* in input 2 is projected to *R'* in the blend: The mitre/dignity makes it harder for the Pope to box. *Mitre* and *dignity* in input 2 are both projected to the same element in the blend, and, crucially, they have no counterpart in input 1. The blend gets an organizing frame from input 1 but also the frame-level relation *R* from input 2. This is what makes it double-scope.

In the blend, we find all the elements of the frame of *boxing* plus the heavy and unwieldy mitre on one boxer's head. It turns out that having a heavy object on the head is an impediment to fighting, and so we have a very natural and automatic pattern completion of the blend, leading to the new frame of *boxing as impeded by heavy headgear*. This frame is an extension of the organizing frame of input 1, not of input 2. This is what makes it asymmetric.

Recall that, in Digging Your Own Grave, the cross-space mapping connected incompatible counterpart relations, such as direction of causality, and that to project causal direction to the blend, it was therefore necessary to choose one rather than the other of these counterpart relations. In the Pope example, because relation *R* in input 2 has no counterpart relation in input 1 (and, *a fortiori*, no incompatible counterpart relation), it can be projected to the blend (appropriately extended by completion) and we need not choose between incompatible counterpart relations.

Nonclashing Double-Scope Networks. Of course, the two organizing frames of a double-scope network are not required to clash. On occasion, they can both contribute to a blend that incorporates both of them. For example, if in a particular corporate community traveling business partners are typically lovers, we can develop the traveling business partners/lovers frame, with emergent structure, which can become familiar and used routinely in the culture. Similarly, if during a Cold War all symphony orchestras that one superpower sends to tour the other contain spies, we can develop the *second violinist spy* frame, a double-scope whose input organizing frames have no frame-level clash.

CHAPTER 7
ZOOM OUT

TOBLERONE

Some networks have what appear at first sight to be extremely tenuous cross-space mappings between the two frames organizing the inputs, but they lead ultimately to extraordinarily rich emergent meaning and global insight. The construction of the network produces strong and essential vital-relation connections between inputs that at first seemed only tangentially related, if at all.

As an example, consider the following ad for Toblerone, a pyramid-shaped chocolate candy that comes in three sizes—small, medium, and large. This ad shows the famous Egyptian pyramids of Giza, two of them small, one medium, and one quite large. Much smaller, the ad shows a group of four Toblerone chocolates in the identical configuration. The caption reads "Ancient Tobleronism?" The ad also includes the motto "Toblerone: Inspires the World." With a wink, it suggests that such a match of shape and proportion between the Giza pyramids and Toblerone chocolates can hardly be an accident. Surely some profound causality connects them! In the blend, Toblerone inspired the world in antiquity to construct these monuments. In the *monuments* frame, monuments are built in honor of, and often in the likeness of, great things. Just so, in the blend, the pyramids have been built to honor and resemble Toblerone. The emergent structure here is that Toblerone has been around throughout history and is the great wonder of the world that inspired all the others.

This integration network approximates a single-scope network: The frame of *great things* and *monuments* organizes the blend, in which Toblerone now fills the role for "great thing." Little if any of the *chocolate-eating* and *chocolate-making* frame needs to be brought in. But the network establishes deep vital relations of causality, intentionality, and time between the inputs: Toblerone now causes the pyramids; it inspires the pyramid-builders, who intend to honor Toblerone. Crucial outer-space vital relations are the main emergent structure of the (ironic) network. The two main elements, the pyramids and the chocolate, though linked by resemblance, are not counterparts in the frames of the inputs: Toblerone is the counterpart not of the monument but, rather, of the great thing that inspired the monument. The emergent vital relations of Causality, Intentionality, and Time are intuitively in the wrong direction, since we must now have Toblerone preceding and being more important than the pyramids. This example, which arises entirely by means of the general structural and dynamic principles of integration networks, shows how far one can go from the prototypes with these principles.

The Toblerone network looks frivolous, but building a rich emergent structure out of a tenuous connection between inputs is a deep principle of science. The connection between an apple and the moon seems at first blush to be only the most tenuous analogy of shape and some motion, but in the blend of Newtonian physics, the apple and the moon are the same and exemplify the laws of the universe. Similarly, the psychoanalyst searches for seemingly "accidental" identities and analogies running across childhood experience, dreams, and adult behavior ("You say you always wear red shoes to the circus. What was the color of your hat, did you say, when you visited your grandmother on her deathbed?

You said the circus was a 'celebration of life.' Of course, you mean 'of death.'"). The psychoanalytic case of building global insights out of weak connections between spaces depends upon building vital relations of Identity, Cause-Effect, Time, and Intentionality between the earlier and the later spaces, on the basis of the emergent structure in the blend. As in the Toblerone network, the outer-space vital relations are emergent. Both Toblerone and the psychoanalytic blend develop theories in order to create the missing vital relations between inputs. In the Toblerone network, the theory is ironic, since it runs entirely contrary to our knowledge. In the psychoanalytic network, the theory is meant to convey the deepest principles of reality, and the specifics of the network explain inexplicable or poorly understood behavior.

DIVERSITY OF PRODUCTS FROM UNITY OF PROCESS

Now consider a straightforward example from geometry. A hyperbola, an ellipse, a parabola, and a circle—all are clearly different shapes. Intuitively, they fall into different categories. Mathematically, they can be given different precise geometric characterizations, specific to each kind of curve. And again, intuitively, a straight line or even a single point seems to be maximally different from conic curves. Yet all these geometric shapes can be viewed as variations on a single theme: the two-dimensional planar section of a three-dimensional (quadric) cone. Keep the plane perpendicular to the axis of the cone, and you get all circles; tilt the plane to obtain ellipses, tilt a little more for parabolas, and tilt again for hyperbolas. Move the plane parallel to one of the sides of the cone until the parabola condenses into a straight line. Move the plane toward the vertex of the cone until the ellipses or circles shrink into a single point. All of these different shapes are conic sections and fit on the same conceptual continuum. On this continuum, we find prototypes: the oval of an ellipse, the parabola of the cannonball shooting through the air, the circle of a beer coaster, the asymptotic hyperbola. And we find limiting cases: the straight line, the point, the circle itself relative to ellipses, or the parabola as a limiting case of ellipses stretched out to infinity.

A theory of conic sections offers a way to see a useful mathematical unity behind the diversity of shapes and figures, without reducing this diversity to a single prototype: We do not say that these shapes are all more or less circles simply because, when viewed from this perspective, they belong to the same continuum on which the circle is a prototype. Similarly, conceptual integration networks offer a way to see unity behind the diversity of particular manifestations of meaning constructions. We start, as in the conic-section case, by formulating the general characterization evoked above (cross-space mapping, projection into a blend, emergent structure by completion and elaboration, etc.). Then we look at specific variants of this general structured dynamic process.

Eight

CONTINUITY BEHIND DIVERSITY

Every minute now should be the Father of some Stratagem.

—*William Shakespeare*

SIMPLEX, MIRROR, SINGLE-SCOPE, and double-scope networks turn out to be powerful, widely deployed tools of the imagination. They look and feel different for all the reasons we have shown, but there is a deep continuity across them. They are not four separate and unrelated species that exhaust the world of blends but, instead, are prominent points that stand out on a continuous landscape. Here, we will take a walk along that landscape to show, step by step, how we get from one to another, encountering intermediate cases as we go.

What could be the consequences of this inquiry? More specifically, what could be the importance of an underlying continuity? For theorists of conceptual integration, the regularities of the operation and the corresponding regularities that operation creates in its products are at the heart of the scientific enterprise. For linguists and cognitive scientists who are already directly interested in how meaning is constructed and how language prompts for meaning, we offer explanations of these constructions and prompts. For those interested in creativity and the construction of meaning in other areas—such as reasoning, invention, or the cultural development of new frames of knowledge—it will be useful and, we hope, illuminating to see how the continuities manifest themselves in those areas and how they extend across language phenomena as well.

But even more is at stake. The findings of this chapter lay the ground for a fundamental hypothesis about the origins of language, which we begin to present in the next section of the book. The origin of language is an enthralling series of multidisciplinary puzzles. For the evolution of species, we find intermediate fossils that have rudimentary structures, and we find species living now that have similar organs but with varying degrees of complexity—different levels of "wing," for example. But for languages, there are no intermediate fossils, and every human language we know of, living or dead, is highly complex. Among the world's full adult languages, there are no simple languages, or languages simpler than others. Even rudimentary pidgin codes that serve as *linguae francae* turn almost immediately into creole languages of great complexity if there are any children around learning those codes as native languages. We also see that many species have vocal skills but no language, and we see that human language can come in different modalities—spoken or signed—with equal levels of grammatical complexity. Moreover, there are very many separate human languages, not just one, and they accomplish their tasks in ways that often seem surprisingly different. Human languages change over cultural time but do not as a result acquire increased complexity. Indeed, it seems that, at any given time, languages present the same kind of diversity and have the same degree of complexity. We will consider these puzzles in the next chapter on the basis of what we find in this one.

Let's begin with an example from the last chapter: the simplex network in which Paul is the father of Sally. The generic space has two people and no relations. The framing input is the *father-child* subframe of our more general kinship frame. The values input consists only of two people with no relation between them, *Paul* and *Sally*. Such a network creates a compression of roles and values in the blended space. This simple network is only the beginning of a long gradient of increasing complexity. Let us look at a gradient of networks all using the *father-child* input and, crucially, the word "father." One result will be that the word "father" will seem to have many different meanings. But in fact the word is always playing the same role, inviting us to use our potential to construct networks that have *father-child* as one input. Consider:

> Zeus is the father of Athena. She was born out of his head, fully clad in armor.

Now, from the kinship space, we bring in general schemas of human progeneration, such as the offspring's coming out of the body of a parent, but we bring in from our knowledge of divinity the possibility of unusual birth. We explicitly build in the blend, on the warrant of the second sentence, an unusual kind of progeneration that involves neither a mother nor an infant.

The divinity in the *Zeus-Athena* input allows for many wonderful blends, each of which contains a *father* and a creative method of progeneration. For instance, Zeus is also the father of Aphrodite—this time, by virtue of having castrated Chronos and cast his genitalia into the ocean foam, whence Aphrodite was born.

The Zeus cases do not entail figurative speech or analogy. Zeus is still felt to be quite clearly the father of Athena, and of Aphrodite. Family structure is inferred, as are the sentiments and emotions that come with it.

Now consider:

> Joseph was the father of Jesus.

In this blend, we do not project the usual structure of the father's role in procreation or the nonvirginity of the mother. But we can project family structure and emotions. We also know that the communities of people in which they lived largely took Joseph to be Jesus' father in every way. Again, this use of "father" is not felt to be metaphoric or analogical.

Now consider a neighbor who takes care of Sally for the day while Paul is away, carrying out fatherly duties such as making her lunch, accompanying her to school, and reading bedtime stories. That neighbor can say to Sally: "I'm your father for today." As in the Zeus and Joseph blends, some family structure and genealogy is projected. As in the Joseph blend (but not the Zeus blends), progeneration is not projected. Many of the typical aspects of the father-offspring relationship are projected (routines, care giving, responsibility, affection, protection, guidance, authority, and so forth). Meaning compositionality is no longer an option to account for this case. Too many properties felt as central are missing. We have moved along the continuum from the pole of simplex networks. But clearly, we have not reached a point on the continuum that would be felt intuitively to be metaphorical. Fatherhood is not a metaphor for what the neighbor is doing. In fact, although some analogy has now contributed to the mapping, the function of this blend is stronger than mere analogy. The neighbor really is filling the role of the father in relevant respects, not just doing something "similar" to what the father does. The flexibility of blending with selective projection and contextual elaboration allows for this intermediate kind of situation, which doesn't fit a usual characterization.

These "father" networks reveal some fascinating principles of how words work. When a word is attached to one of the inputs, it too can be projected to the blend, to pick out meaning there. We will call this principle "word projection." For example, in the Zeus and Joseph cases, "father" is projected to the blend from the *father-ego* input but now picks out *new* meaning in the blend. Because the blend and the framing input remain connected, we can refer to birth as "leaving Zeus's head" in much the same way that we refer to birth as "leaving the womb." Many similar expressions, each using words that already apply to the inputs, can be fashioned that pick out meaning only in the blend. We can also refer to Joseph as "Jesus' mortal father," giving "mortal father" a contrastive rather than redundant meaning, which is likewise inappropriate for the *father-ego* input.

Consider these additional examples linked to "father":

The Pope is the father of all Catholics.
The Pope is the father of the Catholic Church.
George Washington is the father of our country.

The first example still has people in both inputs. From the "kinship" input that provides the word "father," we project not progeneration at all but, instead, authority, size of the family, responsibility, leadership, social role. From the second input, we project specific properties of Catholicism.

"The Pope is the father of the Catholic Church" arguably projects the role of a child to a single social entity (the Church). The blend reflects a type of sociocultural model—specifically, one in which a social entity (church, nation, community) is the "child" of its leader. The word "father" is now felt to have a different meaning, but not a particularly metaphoric one. The same sentence can also be understood to blend the role of the Pope in the institution with the role of father in a family.

With the George Washington sentence, we go a bit further by highlighting the causality in time between parent and child and between founder and nation. This abstraction increases the perceived difference between the two inputs and their domains. The impression of metaphor is undoubtedly stronger. That subjective impression reaches a higher point when the two domains are even more explicitly distinguished, as in "Newton is the father of physics." Physics, unlike church and country, does not even stand in metonymic relation to people and groups of people. Yet Newton and Washington as adult men have all the biological features of possible fathers plus some of the stereotypical social ones (authority, responsibility, and so on). The conceptual integration networks directly bring in frame structure from both inputs.

Even more subjectively metaphorical is an expression like Ezra Pound's "Fear, father of cruelty," in which the two domains (emotions/qualities and people/kinship) have no literal overlap at all, and the projected shared schema, causality, is correspondingly abstract. Finally, Wordsworth's acrobatic metaphor "The Child is father of the Man" comes around almost full circle by using background knowledge (children grow into men) to create emergent structure in the blend that maps kinship to the human condition in an unorthodox way. The oddness of its counterpart connections and the extensive drawing on the frames of both inputs to create a new organizing frame for the blend help make Wordsworth's line feel figurative. But the syntax and mapping scheme of "The Child is father of the Man" are the same as the syntax and the mapping scheme of "Paul is the father of Sally."

The blends we have been talking about are often constructed using language. The reason language can prompt for blends that result in the same word's being used to pick out different meanings is that language does not represent meaning directly; instead, it systematically prompts the construction of meaning. All of the "father" examples are examples of the familiar XYZ construction ("X is the Y of Z") whose purpose is to systematically prompt for blends.

We feel that the word "father" has different meanings because "father" is in each case attached to one of the inputs, blending as a conceptual operation applies to those inputs, and "father" comes to pick out elements in the blend and to participate in phrases that pick out structure in the blend but not in the inputs. The meanings of "father" are not properties of the word "father" but byproducts of the operation of conceptual integration and of the fact that words, like anything else attached to inputs, can be projected to blends. The cognitive operation of conceptual blending, with its mechanisms of selective projection and elaboration, is not restricted to linguistic examples. But a mind that can do blending and that also knows language will inevitably develop meanings for words through blending. If words show up in inputs, they can be projected like any other element of an input. This outcome will change their domain of application, unnoticeably in most cases, but noticeably when the emergent meaning to which they apply in the blend seems remarkably distant from the domain of the input from which they came. When we notice this distance, we call it by one of many names: extension, bleaching, analogy, metaphor, displacement. "Polysemy"—the fact that a single word seems to have "many meanings"—is a very common phenomenon, a standard by-product of conceptual blending, but noticed in only a fraction of cases.

We may think of language as a system of prompts for integration. Since there are many conceptual structures to be integrated, each with ranges of words attached to them, an expression that prompts for their appropriate integration has to combine words, and language has to have forms to make such combinations possible. Obvious examples are predication ("This beach is safe") and compounding ("likely story," "possible solution," "eligible bachelor," "fake gun"). Consider, for example, "This beach is safe," which we discussed in Chapter 2. A common but misguided way of describing the meaning of this sentence is to say that a particular property, *safe*, is predicated of an object, *beach*, by means of the words "safe" and "beach." On this view, "This house is safe" asks us to apply the same particular property, *safe*, to a different object, *house*. So, "safe" has just one meaning. It would be straightforward to say "The beach is safe" when we want to let a child play there. And in that situation, it would be equally true that "The child is safe." But now we see that the purported property *safe* attributed to the beach in "The beach is safe" and to the child in "The child is safe" would have to be two different properties: something like *not potentially harmful* in the first sentence and *not likely to be harmed* in the second. By the same token, the word "safe" in the sentence "The beach is safe" would have to apply many different properties depending on whether the beach is legally protected from development, has a statistically low number of drownings, is a low-crime area, is owned in such a way that its ownership cannot be taken away from the owner, or is a vacation spot that can be proposed without problem to someone (as in a "safe bet"). On inspection, then, "safe" can seem to mean many different things, but users of the word feel that it is the same word and the same concept.

In order to do justice to the meaning of "safe," we must regard it not as applying a particular property but, instead, as prompting for a particular kind of blend. The blend takes into account the frame of harm and the specifics of the situation referred to in the rest of the expression. It requires us to blend them to create a counterfactual scenario in which there is specific harm and to understand how the present situation is disanalogous to that scenario. In fact, the linguistic expression singles out the disanalogous counterparts. For example, "The beach is safe," meaning that the child won't drown, singles out a counterfactual counterpart beach with riptides and deadly waves, and asks us to understand that the beach in the present situation lacks these dangers. The meaning of "safe" is not an invariant set of properties but, rather, a set of instructions for building a suitable conceptual integration network on the basis of the *harm* scenario.

In the network, the complex vital relations of Disanalogy, Cause-Effect, and Identity involved in making the counterfactual space in which harm happens are compressed in the blended space into the inner-space vital relation of Property, applied to one of the roles in the harm scenario.

Can we find other kinds of linguistic evidence for the claim that a unified operation—conceptual integration—is being used to construct meaning across all these continua? One kind of linguistic evidence would be very strong: If a range of meanings that we view as different—simple framing, analogical, metaphorical—are indeed all instances of conceptual integration, then there might be a single syntactic form that prompts for the construction of all these different meanings. In fact, English, almost too good to be true in this regard, has a construction that is specialized in exactly this way. The construction is the apparently pedestrian but immensely powerful XYZ construction, such as "Necessity is the mother of invention" or "He was the Einstein of the fifth century B.C." In this construction—"X be Y of Z"—X, Y, and Z are nouns or noun-phrases. X and Z identify elements x and z in one input space, and Y identifies element y in a second input space. The copula *be* indicates that x and y are counterparts. And the understander must identify the two relevant domains and set up an implicit element w to be the counterpart of y. In "Vanity is the quicksand of reason," one input space is concerned with human traits such as reason and vanity while the other has to do with traveling and falling into quicksand. The missing element w is the traveler. The structure in the input of travel has a traveler falling into quicksand and therefore failing in his enterprise (perhaps to the point of dying). In the input with reason, we understand that reason is similarly imperiled by vanity. The blend corresponding to the grammatical XYZ construction thus has one element that is reason/traveler and another that is vanity/quicksand.

Yet, quite interestingly, we have seen how that same construction also prompts for simplex networks. "Paul is the father of Sally" has the very same syntactic form as "Vanity is the quicksand of reason." It maps the kinship subframe *father-ego* onto the two people *Paul* and *Sally* and integrates the two inputs into a more richly structured blend. The mapping scheme is exactly the same as in the metaphorical case: In one space, x and z are *Paul* and *Sally;* in the other space, y is the role *father* and the counterpart of x (*Paul*). The missing element w is the *ego* of the kinship frame.

The XYZ construction in English is a general prompt to construct an XYZ integration network. This construction covers not only simplex, single-scope, and double-scope networks but also all the intermediate integration networks ("Zeus is the father of Athena," "I'm your father for today," The Pope is the father of the Catholic Church," and so on).

But what about the frequent cases in which the word "of" seems to mean "a part of," as in "the door of the car" or "the top of the building"? Indeed, these cases are also straightforward instances of the general mapping scheme. The word "top" does not in itself denote a part of a building. Rather, it is part of a more general frame—roughly referring to things that have vertical orientations and are bound in space. So when we point to a location in the building and say "This is the top of the building," we are constructing a simplex XYZ blend, just as in the other cases. As depicted in Figure 8.1, "This" identifies location x in the focus input with the building, "the top" identifies a vertical extremity y in the *whole with parts* framing input, and "the building" identifies element z in the building input. The missing element w is the general notion of a whole vertically oriented thing in the *whole with parts* frame. The whole w is mapped onto z, the building.

Are there XYZ phrases that prompt for mirror networks? Yes, of course! Suppose Paul had a daughter Elizabeth who died at age ten, and then, two years later, another daughter, Sally. The loss of Elizabeth, with whom he had a warm, spontaneous relationship, was traumatic for him and made him stiff and withdrawn with Sally as she grew up. Finally, when Sally is sixteen, and Paul has seen her grow up during a stage that Elizabeth never knew, Paul comes to his senses and suddenly begins to behave with full warmth and spontaneity with Sally. Sally, confused, finally asks him why he is treating her in this new way. He responds, after a long pause, "Because you are my long-lost daughter." This expression is the idiomatic equivalent of "You are the long-lost daughter of me." In one space, organized by the *father-daughter* frame, we have Paul as the value of *father* and Elizabeth as the value of *daughter*. In the other space, organized by the same frame, we have Paul as the value of *father* and Sally as the value of *daughter*. The generic space and blended space are also organized by the *father-daughter* frame, making this a straightforward mirror network. In the blend, father is blended with father, daughter with daughter, Paul with himself, and Elizabeth with Sally. In the blend, Elizabeth is Sally. By making this blend, Paul recovers his long-lost daughter Elizabeth, who died, and also recovers his daughter Sally, who had been lost to him in a different way. There are many crucial vital relations between the inputs, so this is a vital-relation mirror XYZ network. The blend has spectacular emergent emotions and dramatic psychological changes for two people and for the memory of a third.

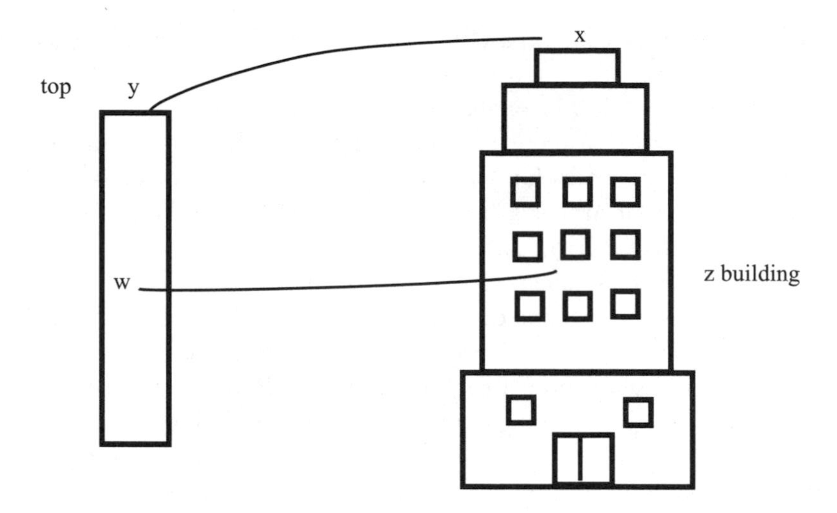

FIGURE 8.1 A SIMPLEX XYZ BLEND

COMPOSITIONALITY

When we see words on a page, do those words stand directly for external realities? No. As we have seen, words and the patterns into which words fit are triggers to the imagination. They are prompts we use to try to get one another to call up some of what we know and to work on it creatively to arrive at a meaning. Blending is a crucial part of this imaginative work, and, as we have seen, blending is not the mere addition of one existing meaning to another to get their sum. Words by themselves give very little information about the meaning they prompt us to construct.

Any integration network we build will end up having a mapping scheme. A simplex network, for example, maps roles in one input onto their values in another, fusing them in the blend. Precisely because it is not merely additive, blending is highly creative. But the capacity to combine and recombine is also a hallmark of human thought. Can blending gain strength by drawing on this capacity for recombination? Yes. Using recombination in conjunction with blending is an exceptionally powerful trick of the imagination.

We have seen various mapping schemes so far—schemes that operate over two inputs, a generic space, and a blend. And we have seen forms of language—"safe," "if," "of," and so on—that prompt us to use some of those mapping schemes. What these examples show us is that blending is a constant mental activity: We blend again and again, building blends out of earlier blends, blends all the way down. Now we ask a crucial question: Is there any regularity to the way that blends compose, or do we have to invent new, idiosyncratic mapping schemes for every compound network? Rather amazingly, no matter how unpredictable creative blending is at every stage, and no matter how various its products seem—simple predication, jokes, metaphors, counterfactuals, mathematical discoveries, analogies, category extensions, event integrations, action blends, and so on and on—it can use the same skeletal mapping schemes again and again and combine them in the same simple ways. And we will see that we combine in simple ways the linguistic expressions that prompt for those skeletal mapping schemes. This is remarkably efficient and elegant: The combination of expressions prompts for a parallel combination of the skeletal mapping schemes they evoke. Composing language forms into more complex language forms thus tells us what kinds of mapping schemes to build, although those language forms say nearly nothing about the kinds of meaning we will construct under their guidance. Meanings themselves are the imaginative products of blending, whether simple or complex, and are not predictable from the forms used to evoke them. The mapping schemes, by contrast, *are* predictable from the language forms used to evoke them. The meaning of the whole is not predictable from the meaning of the parts, but the mapping scheme of the whole is predictable from the mapping schemes of the parts. This compositional aspect of *forms* has been interpreted in the form approaches as telling us the nature of thought itself. But on the contrary, it is only an exceptionally ingenious, useful, and efficient feature of the forms that guide thought.

An XYZ expression like "The adjective is the banana peel of the parts of speech" can mean that unwary writers often slip up on adjectives, that the adjective is a very useful element that helps us get a grip on the noun to which it is attached, that the adjective attracts us but the noun is the real substance, that the adjective is a deceptive surface that hides the essential meaning, and so on. All these different meanings arise from the identical XYZ mapping scheme: X (*adjective*) in the first input is the counterpart of Y (*banana peel*) in the second, Z (*parts of speech*) in the first input is the counterpart of some unmentioned W in the second input. A suitable but unspecified Y–W relationship (*banana peel–W*) is to be projected to the blend and integrated there with the X and Z projections. The language form "X is the Y of Z" directs us to this mapping scheme but does not indicate what we are to do with it. Constructing the meaning involves much more: frames, topologies, general knowledge, context, connections of identity and role and vital relation, and, above all, blending.

For example, we construct meaning for "Sally is the daughter of Paul" on the basis of our full knowledge of human beings, progeneration, family structure, and so on. Using the *same* language form and the *same* mapping scheme, we construct quite a different meaning for "Athena is the daughter of Zeus." In the Zeus-Athena blend, *daughter* is now compatible with the structure *having no mother, born out of the father's head, born as an adult fully clothed in armor,* and so on, so that what look like necessary properties of *daughter* have not been projected to the blend. "France is the eldest daughter of the Church" has the same language form and the same mapping scheme but even more inventive projections to the blend with emergent meaning.

Language forms combine to evoke orderly combinations of mapping schemes. This handy ability can be nicely illustrated by looking at compound XYZ expressions of any length, such as "Elizabeth is the boss of the daughter of Paul" or "Jim is the secretary of the wife of the president of Disney" or "Elizabeth is the boss of the daughter of the secretary of the wife of the president of Disney." To conduct this illustration, we need to back up and look at the backbone of an XYZ expression—namely, *Noun-Phrase of*. We will call this backbone a "Y expression."

Consider, for example, the phrase "The boss of," which is diagrammed in Figure 8.2. It prompts us to call up an input space structured by the *boss-worker* frame, to construct a blended space, to project *boss* from the input to create *boss'* in the blended space, to provide for a w (almost certainly *worker*) in the input space, to project *worker* to create *worker'* in the blend, to project the *boss-worker* relationship in the input to *boss-worker'* in the blend, to construct open-ended connectors from *boss'* and *worker'* in the blend, and to expect the open-ended connector from *worker'* to connect to some element that can be picked out by the noun-phrase that will follow "of."

In general, a Y expression has this very simple form:

Noun-Phrase of

Expressions like "the boss of," "the secretary of," and "a beginning student of" are Y expressions. The Noun-Phrase in a Y expression is typically a role in a common frame, often a relational frame like *father-daughter*, *husband-wife, master-apprentice, president-company*, and so on. A Y expression prompts for blending. Specifically, it prompts us to perform the following operations:

- Call up an input space for the relational frame containing y (the element named by Y).
- Construct a blended space.
- Project the element y to create an element y' in the blend.
- Provide for a w in the input space that will bear an appropriate relationship to y.
- Project that w to create an element w' in the blend.
- Project the y-w relationship onto y'-w' in the blended space.
- Provide open-ended connectors from y' and w' in the blend, as shown in Figure 8.3. These connectors are expected to make connections at some point.
- Expect the open-ended connector from w' in the blend to connect to something picked out by the noun-phrase that will follow "of."

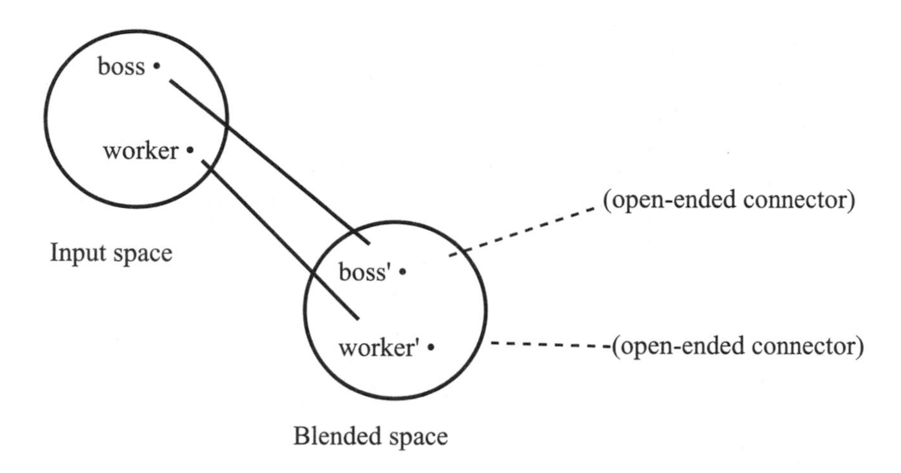

FIGURE 8.2 "THE BOSS OF"

What are these open-ended connectors? For w', what follows "of" will tell us how to attach that open-ended connector. For example, "The boss of Hieronymous Bosch" prompts us to build a space containing the painter Hieronymous Bosch and to connect w' to him. "The boss of the company" prompts us to build a space structured by the *company* frame and to connect w' to the role *company* in that space. For y', the open-ended connector can be directed by explicit linking, as in the simple XYZ expressions we discussed above—for example, "Elizabeth is the boss of Hieronymous Bosch" or "Elizabeth is the boss of the orchestra." Alternatively, it can be explicitly directed by a gesture, as when you point at someone and say "The boss of the orchestra." If there is no explicit linking, the y' element stands on its own as a role.

In the default case, open-ended connectors will attach to elements in the same space. This is exactly what we saw in the original simple XYZ construction, exemplified by "Paul is the father of Sally." "The father of" is a Y expression that sets up the Y-of network diagrammed in Figure 8.4.

In the full XYZ expression "Paul is the father of Sally," *Paul* is explicitly linked to the role *father*. Because "Sally" follows "of," *Sally* is necessarily linked to w' (*child*) in the blend. In this case, our Y-of network is completed as shown in Figure 8.5.

Most important, we can compose Y expressions by letting the open-ended connectors attach to other roles. That is, what follows the "of" in the first Y expression can be another Y expression, for as long as we like: "The doctor of the sister of the boss of Hieronymous Bosch . . . " (or, more generally, "The y1 of the y2 of the y3 of the y4 of . . . "). Earlier, we saw the mapping operations called for by a Y expression. Because these operations have open-ended connectors, they can be repeated, so that an open-ended connector of one Y-of network

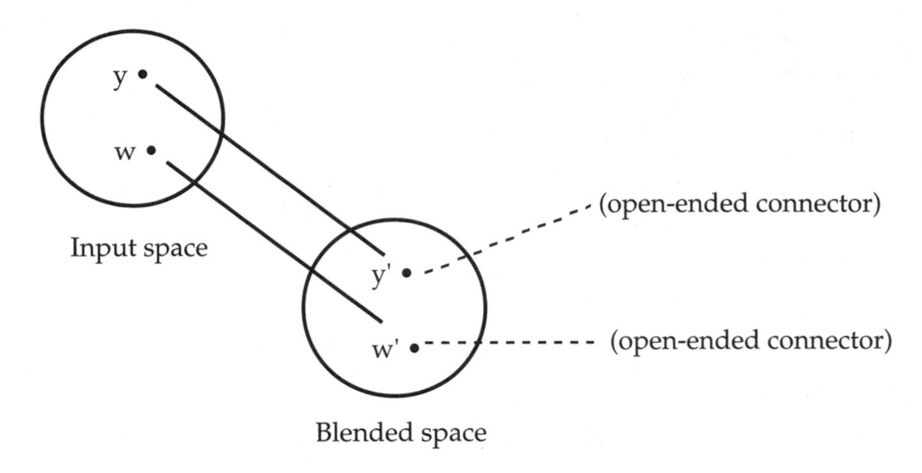

FIGURE 8.3 THE Y-OF NETWORK

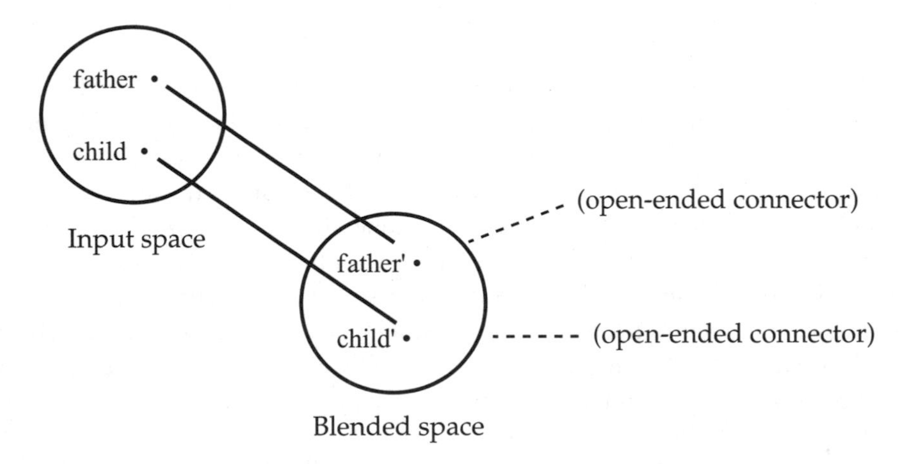

FIGURE 8.4 "THE FATHER OF"

attaches to another Y-of network. Repeating the expression just asks us to repeat the mapping operations. So if we have "The Y1 of the Y2 of," which contains two Y expressions, we are instructed to go through the mapping operations twice. We begin with Y1 to construct the Y1-of network. That gives us an element in the blend w1' with an open connector. Then we connect that network to y2 by attaching that open connector to y2. This is just what the mapping instructions tell us to do: Whatever follows "of" tells us where to connect the w in the blend. In this case, what follows "of" is Y2, so we attach the w1' open connector to y2. Then we proceed to construct the Y2-of network. The composed mapping scheme is shown in Figure 8.6.

"The secretary of the wife of the president of" has three Y expressions, which, as shown in Figure 8.7, prompt us to repeat the mapping operations three times.

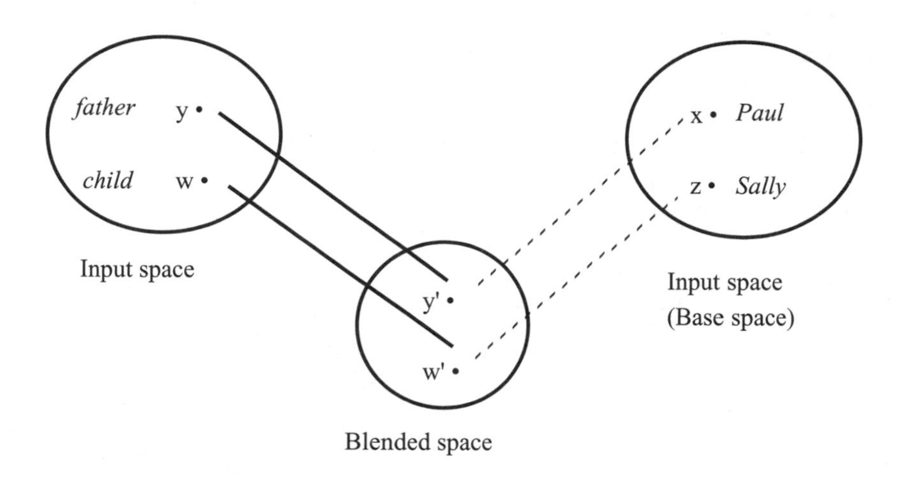

FIGURE 8.5 THE XYZ NETWORK

A general network of this kind is what we call a Y^n network, where n is the number of repetitions. In principle, there is no limit to the number of Y expressions or to the number of mapping schemes. But in practice, we find few examples that are higher than Y-of cubed. One important reason for this is that there is more going on in these networks than simply stringing along the spaces and doing the successive blends. Compression for global insight is a general value for cognition, as we have discussed. In Y networks, new meaning is being constructed in the blends. Therefore, as blends arise, the pressure will be to integrate them if they are compatible. As we build Blend 2, we are trying to blend it with Blend 1, and each time a new blend arises, we try to blend it with the most recent megablend of all the previous blends. At any moment, the megablend of blends is giving the best global insight into the entire network. For example, "the valet of the secretary of the president" prompts for a compressed megablend, as depicted in Figure 8.8. In the successive blends given by the mapping schemes, we get new roles *valet of the secretary* in Blend 1 and *secretary of the president* in Blend 2. In the megablend, compression gives us yet another role, *valet of the secretary of the president*, which can become conventional and have its own emergent structure that is not a composition of the features of the individual roles. We might say, for example, "The valet of the secretary of the president is a hereditary title" or, in another domain, "The wife of the secretary of the president is the most important position in the government." That role is not in any of the intermediate blends or in any of the input spaces to those blends. It provides evidence for the building of the megablend.

Figure 8.8 depicts only the mappings to the megablend, not the richness of its emergent meaning. For example, *a* in the megablend is now a compressed

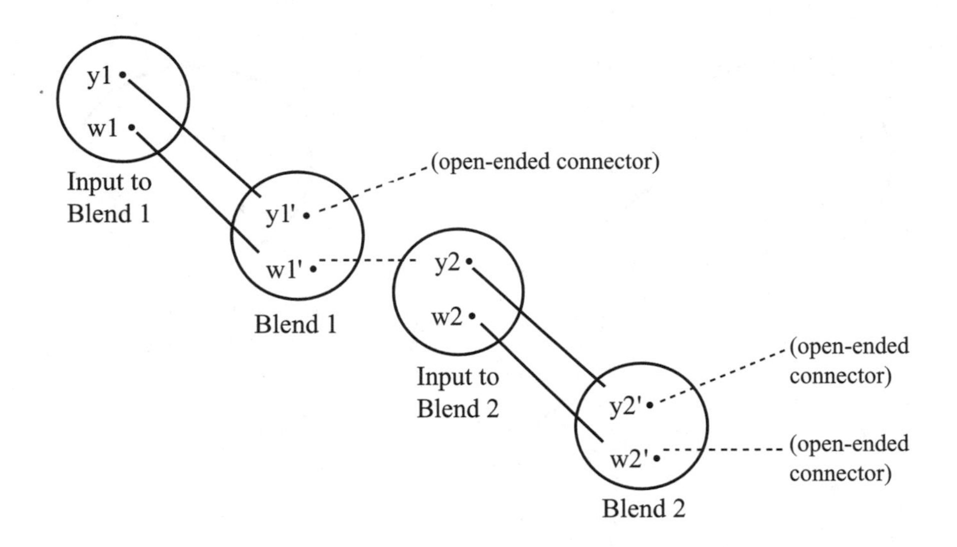

FIGURE 8.6 THE Y-OF SQUARED NETWORK

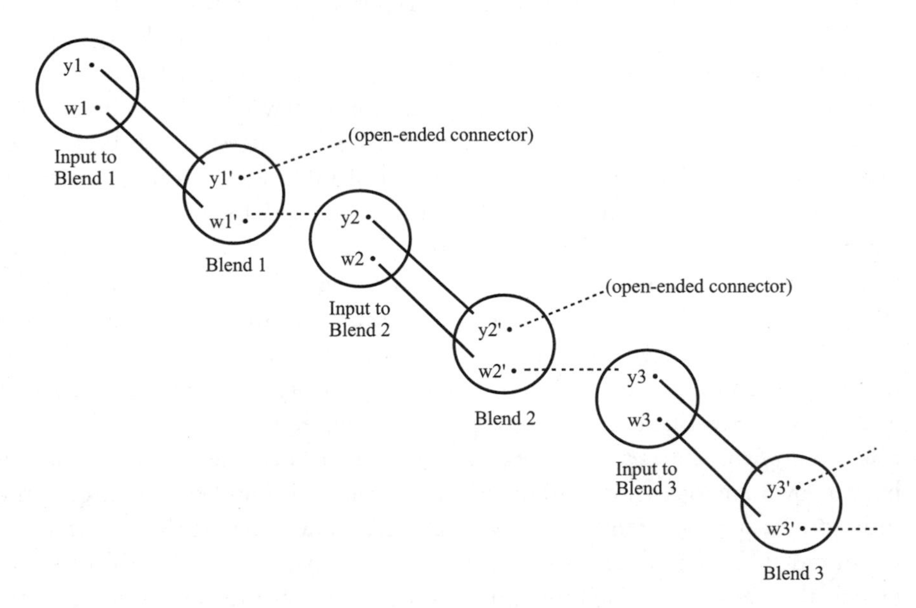

FIGURE 8.7 THE Y-OF CUBED NETWORK

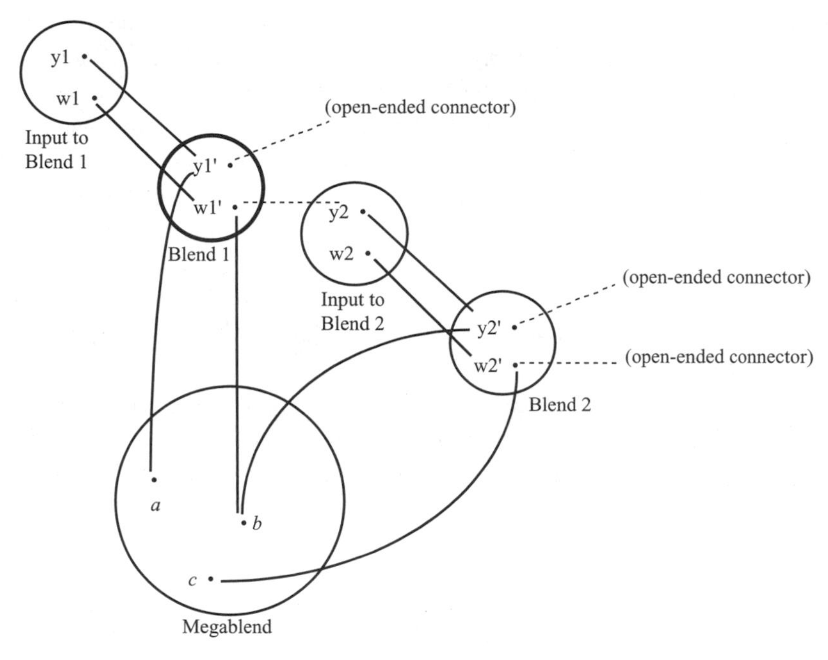

FIGURE 8.8 MEGABLEND #1

role in a world where the president has a secretary and the secretary has a valet, and *a* is that valet. In general, as we move from blend to megablend and from earlier megablends to later ones, the frames become more complex and perform greater compressions. Unconnected roles like *president, valet,* and *secretary* end up integrated into a single emergent frame.

There is a true elegance and simplicity to the composition of the mapping schemes and the emergent global blends. The Y^n expressions are not at all unusual in being guided by these principles of linking, compression, and blending. In Chapter 17 we will tell a more general story, in which grammatical constructions in general are pre-compressed integration networks with open-ended connectors available to be attached. Grammatical form may provide multiple ways to set up the same complex blend. For example, "the valet of the secretary of the president" and "the president's secretary's valet" result in the same ultimate compound integration network, and the mapping operations of both the "of" and the possessive form ("'s") proceed from the general principles of linking, compression, and blending; yet the dynamics of building that network differ in systematic ways. The two expressions prompt us to begin building the same network, but from different starting points and in different directions. "The valet of the president's secretary" gives us another, closely related way to build the network. We emphasize that forms as simple and basic as "the valet of" and "the president's" are fundamentally prompts for intricate blending. Each of these phrases, for example, asks us to build an input, a blend, elements in each, projections from the input to the blend, and two open-ended connectors from the two elements in the blend!

At this point, let us return to the XYZ expressions with which we began and diagram a compound form like "Ann is the boss of the daughter of Max." This expression contains the Y^2 expression "the boss of the daughter of," which builds up the integration network already discussed, including the megablend. In addition, it links the three open connectors to elements in a specific mental space: Ann and Max and an element that is the counterpart of the worker/daughter in the megablend. The resulting diagram, shown in Figure 8.9, looks identical to the previous one, except for the labels and the grounding of the open connectors in a specific space.

This example clearly shows the standard features of conceptual integration as they operate in composed integration networks prompted by composed syntactic cues. The construction of links and spaces systematically follows the syntactic order.

COMPOSITION OF METAPHORIC INTEGRATIONS

The language forms and the corresponding mapping schemes we have seen work identically in metaphoric cases. Metaphoric integrations are typically single-scope or double-scope, and they feel quite different from simplex integrations like "Paul is the father of Sally." They involve quite different projections and constructions of meaning. But the differences in the imaginative construction of meaning are one thing; the language forms and the mapping schemes are another.

We now come to an exciting insight into how form and meaning are driven. The language forms that lead to intuitively literal meanings can also give us intuitively metaphorical meanings that seem to belong to radically different kinds of thinking. Yet those identical forms are prompting for identical mapping schemes to guide those radically different constructions of meaning. And those mapping schemes compose in identical ways, regardless of whether the ultimate meanings are flatly literal, poetically metaphorical, scientifically analogical, surrealistically suggestive, or opaque. Once again, and this time arrestingly, we see that grammar is a set of prompts for guiding us quite precisely in our use of imaginative mental operations. The grammar indicates a kind of path. But what ultimately happens on that path depends on what specifically is encountered and on the imaginative operations conducted along the way. The results may subjectively seem to occupy different realms of thought, even though we are unconsciously carrying out the same mapping schemes.

What have we developed so far? We have shown that the same XYZ language form prompts for the same integration mapping scheme, regardless of the conceptual content. It works across simplex, mirror, single-scope, and double-scope

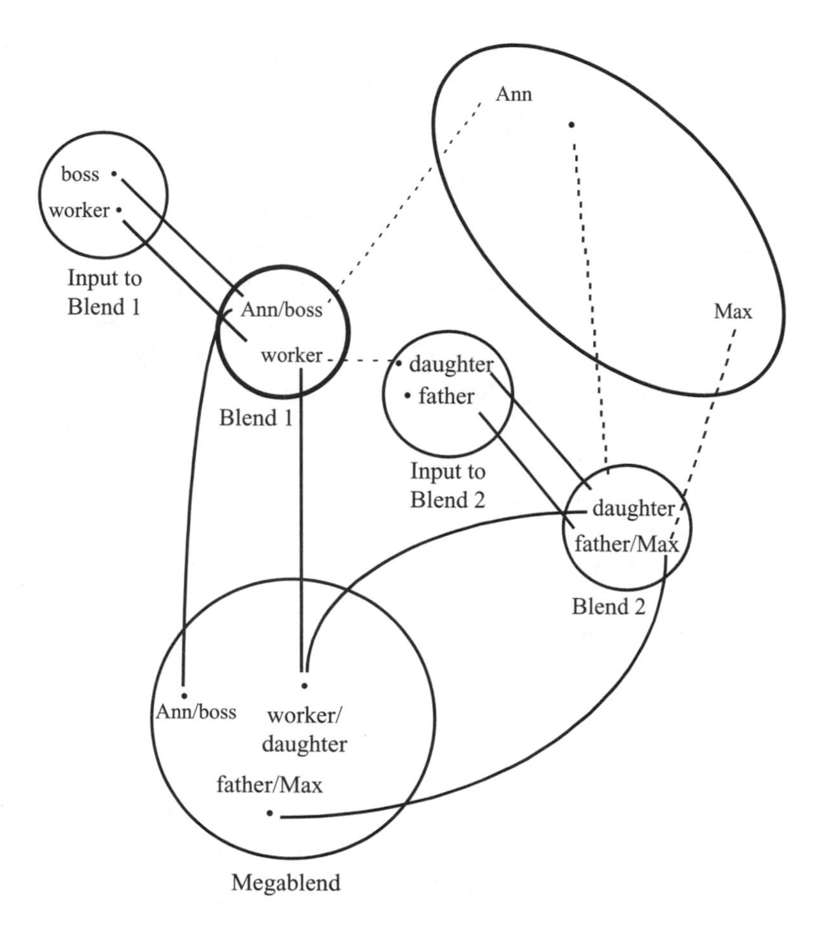

FIGURE 8.9 MEGABLEND #2

arrays as well as every intermediate kind of array: "Paul is the father of Elizabeth," "You are my long-lost daughter," "Zeus is the father of Athena," "Fear is the father of violence," "Vanity is the quicksand of reason," "Causation is the cement of the universe." We have also shown in detail how Y expressions prompt for integration mapping schemes and how those expressions compose syntactically and call for precise compositions of the corresponding mapping schemes.

If what we have developed is correct, then any compound XYZ expression that contains a Y^n expression should prompt for the same composed mapping scheme, regardless of the nouns in the expression. This is an unavoidable prediction of the theory we have proposed, but it is not a feature of any other theory of language, metaphor, logic, rhetoric, or cognition that we know of. Most of the major theories of meaning have assumed that these various kinds of expressions are incommensurable and that notions like compositionality are meaningless for figurative phenomena like metaphor.

This is a major prediction. Is it true? Well, pick any nouns that are acceptable for the XY^nZ construction and splice them in. The resulting sentence may be unintelligible, but our prediction says that you will know the mapping scheme you are supposed to carry out across all those spaces. Some expressions, like "Jean is the sister of the mother of the son of Joseph," will seem conventional. Others, like "My office is the secretary of the understudy of despair," will seem bizarre. But they will all have the same mapping scheme, as we can see by taking one that is doubly metaphoric and going through it. Consider the doubly metaphoric expression "Prayer is the echo of the darkness of the soul." Its grammatical form and its mapping schemes are identical to those in "Ann is the boss of the daughter of Max." And both are XY^2Z expressions. In fact, we have already seen, in Figure 8.9, a diagram of the mapping scheme for "Ann is the boss of the daughter of Max." If we are right, that diagram, with appropriate labels, should directly reflect the complex compound mapping scheme prompted by "Prayer is the echo of the darkness of the soul," which is diagrammed in Figure 8.10.

For this latter example, we have a starting point—a religious space containing things like *prayer*. The first Y expression, "the echo of," prompts a blend where the space with *echo* is one input. One obvious provisional choice for that input is a space where there is an echo in response to a sound. *Echo* in that space is linked by "is" to *prayer* in the religious space, which thereby becomes the second input space to the blend. At this point, the part of the language form that has been provisionally processed is "Prayer is the echo of," and Blend 1 has been set up, complete with its open-ended connector from *sound*. The mapping scheme of the Y expression now instructs us to attach that open-ended connector from *sound* to something picked out by what follows "of"—in this case, "darkness." Again it is up to the imagination to construct an appropriate space containing *darkness*, and again the imagination may make many different choices, any of which could later be revised; but one obvious choice is a space in which there is a locus and a gradient of light in that locus. The Y expression "the darkness of" prompts for another blend network, so we set up Blend 2 accordingly, complete with open-ended connectors from both *darkness* and *locus*. We also blend Blend 1 and Blend 2, giving us, in the megablend, *prayer/echo*, *sound/darkness*, and, so far, *locus*, whose corresponding open connector has yet to be completed. Now the word "soul" completes the second Y expression (and the sentence), telling us to attach the open connector from *locus* to *soul*, which is in the same religious space as *prayer*. The remaining open connector from *darkness* attaches to a new and unnamed element in that religious space.

As a consequence of this complex mapping scheme, we ultimately have one specific space with elements corresponding to X, Z, and all the elements in the blends that have open-ended connectors hanging off them. It may be that this

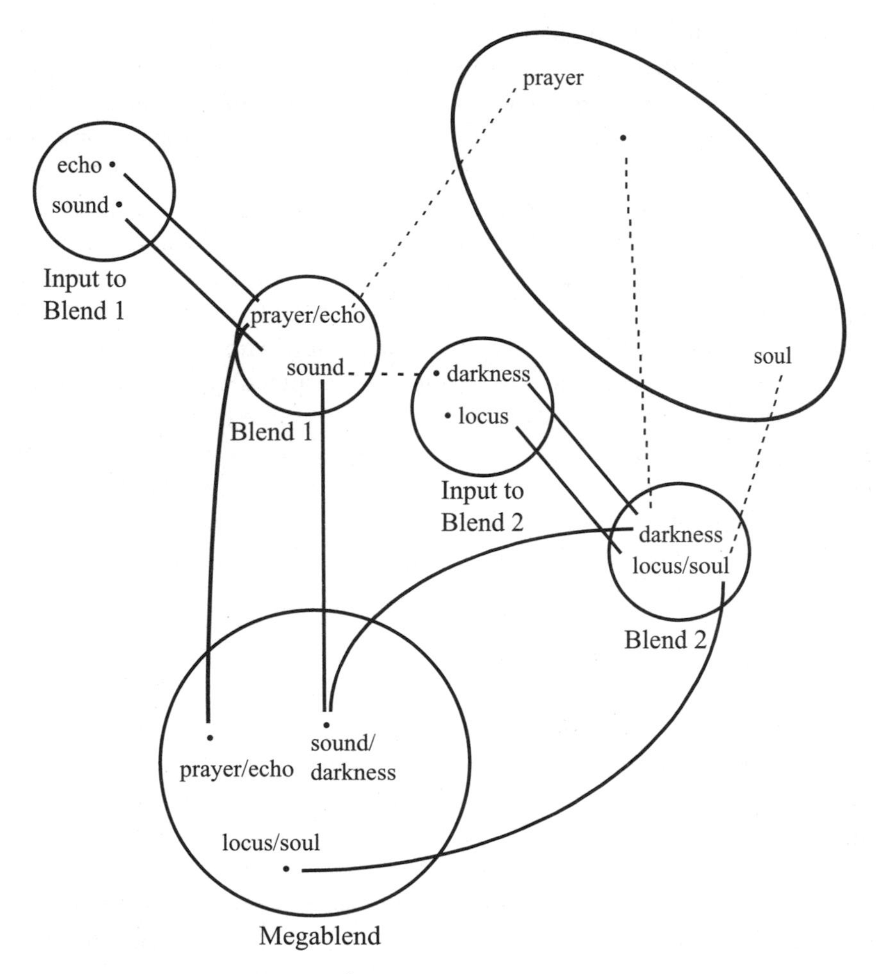

FIGURE 8.10 MEGABLEND #3

specific space is already available in the context (e.g., religion), or that it is mostly available and we need only build some new elements and relations in it, or that we have to create it from scratch by means of this very compound integration network, as in "Ann is the boss of descendants of dinosaurs," which leads us to group Ann, her workers, and the ancestor dinosaurs in one space.

The mapping scheme just outlined faithfully reflects all of the metaphorical blends and mappings prompted by "Prayer is the echo of the darkness of the soul." Moreover, that mapping scheme is exactly the same as the one for "Ann is the boss of the daughter of Max" and, therefore, exactly the same as the one for any XY^2Z expression.

So the prediction is true!

But there is more to say. Blending is the tool of compression *par excellence*, and in this case it delivers rich and indispensable new compressions. The new compressed frame of the megablend gives us a new conception of the causality of the soul and religious expression. In this frame, there is a new role for something that is unnamed but sharply conceived: something that is simultaneously sound and darkness. This new role is darkness within a scale of light; it is sound that has something against which to reverberate. It is part of your soul and part of your spiritual experience. Given this experience and the nature of the soul, it results in prayer. The imagination is free within this emergent frame to develop richer meaning, further associations. In fact, the imagination is free to develop as many differently nuanced and potentially conflicting interpretations of this suggestive phrase as it likes, provided they are all guided by this one mapping scheme. We can use "echo," "darkness," and "soul" to call up any space these words suggest from publicly shared mental spaces or deeply personal experiences. We may prefer some of these interpretations over others; some may be memorable and others nonsense. But they all have the same mapping scheme, and in all of them the megablend is the site for possible global insight.

Does this variety of possible meanings for the doubly metaphoric example imply that the metaphoric mapping schemes are radically unlike those in doubly simplex cases such as "Ann is the boss of the daughter of Paul"? On the one hand, the answer is no: The mapping schemes are not different, and in principle we are free to call up any spaces we like for "boss," "daughter," and "Paul." People's ideas of "boss" or "daughter" might be very different by virtue of personal experience and memories. For some, the sentence might evoke Paul's willingness to accommodate Ann. For others, it might evoke Paul's animosity against someone who gives orders to his daughter or presents a challenge to his authority over her. Such larger frames emerge in the megablend and can vary from context to context. We get one reading if this sentence is said in explanation to someone who has just observed Paul doing something for Ann that he would never do for anyone else; we get another if he is being exceptionally contrary with Ann. But none of this variation changes the mapping scheme. In the same way, "Prayer is the echo of the darkness of the soul" could be said to indicate that prayer is a symptom of psychosis or the most visible and reliable proof of spiritual conversion. But, again, either interpretation has the same mapping scheme.

On the other hand, even as the mapping schemes remain the same, the particular kinds of integration networks within those schemes are different. "Ann is the boss of the daughter of Paul" is a compound of two simplex networks, while "Prayer is the echo of the darkness of the soul" is a compound of two double-scope networks. Those two kinds of networks are profoundly different in all the ways we explained in Chapter 6. For example, in the *boss-daughter* case the intermediate blends have organizing frames that come from only one input, while in the *echo-darkness* case the intermediate blends have organizing frames that draw on the organizing frames of *both* inputs. The Compound Double-Scope XY^{2}Z offers exceptionally rich possibilities for creative clashes. *Prayer* is voluntary, vocal, meaningful, human, and deep, while *echo* is nonintentional, automatic, not necessarily vocal, meaningless, physical, and merely a thin, distorted copy of something else. Similar clashes between *sound* and *darkness,* and between *locus* and *soul,* come immediately to mind. Then there are the clashes between the two intermediate blends as inputs to the megablend. Across all this mismatch we are to find a deeply meaningful compressed role connecting *prayer* to *soul.* We might well have had connections between *prayer* and *soul* already: Prayer "comes from" the soul, prayer is the soul "talking to God," prayer is "evidence of" having a soul, and so on. The clashes in this case lead not to failure but to effective new meaning.

For both "Ann is the boss of the daughter of Max" and "Prayer is the echo of the darkness of the soul," the integration network builds an element in the specific space (the one with Ann and Max or the one with prayer and soul). In the first case this element is someone who is both a worker and a daughter but who is unnamed and whose existence we need not have recognized at the outset. In the second it is an element related to prayer and soul for which we have no common name and whose existence we likewise need not have recognized at the outset. The religious element in this specific space may be profound and distinct but ineffable, and the compound expression allows us to prompt for it as such.

Had we set up the XY^{2}Z diagram in Figure 8.10 solely to depict the mapping scheme of the doubly metaphoric definition of *prayer*, it would have seemed plausible, because everyone recognizes that the interpretation of the mysterious and suggestive metaphor involves elaborate and creative mappings. The same diagram might have seemed implausible as a representation of what is going on when we make sense of "Ann is the boss of the daughter of Max," because that is not how logic approaches usually analyze such a case. But the blending explanation, which applies to both, is more accurate, more parsimonious, and more general than the traditional logic explanations. It is more accurate because the logic explanations miss the many compressions of the megablend, including the emergent frame with its compressed role. In fact, logic notations are just labels for the input frames. They presuppose psychological access to those frames and say nothing about emergent new frames. The blend explanation is more parsimonious because the same blending and mapping schemes account for many superficially different meaning phenomena, generally described as literal, metaphoric, analogical, categorical, and so on. The general explanation does away with a jumble of highly specialized operations, each tailored to deceptive surface features of a phenomenon. Finally, the blending explanation is more general because it accounts in a deep way for the harmony of form and meaning over a vast and open-ended array of data. That harmony consists in the pairing of the form and the mapping scheme, and in the way the composition of the forms is paralleled by the composition of the mapping schemes.

CHAPTER 8
ZOOM OUT

SUPERNATURAL FATHERS

We have discussed XYZ blends that have *father* in one of the inputs.

Question:

- How far out can a father be pushed and still be a father?

Our answer:

Literary works frequently prompt for highly intricate blending. Milton's portrayal of Satan as father in the second book of *Paradise Lost* is an extended display of double-scope blending.

The commonplace notion of Satan is already a blend for which a conceptual domain has been elaborated. Satan is a blend of individual human being—thinking, talking, desiring, intending—and theological ontology. In the theological space, he has eternal features such as evil as well as nonhuman powers and limitations. He is anthropomorphic but has theological features and nonhuman conditions. The blended domain for Satan is quite elaborated: He has like-minded colleagues in the form of a cohort of devils, together they form an intricate hierarchical organization of social groups, and so on. This blended domain is entrenched both conceptually and linguistically. Consequently, although the blend is in some ways double-scope, expressions like "The devil made me do it" or "Get thee behind me, Satan"—or even expressions based on further blending, such as the reference to a child as a "little devil"—do not feel especially figurative.

Milton extends this blend in ways that seem strikingly figurative and allegorical. His theological space includes evil, disobedience, sin, death, and their relations, as well as the psychology of the sinner confronted with spiritual death. His kinship space includes progeneration and kinship relations, especially the role *father*, and he adds a preexisting blend of the birth of Athena from the brow of Zeus. In short, he gives us two double-scope theological blends: Satan as the father of Sin and Satan as the father of Death.

In Milton's blend, Satan conceives of the concept of sin; a fully grown woman, Sin, leaps from his brow. Satan is attracted to sin/Sin: he has sex with her. Although he does not know it at the time, his involvement has a consequence—namely, death; in the blend, Death is the male offspring of Satan's incestuous involvement with Sin. Death rapes his mother, causing her to give birth to a small litter of allegorical monsters.

After Satan has been sent to Hell and has decided to try to escape, he meets two characters at the gates of Hell who have been stationed there to keep him in. They are Sin and Death. He does not recognize them.

The mental spaces that contribute to this blended story—the kinship space and the theological space—correspond in some ways but not others. Milton draws from both of them, selectively, to create a double-scope blend. For example, he takes exclusively from the kinship space Sin's intercession between Death and Satan—father and son—when they are on the brink of terrible combat.

He also takes many central features exclusively from the theological space. In the theological space, there is a sinful cast of mind that does not recognize physical and spiritual death as the result of sin and that is at last appalled when it must acknowledge these consequences. Hence, in the blend, Sin is surprised to have conceived Death, and she finds her son odious. Next, in the theological space, mortality and spiritual death overshadow the appeal of sin and are stronger than sin; acknowledging death devalues sin; willful, sinful desires are powerless to stop this devaluation. Hence, in the blend, Sin is powerless to stop her horrible rape by Death. In the theological space, the fact of spiritual death brings ceaseless remorse and anguish to the sinful mind, and the torments of Hell bring eternal punishment. Hence, in the blend, the rape of Sin by Death produces monstrous offspring whose birth, life, actions, and relationship to their mother are impossible for the domain of human kinship:

> *These yelling Monsters that with ceaseless cry*
> *Surround me, as thou saw'st, hourly conceiv'd*
> *And hourly born, with sorrow infinite*
> *To me, for when they list, into the womb*
> *That bred them they return, and howl and gnaw*
> *My Bowels, their repast; then bursting forth*
> *Afresh with conscious terrors vex me round,*
> *That rest or intermission none I find.*

Milton creates subtle correspondences between the kinship space and the theological space. For example, he blends the unusual scenario of disliking a child with feeling horror at the fact of death. He blends the unusual scenario of a son raping a mother with the effect of death on sin. And perhaps most ingeniously, he blends the unusual medical frame of traumatic vaginal birth that physically deforms the mother and so makes her less attractive with the way sin becomes less attractive once death is acknowledged as its outcome:

> *At last this odious offspring whom thou seest*
> *Thine own begotten, breaking violent way*

Tore through my entrails, that with fear and pain
Distorted, all my nether shape thus grew
Transform'd.

Although Milton's portrayal of Satan as a father is double-scope, it preserves considerable structure associated with *father* and *birth*. Consider first the paternity of Death. The "father" has human form and speaks human language, is excited by feminine beauty, and has anthropomorphic sex with an anthropomorphic female in a typically human scene. There is a birth through a vaginal canal; the son inherits attributes of both father and mother; father and adolescent son have a conflict over authority. Now consider the paternity of Sin. Here, too, the father has human form and speaks human language. There is an offspring in human form, who emerges from a container-like body part and develops into a sexual being.

Readers of Milton never stop to ask themselves whether or in what sense Satan is the father of Sin and Death. Milton gives them the answers before they can ask those questions. Although many of what might have been considered criterial features or truth-conditional features of *father* are abandoned in these two double-scope blends, which are themselves connected by vital relations and compressed into one huge megablend, Milton's imaginative work has provided a coherent global insight in which the ascription of fatherhood to Satan is straightforward and true. This is particularly amazing if we recall that turning Satan into a father was originally the great challenge that motivated the creation of this network. God the Father has a son, Christ. The other member of the Holy Trinity is *Spiritus Sanctus*, the Holy Ghost, the breath of eternal life. Milton intends to give us a picture of the infernal as the opposite of the divine. He needs to create an Infernal Trinity, in which Satan is the counterpart of God. How to make Satan a father is a conundrum, which Milton solves so well that it seems never to have been a question at all.

WAYS OF THINKING ABOUT COMPOSITIONALITY

A lot has been said about compositionality in logic, semantics, artificial intelligence, and philosophy of language.

Question:

- What is different about the compositionality story here?

Our answer:

First, we must distinguish compositionality in our general sense from narrow, truth-conditional compositionality. In a logical system, there must be a compositional syntax and a compositional semantics such that any syntactic combination of propositions that is itself a proposition encodes a computation of the truth value of the whole from the truth values of the parts. For example, the connective "v" (inclusive *or*) in propositional logic is compositional because there is a truth function that yields the truth value of $p \vee q$ given the truth value of p and the truth value of q.

Some form of compositionality is always assumed in semantic theories because it appears that people can construct meanings in consistent ways on the basis of syntactic combinations—in particular, they understand one another. Compositionality accounts naturally for the generative nature of meaning: An unlimited number of meanings can be constructed from combinations of a finite number of forms.

The most basic view of compositionality for semantics is the one that holds for logical systems: Linguistic expressions in isolation have truth conditions that are computable from the truth-conditional properties of their parts. For example, the conditions under which an expression like "brown cow" will be true of something can be computed from the conditions under which "brown" is true of something and "cow" is true of something. Many versions of formal semantics—for instance, Montague grammar—have attempted to maintain this basic view of compositionality. But there is a general consensus that this rudimentary view is refuted by empirical evidence. The examples discussed throughout this book are part of that empirical evidence.

A looser view of compositionality permits elements from context to come into the computation of truth conditions, in addition to what is given by the truth conditions of the parts of the expression. An obvious case is an expression like "I am here," where the deictics *I* and *here* have to be given reference in context. A less obvious case is an expression like "The newspaper is on Main Street," where we need to know from context what newspaper is referred to, but also whether the understood reference is to the editorial offices, a copy of the paper, or the momentary location of the action reporters covering the current scene.

In this looser view of compositionality, the purpose of specifying the context is to add needed referential assignments to the computation of truth conditions, not to remanufacture the truth conditions of stable parts. But in the blend for "Athena is the daughter of Zeus," *daughter* is now compatible with the structure *having no mother, born out of the father's head, born as an adult fully clothed in armor, immortal,* and so on, so that what look like truth conditions of *daughter* have not been projected to the blend. Specifying the context to the degree that it would specify all the appropriate truth conditions in this case would require specifying the framing of the inputs, the counterparts in the cross-space mapping, the details of the selective projection, the degree of matching topology, and all the other parts of the network model of conceptual integration.

Views of truth-conditional compositionality—whether of the narrow or looser species—are inadequate to account for the variety of meaning constructions people actually achieve. But this is not to say that meaning construction is not compositional in other fundamental ways: People do construct meanings in consistent ways on the basis of syntactic combinations. As we have seen in this chapter, the consistency is accounted for by the fact that there is compositionality at the level of the general mapping schemes themselves and at the level of the syntactic forms that prompt for those schemes.

The general pattern of this two-level compositionality is this: A given syntactic form—such as an "of" construction—prompts for the creation of structure in a blend and also provides a way of referring to that structure. That blend then becomes available for further mapping under prompting from additional syntactic constructions. The mapping schemes themselves compose, as do the grammatical constructions.

Although Fregean compositionality has an intuitive appeal as a model of how we construct complex meanings in response to complex syntactic prompts, it cannot account for actual construction of meaning. The network model applies uniformly over the cases of supposed Fregean compositionality and the cases that are clearly beyond the Fregean model. In a simplex network in which, additionally, the values in one input are prototypical for a rigidly categorical frame in the other ("Paul is the father of Sally"), the blending mapping scheme is truth-conditionally equivalent to Fregean compositionality. In all the other cases—involving looser frames, mirror, single-scope, or double-scope networks—the network model continues to apply, but Fregean compositionality fails on principle since it presupposes the very features of rigid categorization that these examples do not have. Even if a mirror network happens to have a rigidly categorical shared frame and absolutely prototypical values for its roles (e.g., "You are my long-lost daughter"), that integration network is not captured by Fregean compositionality.

Rather surprisingly, the network model is less dependent on reference to context than the "looser" version of Fregean compositionality we discussed above, which attempts to preserve Fregean truth-conditional compositionality by adding discourse information to the computation of truth conditions. The network model shows a relationship between syntactic composition and composition of mappings that holds independent of reference to domains, situations, and contexts. The compositional mapping schemes we discovered in the case of "Ann is the boss of the daughter of Max" and "Prayer is the echo of the darkness of the soul" were essentially identical, and identically based on syntactic composition of "of" constructions, even though their conceptual domains and contexts were radically different. Of course, context is all-important—but for the imaginative elaboration of the blends, not for the mapping scheme.

It is also worth reemphasizing that the general construction of the mapping scheme follows the order of the syntactic construction quite closely, thus suggesting a plausible model of the way in which such meanings are actually processed. We saw that the same ultimate blend could be constructed from many different starting points so that different languages can use different syntactic constructions, and in fact the same language can have more than one way to prompt for the same mapping scheme. "The valet of the secretary of the president" and "The president's secretary's valet" prompt for the same ultimate mapping scheme but start at different points.

In this chapter we have also studied the way in which a composition of syntactic cues prompts for a composition of mapping schemes, and we have used as our showcase illustration the syntactic composition "Noun-Phrase 1 is the Noun-Phrase 2 of Noun-Phrase 3," where Noun-Phrase 3 has the form Noun-Phrase 4 of Noun-Phrase 5. That is, we have used the syntactic part of the XY^2Z construction. The points we have made apply across all syntactic compositions. Indeed, syntactic compositions as different as "Language is the fossil poetry of the soul," "Las Vegas is the American Monte Carlo," "Paul's daughter is the president of Disney," "Social movements are at once the symptoms and the instruments of progress," and "As poetry is the harmony of words, so conversation is the harmony of minds" all compose syntactic cues to prompt for composition of mapping schemes.

GREAT-GRANDFATHER OF

We have seen actual Y^n expressions for n as large as 3.

Question:

- If the mapping scheme can in principle be compounded indefinitely, why don't we routinely see Y^n expressions for n greater than 3 or 4?

Our answer:

The "Y of" language form provides the ability to compound indefinitely or, more important here, to compound five or six times and still give a perfectly memorable and pronounceable sentence. But as a matter of on-line processing, the megablend must maintain active connections across the whole network to give us a sense of overall global insight. Our hypothesis is that the megablend and its connections run into limits on working memory. The most obvious way around these limits is to build and make conventional a subscheme of a compound network. Consider a Y^3 network in which the compressed role in the megablend is *Father of Parent of Parent of.* The English language has a name for this compressed role: "great-grandfather." That megablend and the word for its compressed role are available to us as a chunk from long-term memory and so they place less demand on working memory than would a fresh Y^3 network such as "The father of the mother of the mother of." By providing a compression that is available as a unit from long-term memory, with an attached word for the compressed role, *great-grandfather* makes it relatively easy to construct on-line a megablend for "The mother of your father's great-grandfather."

FREEDOM AND LIMITS ON INTERPRETATION

We have pointed out that different interpretations are possible for the same XYZ expression.

Questions:

- What are the sources of these differences, if the mapping schemes are the same?
- Are there any limits?

Our answers:

First, we should be careful to note that grammar pairs forms with mapping schemes, and that the same grammatical form can be paired with more than one mapping scheme. For example, "Y of" is a form that is paired with the Y-of network mapping scheme we have discussed. But that same form is paired with a different mapping scheme, in which y is the counterpart of what is picked out by the noun following "of," as in "the state of Alabama." Charles Fillmore gives this example: "One needn't throw out *the baby of personal morality* with *the bathwater of traditional religion.*" These counterparts need not be metaphoric, as we see from examples like "the nation of England," "the island of Kopipi," "the stigma of cowardice," "the feature of decompositionality," "the condition of despair." In this chapter, we have not talked at all about the "of" expressions that prompt us to use this different mapping scheme. But of course it is possible to find expressions that are ambiguous. "The fire of love" could be read either way: Love is metaphorically fire, or fire is the metaphoric counterpart of something related to love, such as sexual passion or anguish. One source of difference in readings, then, comes from the different mapping schemes associated with the same language form. But this is not our focus here.

Suppose we are concerned with just one mapping scheme. We already saw that the imagination has wide latitude in recruiting, projecting, and blending additional background knowledge, context, and memories in order to develop a full meaning on the basis of a particular mapping scheme and a choice of particular domains and counterpart elements. But the mapping scheme itself leaves latitude in the selection of domains and counterpart elements. Consider "Vanity is the quicksand of reason." The mapping scheme prompts us to find an unnamed w in the space with *quicksand* and an unnamed relationship between *quicksand* and that element w. It seems straightforward to select *traveler* as w, since quicksand is a potential trap for the traveler just as vanity is a potential trap for reason. And given that choice of elements for the mapping scheme, it seems automatic to conclude that vanity is something reason would want to steer clear of. But Oscar Wilde might have offered the reading that only the virtue of vanity can keep reason from wandering aimlessly in a barren, uncharted desert; that vanity provides clarity of vision, excitement, purpose. Crucially, the mapping scheme itself does not even guide us to choose *traveler* as w. Instead, the imagination has full latitude to choose, for example, *desert* as w, since quicksand is *part of* the desert, and *part of* is a very common and basic relationship. Under that choice, we can form a blend in which *vanity* is part of *reason*. For example, we might conceive that there can be no reason without vanity, even though vanity is the weakest and least reliable part of reason.

Does this freedom of the imagination mean that anything goes? On the contrary, finding one or two such integrations out of the potential infinity of connections afforded by the world is extremely difficult and anything but random. There are several different influences on what connections will be made and what resulting networks will be satisfactory. In order to use a connection, such as the connection between *quicksand* and *traveler*, it has to be made active, and it cannot be made active unless it first exists or is constructed. Different connections that might be made active vary on their degree of entrenchment and on how easy it is to activate them in any given moment. Suppose that you hear "Vanity is the quicksand of reason." It launches the XYZ mapping scheme, and your brain is now looking for a w. Suppose also that your brain has complete latitude and tries *bacteria*. What happens at that point in pursuing the elaboration of the integration network? We need a frame in the *quicksand-bacteria* input space that will project to the blend and contribute to its emergent meaning. A minimal requirement on that frame is that it must contain the elements *quicksand* and *bacteria* and a relationship between them. But for most people, such a frame is not available because they have no conventional public knowledge or personal memories of such a frame. So even if the brain does try out *bacteria*, nothing comes of it, no blend is formed, and there is no conscious memory of the attempt. Alternatively, suppose we bring into existence the appropriate frame just before "Vanity is the quicksand of reason" comes along. Then this frame will be activated and probably tried out. In that case, *bacteria* might prove a superbly successful candidate for the missing *w*. This prediction is easily confirmed by interpreting "Vanity is the quicksand of reason" within the following context:

> Did you know that some bacteria can live only in quicksand and depend on it for everything? For them, quicksand is not a trap, it is what they need to live at all. Well, for some people, vanity is the quicksand of reason. Their vanity gives them the self-confidence to think well.

In any theory of meaning, activation does not come for free. The existence of frames, knowledge, experience, scenarios, and memories does not come for free. Ease of activation and degree of entrenchment by themselves impose very strong constraints on the imagination and the use of language. Linguists, logicians, and, for the most part, even psychologists tend to focus on the entrenched cases, which are already built and usually easy to activate. When only the rigid and entrenched patterns are used, meaning becomes predictable based on the mapping schemes and those patterns. This is probably why linguists, logicians, and analytic philosophers of language have often incorrectly excluded inventive, figurative, creative, and literary examples from their domain of inquiry. The mistaken view was that only predictable composition of meaning can be scientifically tractable and important, and that only predictable composition of meaning can support genuine rational thought as opposed to the glinting ephemera of whimsy. As we discover repeatedly, the power of thought—whether rational or whimsical, emotional or practical—lies in the same basic mental operations. To focus exclusively on the entrenched cases is to be blind to the way we think. It obscures not only our general operations of thought but nearly all of what is happening in the entrenched cases themselves. The action on stage in the entrenched cases is possible only because of much greater activity in the wings of the imagination.

The latitude of the imagination can be pointed out for even the most entrenched cases. Take, for example, "Mary is Paul's mom." The strong default mapping is that *Paul* is the counterpart of *son* in the input with the kinship frame. That default is a choice the imagination makes, even though it is the easiest choice because of activation and entrenchment. But it is defeasible, and other choices, requiring different activations, can be made using the same principles of thought. Suppose in a nuclear family with children, Paul is the male parent and there is some scheme of divided labor or psychological roles with respect to the children. In this scenario, Paul's nuclear family has a role *mom*, and Mary fills that role for Paul; she is, as it were, *mom of the household that has Paul as male parent*, and we might say "Mary is Paul's mom" to mean this. Whether we choose as a counterpart of Paul *son* or *male parent in a household,* we have the identical general XYZ mapping scheme. But the two readings have different actual mappings, because different counterparts have been selected for the cross-space mapping, and of course different integrations have been performed in the blend.

Part Two

HOW CONCEPTUAL BLENDING MAKES HUMAN BEINGS WHAT THEY ARE, FOR BETTER AND FOR WORSE

Nine

THE ORIGIN OF LANGUAGE

Every child is born a genius.
—*Buckminster Fuller*

LANGUAGE IS A MYSTERY because it is a singularity: Only human beings have grammar of the sort we find in natural languages. But language is not the only singularity: The considerable efforts of animal psychologists have uncovered no evidence that other species can reach very far in conceiving of counterfactual scenarios (such as those underlying pretense), metaphors, analogies, or category extensions. Even the most impressive nonhuman species have very limited abilities to use tools, let alone design and make them. Human beings have elaborate rituals that constitute cultural meaning without being tied to immediate circumstances of feeding, fighting, or mating, but the nearest that other species can come to "rituals" are instinctive displays directly tied to such immediate circumstances. As Merlin Donald puts it, "Our genes may be largely identical to those of a chimp or gorilla, but our cognitive architecture is not. And having reached a critical point in our cognitive evolution, we are symbol-using, networked creatures, unlike any that went before us."

As we discussed at the beginning of Chapter 8, it is hard to get a reasonable explanation of language that sees it as the product of gradual steps, each producing a grammar by making the previous grammar a little more complicated. Such an explanation would have to propose that a group of human beings began with a very simple grammar that gradually, generation after generation, grew ever more complex, until it reached the level we see in the languages in the world today. However, such an explanation runs up against the fact that we can point to no simple languages, or even ones that are simpler than others. There are many evolutionary developments for which it is relatively easy to see a gradual path: We can see gradual steps by which early mammals plausibly evolved into primates or cetaceans. But we do not see any gradual path in mammalian history for the development over many generations of ever more complex grammars.

EXISTING THEORIES

The hunt is on for the origin of language. What could have caused this singularity to come into existence? One line of thinking looks at language as a very specific human production and asks how it could have arisen. The language faculty is viewed as distinct from other human capacities, and so the correlation with the other human singularities has no theoretical place: Those other abilities are distinct from language and call for other explanations. This line of exploration has room for many different kinds of theories.

Nativist theories—Chomksy is the preeminent name here—place the distinctiveness of language in specific genetic endowment for a specifically genetically instructed language module. Under that view, there is minimal learning involved in acquiring a language. Most of the language module is already in place. Which language actually gets spoken—Chinese, Bantu, English—is relatively superficial: Very thin exposure to a given language sets parameters that control the output of the language module—that is, gives us one language instead of an alternative language. But it is not clear what, in the evolution of the human brain, could have been the precursor of the language module. Nor is it clear what pressures from natural selection would have produced such a module, given that we find no intermediate stages. This is why many nativists have embraced the view that a sudden, dramatic, perhaps unique event in human evolutionary history produced in one leap a language module resembling nothing like the brain's previous resources.

Other nativist views of language see it as having arisen by gradual natural selection. For example, Stephen Pinker and Paul Bloom argue that "there must have been a series of steps leading from no language at all to language as we now find it, each step small enough to have been produced by a random mutation or recombination."

Although viewing language as a specific production distinct from other human capacities is associated with nativist theories of language as modular, neither nativism nor modularity is crucial for the view of language as a distinct capacity. A radical associative theorist who sees cognition as developing during childhood through the forming, strengthening, and weakening of connections between neurons might easily view language as a very special set of operations that arise in a network. On this view, language would be essentially distinct from other capacities of the network, even though all of them would share a common denominator in the basic associative operations. While the language faculty would not be localized in any area of the network, it would still be distinct operationally.

Some associative theories emphasize the role of evolution in developing powerful learning mechanisms that perform statistical inferences on experiences. In these views, the brain has evolved rich, specific architectures for statistical extraction, and language is one of the things that can be learned through those domain-general processes of statistical inferencing. Language is intricate and depends upon the evolution of those learning abilities, but the way we learn it is not specific to language. The language the child hears—far from impoverished—is adequate for the purpose of converging on grammatical patterns by doing statistical inferencing. The evolutionary story here is that the brain evolved learning abilities with some bias for learning things like language, but did not evolve "language" or a neural "language module." As Terrence Deacon writes, "The relevant biases must be unlike those of any other species, and exaggerated in peculiar ways, given the unusual nature of symbolic learning." It remains a challenge, of course, to explain how those particular learning abilities and biases for language could have evolved, and it is still not clear under this explanation why we have no evidence of intermediate, simpler forms of language.

One line of thinking, associated with theorists like William Calvin and Derek Bickerton and Frank R. Wilson, tries to find preadaptations for language—such as the development of the hand or of reciprocal altruism—that could have put in place some of the computational ability that language needs. In this way of thinking, there were gradual steps toward language, but the early steps did not look like language because they weren't. They simply had some powers that made sophisticated language possible later on.

There are also co-evolutionary proposals, including an influential recent proposal by Terrence Deacon. Language, he argues, is not an instinct and there is no genetically installed linguistic black box in our brains; language arose slowly through cognitive and cultural inventiveness. Two million years ago, australopithecines, equipped with nonlinguistic ape-like mental abilities, struggled to assemble, by fits and starts, an extremely crude symbolic system—fragile, difficult to learn, inefficient, slow, inflexible, and tied to ritual representation of social contracts like marriage. We would not have recognized it as language. But language then improved by two means. First, invented linguistic forms were subjected to a long process of selection. Generation after generation, the newborn brain deflected linguistic inventions it found uncongenial. The guessing abilities and intricate nonlinguistic biases of the newborn brain acted as filters on the products of linguistic invention. Today's languages are systems of linguistic forms that have survived. The child's mind does not embody innate language structures. Rather, language has come to embody the predispositions of the child's mind.

The second, subordinate means by which language improved, in Deacon's view, had to do with changes in the brain. Crude and difficult language imposed the persistent cognitive burden of erecting and maintaining a relational network of symbols. This demanding environment favored genetic variations that rendered brains more adept at language. Language began as a cognitive adaptation, and genetic assimilation then eased some of the burden. Cognitive effort and genetic assimilation interacted as language and brain co-evolved. In Deacon's view, language was "acquired with the aid of flexible ape-learning abilities." It was grafted onto an ape-like brain. Language, then, is not walled off from other cognitive functions such as interpreting and reasoning. Grammatical form is not independent of conceptual meaning. There is no linguistic black box and there was no genetic installation of language.

A RANGE OF ARRESTING HUMAN SINGULARITIES

Three of the biggest singularities that seem to have entered explosively on the human stage around the same time in human prehistory are art, religion, and science. As Stephen Mithen writes in "A Creative Explosion?":

> Art makes a dramatic appearance in the archaeological record. For over 2.5 million years after the first stone tools appear, the closest we get to art are a few scratches on unshaped pieces of bone and stone. It is possible that these scratches have symbolic significance—but this is highly unlikely. They may not even be intentionally made. And then, a mere 30,000 years ago, at least 70,000 years after the appearance of anatomically modern humans, we find cave paintings in southwest France—paintings that are technically masterful and full of emotive power.

Mithen makes the same claim for religion and science, and naturally asks what could have led to these singularities. His answer is that human beings suddenly developed a totally new capacity for "cognitive fluidity"—that is, a capacity for the "flow of knowledge and ideas between behavioral domains," such as "social intelligence" and "natural history intelligence." Although he has no theory about the principles of "cognitive fluidity" and views it as a higher-order operation used for the special purpose of putting different domains together, Mithen's general notion bears some resemblance to our idea of conceptual integration, and even his diagrams look strikingly like Figure 3.1, which we used to illustrate the Buddhist Monk riddle. What caused cognitive fluidity? In Mithen's view, there must have been some singular, explosive evolutionary event that produced a quite different sort of brain.

Mithen observes that time and again, theorists have confidently if vaguely located the exceptional cognitive abilities of human beings in their capacity to put two things together. Aristotle wrote that metaphor is the hallmark of genius. And as we saw in Chapter 2, Koestler proposed that the act of creation is the result of "bisociating" different matrices.

The prehistoric picture we are left with is one of mysterious singularities: explosions, some perhaps simultaneous, in new human performances. We also have, for all these singularities, the problem that there is essentially no record of intermediate stages between the absence of the ability and its full flowering. And this prehistoric story has, at least for language, its contemporary parallel: We find no human groups, however isolated, that have only rudimentary language. We find no primates with rudimentary language. At first glance, this seems a completely abnormal situation. We find no parallels in evolution of species, for example—no complex organisms that leap without precursors out of the slime. What kind of theory do we need, then, in order to account for such a strange and unprecedented picture?

WHAT SHOULD A PROPER THEORY OF THE ORIGIN OF LANGUAGE LOOK LIKE?

As Darwin noted, evolution's main trick seems to be gradual change, so an adaptationist account is obliged to show that each step would have been adaptive. Evolution is never allowed to think "Well, if I could get to stage ten, it would be good, so give me a break while I go through the first nine." Other things being equal, we prefer an evolutionary account that shows continuity of change rather than a spectacular singularity. Even "punctuated equilibrium" theories propose only relatively minor jumps—not jumps that produce an eye or language out of nothing.

But we face a problem: How do we explain the emergence of arresting human singularities out of relatively continuous changes in brain and cognition?

To think about this question, we must put aside two major fallacies. The first is the fallacy of Cause-Effect Isomorphism. Compressing cause and effect is indispensable to cognition, but it often has the bad consequence for scientific thinking that, recognizing an effect, we conceive of the cause as having much the same status as the effect. If the effect is dramatic, we expect a dramatic causal event. If the effect is unusual, we expect an unusual causal event. This way of thinking is so common that popular science accounts routinely offer entertaining demonstrations of the way in which unusual cases come from boring, routine causes. This is always the story for popular accounts of evolution or chaos theory: The beetle whose abdomen has grown into a covering costume that makes it look like a termite so it can live in termite nests arose through the most routine operations of gradual natural selection.

The Cause-Effect Isomorphism Fallacy leads us to think that a discontinuity in effects must come from a discontinuity in causes, and therefore that the sudden appearance of language must be linked to a catastrophic neural event. The only evidence we need against this fallacy is the straw that broke the camel's back, but in fact we see such evidence everywhere in science. Under heat, there is a smooth continuity of causation as ice goes abruptly from solid to liquid. The change from solid to liquid is a singularity, but there is no underlying singularity in the causes or in the causal process. One more drop of water in a full cup causes a lot of water to flow suddenly out onto the table, not just the one drop that was added. One more gram of body fat can make it possible for you to float on your back without effort in the middle of the South Pacific; a gram fewer and you sink down. In this last case, a life-and-death singularity arises from smooth continuity in causes and causal operations. Nothing has changed about the principles of hydrodynamics or buoyancy when you drown.

So, in principle, the sudden appearance of language is not evidence against evolutionary continuity. Singularity from continuity is a normal occurrence. The only remaining question is, Can biological evolution also work this way? Are there specific evolutionary processes that give us remarkable singularities out of causal continuities? Here, we encounter a second major fallacy, the Function-Organ Isomorphism Fallacy. This is the well-known idea that the onset of a new organismic function requires the evolution of a new organ. Under this fallacy, because opossums hang from trees by their tails, tails are organs for performing the function of hanging from trees; because people speak with their tongues, tongues are organs for talking. But biologists routinely point out that as an organ evolves it may acquire new functions or lose old functions or both. There were intricate mammalian tongues before there was language; the tongue did not need to be invented afresh. And the ancestors of opossums had tails before opossums hung from trees. The continuous evolution of an organ does not necessarily correlate with a continuous evolution of a function. Functions can be singularities while the evolution of an organ is continuous. Like the body in the water, an organ may need only the tiniest increment of change to subserve a striking new function, such as floating.

We see this in the case of the proposed theories of how dinosaurs evolved into birds. Nobody proposes a theory of discontinuity for the organ (wings) but nobody proposes a theory of continuity for the function (flight). Rather, according to such theories, wings came gradually: Scales seem to have developed slowly into feathers, feathers provided warmth, the existence of longer arms and feathers made it possible to flap for a little extra ground speed and so the arms got longer and more feathery. Flight in all these theories came at one fell swoop: At a critical point, the organism could become truly airborne, and so could now fly after the dragonfly and gobble it up. It occurs to no one to propose that flight—a spectacular singularity of function—came about because an organ for flight suddenly evolved from scratch. It also occurs to no one to propose that since modern birds fly higher than one hundred feet, there must have been intermediate stages in which birds could attain heights of first one foot, then many generations later two feet, and so on up to one hundred feet. Being truly airborne is all-or-nothing, and so the behavior of flying is also basically all-or-nothing. An organism either flies or does not.

In thinking about the origin of language, we must put aside the fallacies of Cause-Effect Isomorphism and Function-Organ Isomorphism. Language is not an organ. The brain is the organ, and language is a function subserved by it, with the help of various other organs. Language is the surface manifestation of a capacity. It is a singularity of function, and so nothing prevents it from having arisen from a basically continuous and adaptive process of evolution. The function can have arisen recently in human evolution even though the continuous changes that brought it about can have been working for many millions of years. The causes are very old but one particular effect showed up just yesterday. This is what we propose.

The best theory of the origin of language would have the following features:

- A recognition of the singularity of language. There is no phylogenetic evidence of sustained intermediate stages, and no evidence of present human languages that are rudimentary.
- Rejection of an extraordinary event as responsible for the extraordinary capacity. In other words, no Cause-Effect Isomorphism.
- A continuous path of evolutionary change over a very long period as the cause of language, since that is how evolution almost always works.
- A path that is a plausible adaptive story: Each change along the path must have been adaptive in itself, regardless of where the path ultimately led.
- Hence a continuous evolutionary path that produces singularities.
- A model of what mental operations developed along that path, and in what order.
- An explicit account of what continuous changes produced what singularities, and how they did so.
- Robust evidence from many quarters that human beings actually perform the mental operations on that hypothetical path.
- Intermediate steps not for the function of language itself but for the cognitive abilities that finally led to the precipitation of language as a product.
- Evidence in the anatomy or behavior of today's human beings pointing to the history of these steps, just as anatomical evidence in today's human beings points to our once having had tails.
- Other things being equal, a parsimonious way of explaining the emergence of many related human singularities as products that arise along the same continuous evolutionary path.

THE CENTRAL PROBLEM OF LANGUAGE

The world of human meaning is incomparably richer than language forms. Although language has been said to make an infinite number of forms available, it is a lesser infinity than the infinity of situations offered by the very rich physical mental world that we live in. To see that, take any form, such as "My cow is brown," and try to imagine all the possible people, cows, and shades of brown to which it might apply, as well as all the different uses of the phrase as ironic or categorical or metaphoric, including its use as an example in this paragraph.

A word like "food" or "there" must apply very widely if it is to do its job. The same is true of grammatical patterns independent of the words we put in them. Take the resultative construction in English, which has the form A-Verb-B-Adjective, where the Adjective denotes a property C. It means *A do something to B with the result that B have property C*, as in "Kathy painted the wall white." We want this construction to prompt for conceptions of actions and results over vast ranges of human life: "She kissed him unconscious," "Last night's meal made me sick," "He hammered it flat," "I boiled the pan dry," "The earthquake shook the building apart," "Roman imperialism made Latin universal." We find it obvious that the meaning of the resultative construction could apply to all these different domains, but applying it thus requires complex cognitive operations. The events described here are in completely different domains (Roman imperialism versus blacksmithing) and have strikingly different time spans (the era in which a language rises versus a few seconds of earthquake), different spatial environments (most of Europe versus the stovetop), different degrees of intentionality (Roman imperialism versus a forgetful cook versus an earthquake), and very different kinds of connection between cause and effect (the hammer blow causes the immediate flatness of the object, but eating the meal at one time causes sickness at a later time through a long chain of biological events).

This very simple grammatical construction allows us to perform a complex conceptual integration that, in effect, compresses over Identity (e.g., Roman imperialism), Time, Space, Change, Cause-Effect, and Intentionality. The grammatical construction provides a compressed input space with a corresponding language form. It is then blended in a network with another input that typically contains an unintegrated and relatively diffuse chain of events. So, if it is our job to turn off the burner under the pan containing zucchini in boiling water, and we forget about it and all the water evaporates, we can say, confessionally, "No zucchini tonight. I boiled the pan dry. Sorry." In the diffuse input, the causal chain runs from forgetting to the invariant position of the burner knob, to the flow of gas, to the flame, to the temperature of the pan, to the temperature of the water, to the level of the water, to the dryness of the pan. The agent performs no direct or indirect action on the pan at all. But in the blend, the compressed structure associated with the grammatical construction is projected together with some selected participants from the diffuse chain of events in the diffuse input. In the blend, the agent acts directly on the pan. Moreover, although the boiling of the water is an event and its cause was something the agent did or did not do, there is cause-effect compression in the blend such that in the blend, although not in the input spaces, *boiling* is an action the agent performed on the pan.

As this example shows, the simplest grammatical constructions require not only high abstraction over domains but also complex double-scope integration. This is a very general feature of grammatical constructions, as we will demonstrate with considerable additional evidence in Chapter 17.

Paradoxically, language is possible only if it allows a limited number of combinable language forms to cover a very large number of meaningful situations. The previous chapter described an example of how that happens. We saw how a simple form, "Y of," prompts for an integration mapping scheme that efficiently covers vast ranges of meaning. We also saw in some detail how that form combines to prompt for ever-larger mapping schemes.

There is every reason to think that some species are able to operate efficiently in separate domains of, say, tool use, mating, and eating without being able to perform these abstractions and integrations. If that is so, then grammar would be of no use to them, because they cannot perform the conceptual integrations that grammar serves to prompt. But couldn't they just have a simpler grammar? The only way they could have a simpler grammar and yet have descriptions in language for what happens would be by having separate forms and words for everything that happens in all the different domains. But the world is infinitely too rich for that to be of any use. Trying to carry around "language" of that size would be crippling. The evidence does not suggest that primates have compensated for lack of language by developing, for example, one million special-purpose words, each conveying a special scenario. On the contrary, although primate species are capable of specific "vocalizations" (e.g., in response to a potential predator), the best efforts to teach words to chimpanzees cannot get them past a vocabulary of about two hundred items. Having a handful of vocalizations is clearly a help, but evolution has found no use in trying to extend that strategy very far. The extraordinary evolutionary advantage of language lies in its amazing ability to be put to use in *any* situation. We will call this crucial property of language "equipotentiality." For any situation, real or imaginary, there is always a way to use language to express thoughts about that situation. The key to the amazing power of the equipotentiality of language, which we take for granted and use effortlessly in all circumstances, is *double-scope conceptual integration.*

GRADIENTS OF CONCEPTUAL INTEGRATION AND THE EMERGENCE OF LANGUAGE

On independent grounds, we must grant that human beings today have powerful and general abilities of conceptual integration. In particular, double-scope networks are the kind of mental feat that human beings perform with the greatest of ease but that other species are unable to achieve. Thus far we have seen the crucial role of double-scope networks in grammatical constructions, the invention of scientific and mathematical concepts, religious rituals, counterfactual scenarios, persuasive representations, and vital-relation compressions.

We have also seen that networks of conceptual integration fall along gradients of complexity. At the top end are networks whose inputs have clashing organizing frames and blends that draw on both of those frames, the double-scopes. At the bottom end are simplex networks with conventional frames and ordinary values for their roles. In Chapter 8 we surveyed a continuum of complexity from the most rigid and basic simplex networks to topology-clashing double-scopes. Along the way, we discovered other prototypes: mirror and single-scope networks, and saw that these types of network are not separate categories but products of the basic mental operation of conceptual integration. They stand out from the great range of blending networks.

Our hypothesis for the origin of language is as follows:

- Double-scope conceptual integration is characteristic of human beings but not other species and is indispensable across art, religion, reasoning, science, and other singular mental feats that are characteristic of human beings.
- The hallmark virtue of advanced blending capacity is its provision of efficient, intelligible, strong compressions across ranges of meaning that would otherwise be diffuse and unmanageable. There are many scenes that are immediately apprehensible to human beings: throwing a stone in a particular direction, breaking open a nut to get the meat, grabbing an object, walking to a visible location, killing an animal, recognizing a mate, distinguishing friend from foe. Double-scope blending gives us the supremely valuable, perhaps species-defining cognitive instrument of anchoring other meanings in a highly compressed blend that is like these immediately apprehensible basic human scenes, often because such scenes are used to help frame the blend.
- The development of blending capacity was gradual and required a long expanse of evolutionary time: Basic blending is evident as far back as the evolution of mammals.

- Each step in the development of blending capacity was adaptive. From very simple simplex blends to very creative double-scopes, each step of the capacity would have been adaptive because each step gives increasing cognitive ability to compress, remember, reason, categorize, and analogize.
- There is ample evidence of intermediate stages in the development of blending capacity. Some species, for example, seem able to do only simple simplex networks, while others seem able to do slightly more unusual simplex networks.
- There is also ample evidence of intermediate stages in human beings, in the sense that although we can do double-scope blending, we can of course still do simplex blending.
- A special level of capacity for conceptual integration must be achieved before a system of expression with a limited number of combinable forms can cover an open-ended number of situations and framings.
- The indispensable capacity needed for language is the capacity to do double-scope blending.
- The development of double-scope blending is not a cataclysmic event but, rather, an achievement along a continuous scale of blending capacity, and so there is no Cause-Effect Isomorphism in the origin of language: The cause was continuous but the effect was a singularity.
- Language arose as a singularity. It was a new behavior that emerged naturally once the capacity of blending had developed to the critical level of double-scope blending.
- Language is like flight: an all-or-nothing behavior. If a species has not reached the stage of double-scope blending, it will not develop language at all, since the least aspects of grammar require it. But if it has reached the stage of double-scope blending, it can very rapidly develop a full language in cultural time because it has *all* the necessary prerequisites for a full set of grammatical integrations. The culture cannot stop at a "simpler" language—for example, one that has only the Subject-Verb clausal construction. A grammatical system, to meet the crucial condition of equipotentiality, must possess a full set of possible integrations and corresponding forms that can combine to give expressions suitable for any situation. Therefore, language will automatically be multiply double-scope and complex. And there will be no stopping the development of language from achieving that level, since the engine of double-scope blending that produces equipotentiality will be fully in place.
- The story of the origin of language does have room for intermediate stages, in the capacity: Human beings still have the capacity to do simple forms of blending. But no intermediate stages will be found in languages themselves because full grammar precipitates quickly as a singular product

of the blending capacity once it reaches the critical stage. By "quickly" here we do not mean instantaneously but within cultural rather than evolutionary time.

- The hallmark virtue of language is its ability to use grammatical patterns, suitable for basic human scenes, to capture and convey much less tidy meanings. This is done through the massive compression offered by double-scope blending, which can achieve blends that fit the grammatical patterns associated with those basic human scenes. (In Chapter 17 we will survey the ways in which the basic constructions of a grammar rely on double-scope blending.) Language, in the strong sense, must be equipotential. It must be serviceable for the innumerable new situations we encounter. But the only way it can be equipotential is for the human mind to be able to blend those new situations with what we already know to give us intelligible blends with attached grammatical patterns so those existing grammatical patterns can express the new situations. To say something new, we do not need to invent new grammar—and a good thing, too! Rather, we need to conceive of a blend that lets existing grammar come into play. Only in this way can an individual with a small, relatively fixed vocabulary of words and basic grammatical patterns cope with an extremely rich and open-ended world.
- If we follow the view of Stephen Mithen—according to which other singular explosions in human capacity and society, such as tool design, art, religion, and scientific knowledge, were the result of "cognitive fluidity"—then it is plausible to conclude that all these spectacular changes in human performance came about once the continuous improvement of blending capacity reached the critical level of double-scope blending. Mithen explicitly places the origin of language far before the development of "cognitive fluidity." For him, language is an input to "cognitive fluidity." For us, by contrast, it is the most impressive behavioral product of double-scope blending.

In summary, continuous improvement of blending capacity reached the critical level of double-scope blending, and language precipitated as a singularity. But why should double-scope blending have been the critical level of blending that made language possible? The central problem of expression is that we and perhaps other mammals have a vast, open-ended number of frames and provisional conceptual assemblies that we manipulate. Even if we had only one word per frame, the result would be too many words to manage. Double-scope integration, however, permits us to use vocabulary and grammar for one frame or domain or conceptual assembly to say things about others. It brings a level of efficiency and generality that suddenly makes the challenging mental logistics of expression tractable. The forms of language work not because we have managed to encode in them these vast and open-ended ranges of meaning, but because they make it possible to prompt for high-level integrations over conceptual arrays we already command. Neither the conceptual operations nor the conceptual arrays are encoded, carried, contained, or otherwise captured by the forms of language. The forms need not and indeed cannot carry the full construal of the specific situation but, instead, consist of prompts for thinking about situations in the appropriate way to arrive at a construal.

Our proposal would explain the apparent discontinuity of the appearance of language: No "fossils" of early simple language have been found because there were none. The appearance of language is a singularity much like the rapid crystallization that occurs when a dust speck is dropped into a supersaturated solution. When a community graduates to double-scope integration at the conceptual level, any local problem of expression that is solved by a specific double-scope integration gives the pattern for solving the general problem of expression, making that general problem tractable, and resulting in the complex singularity of the appearance of a system that uses a limited number of combinable forms to cover an open-ended number of situations. It "covers" these situations not by encoding construals of them (e.g., through truth-conditional compositionality) but, rather, by using a limited number of forms to prompt for on-line inventive integrations that are full construals.

THE ORIGIN OF COGNITIVELY MODERN HUMANS

Here are some fascinating individual truths about evolution and the origin of modern human beings that have been widely, if disparately, recognized but never before combined into a single coherent story:

- Biological evolution happens gradually.
- Human language appears, in evolutionary terms, very suddenly in recent prehistory.
- Art, science, religion, and tool use also appear very suddenly in recent prehistory.
- Human beings differ from all other species in having these behavioral singularities, and their performances in these areas are extraordinarily advanced.
- Anatomically modern human beings arose 150,000 years ago.
- But behaviorally modern human beings date from around 50,000 years ago. That is, evidence of advanced modern behavior in tool use, art, and religious practices appears in the archeological record around 50,000 years ago.
- There is no evidence of "simple" languages in other species.

- There is no evidence of "simpler" languages in other human groups.
- Children learn complex languages remarkably easily. But they go through what look like intermediate stages.

None of the previous theories puts all these truths together, and the theories that do exist conflict with one another, sometimes in extreme ways.

Some theorists propose that a dramatic biological event produced dramatically different human beings who had language. Chomsky proposes such a dramatic biological event for language. Mithen, by contrast, proposes a neurological "big bang" for cognitive fluidity but *not* for language. For Mithen, the earliest anatomically modern human beings already have language, but it takes them another hundred thousand years to get art, religion, science, and elaborate tool use, and when they do get those performances, it happens overnight. That change in behavior is triggered by an exceptional, singular change in the human brain that is highly adaptive. For Mithen, that dramatic biological change is unrelated to the origin of language but instead produces remarkable human creative abilities. Language, already available, latches on to these new abilities. It is a beneficiary of cognitive fluidity but is not in itself creative under this account. For Chomsky, the dramatic biological event has only syntax as its direct product; he is also skeptical of accounts that adaptation played any role in the appearance of language. Both Chomsky and Mithen look at a singular result or results and explain them by postulating a singular biological cause. In this way, they deal efficiently with the absence of intermediate stages: The full results followed quickly from the causes. For Chomsky, the singular result of the dramatic biological change is language, which appears explosively on the human stage. For Mithen, art, science, and religion appear explosively on the human stage, but not language. These theories do not come without cost, however. They go against the principle of gradualism in evolution. Chomsky even seems to go against natural selection. Both he and Mithen pull a speculative, catastrophic, indeterminate, but all-powerful biological event out of a hat. Their explanations have built-in limits beyond which they cannot be pushed. Chomsky would need an extra theory to account for all the other human singularities, and Mithen would need an extra theory to account for language. Their theories are driven by Cause-Effect Isomorphism. Chomsky adds Function-Organ Isomorphism of the strongest possible sort. Since these isomorphisms give us compressions and hence global insights, they are seductive.

Other theories, such as those of Terrence Deacon on the one hand and Steven Pinker and Paul Bloom on the other, propose gradual evolutionary or co-evolutionary development of language ability. Both of these theories avoid the trap of proposing dramatic biological causes, but they face the problem of explaining why there are no surviving intermediate stages: Both propose that there were intermediate stages, but that the people who had them are gone and left no trace of those stages. Pinker and Bloom additionally face the difficulty of explaining the other human singularities; their theory, like Chomsky's, is directed exclusively at the origin of language and not at the development of certain forms of conceptual thought. Deacon is the one theorist on our list who leaves ample room for relating the origin of language to the origin of other cultural behaviors. He proposes the gradual, adaptive evolution of a relational ability that underlies a range of human performances. Those performances then co-evolved with that mental and biological capacity. From our point of view, Deacon has the right overall frame for the origin of language, but his theory is missing an explanation of the mental operations underlying this relational ability. The findings we present in this book were not available when Deacon was developing his views. More generally, the notion that human mental feats as disparate as simple framing, counterfactual thinking, and event integration could proceed from the same cognitive ability and lie on a common continuum was unavailable in the cognitive neuroscience community. We have seen that conceptual blending is a good candidate for a continually evolving mental ability that could produce the singularity of language. This opens up possibilities that Deacon could not have considered. Another consequence of our findings is that language would have precipitated much more quickly than Deacon proposes, over a span of thousands of years rather than millions. But on the other hand, the evolution of the cognitive capacity that yields language as a singularity could have begun long before there were human beings, hominids, or even primates.

Still other theories, such as the *Lingua ex Machina* proposal of William Calvin and Derek Bickerton, offer a preadaptation story. According to these theories, evolution labored long to produce abilities that ended up subserving syntax. These theories thereby avoid postulating a singular cataclysmic cause. On the contrary, they are gradualist stories. Calvin and Bickerton also go into the details of what those evolved capacities were (e.g., the ability to throw a projectile, the ability for reciprocal altruism) and what computational abilities they could provide to syntax. There is certainly nothing wrong with thinking that preadaptations played an important role in making the origin of language possible. Indeed, the capacity we invoke, conceptual blending, far from being limited to language, extends to action, reasoning, social interaction, and so on. The emergence of conceptual blending would have been favored by preadaptations. Where we differ with Calvin and Bickerton is that they propose that evolution delivered an ability for grammar while we propose that evolution delivered an ability for conceptual blending which, once it reached the stage of double-scope integration, had grammar as a product.

None of the proposals we have seen explicitly links all the singularities—language, science, religion, the arts—as deriving from a common cause. But there are other accounts that do see that linking as a priority. For example, Richard Klein, in *The Human Career*, offers the hypothesis that a dramatic mutation produced neurological change about 50,000 years ago, and that this neurological change gave human beings some signal capacity such as language. Once that particular capacity was in place, it led to the development of advanced tool use and the invention of art and perhaps other abilities, and these neurologically advanced human beings spread throughout the world.

Our proposal for the origin of language has ample room for full linkage across the singularities in human performance that arose around 50,000 years ago, but it does not require any one of them to have been the cause of the others. On the contrary, there is a deeper, underlying cause—namely, the continuous development of blending capacity until it arrived at the critical point of double-scope blending, and all these staggering new performances of human beings proceed from that capacity as products developed in parallel. On our view, these new performances reinforced each other in cultural time. The evolutionary achievement of double-scope blending still needs cultural time in which to bear all its fruit. The visible products of the new cognitive capacity are all social and external—art, religion, language, tool use. There is every reason to think that once the capacity was achieved and the cultural products started to emerge, they reinforced each other. Language assisted social interaction, social interaction assisted the cultural development of language, and language assisted the elaboration of tool use, as the tree of culture put forward these exceptional new products. Language and art became part of religion, religion part of art, language part of the technology of tools, all intertwined. Certainly this is the picture we see when we look at human beings today. We will consider more evidence for this intertwining in the next chapter, when we look at blending and material culture.

We agree with Klein that the singularities are linked, but this does not imply that one of them caused the others. They are all products of the underlying evolution of the capacity for double-scope blending. There is another aspect of Klein's work, however, that is crucial to our account. He places the origin of language near in date to the origin of the other singularities. Why would a theorist like Mithen, who saw cognitive fluidity as the "big bang" of human evolution, not have considered language as part of the constellation of singularities like art, science, and religion that resulted from that big bang? The answer is simple: He assumes that language falls out of a combination of big brains and modern vocal apparatus. Mithen writes: "During the last few years the argument that both archaic *H. sapiens* and Neanderthals had the brain capacity, neural structure and vocal apparatus for an advanced form of vocalization, that should be called language, is compelling." This account would place the origin of language in the range of 100,000 to 400,000 years ago, and perhaps as much as 780,000 years ago. Therefore, language must have arisen, on Mithen's view, at least 50,000 years before the explosion of art, science, and religion in the human record.

Yet Mithen himself takes the view that human beings about 50,000 years ago developed striking new mental abilities that did not require a change in brain size or in anatomy. We think that is exactly right, but that language was part of the suite of products that flowed from that evolution. This unifying hypothesis receives strong support from recent archeological and genetic studies that were not available to Mithen.

Klein provides archeological evidence that there are two distinct types of modern human beings—anatomically modern and behaviorally modern. The anatomically modern humans have our anatomy, but not our characteristic behaviors. The behaviorally modern humans have both. The anatomically modern human beings, dating from about 200,000 years ago, at some point cohabited with more archaic human beings, such as Neanderthals. The behaviorally modern human beings originated much more recently—about 50,000 years ago—and dispersed eastward from Africa, ultimately supplanting all other human beings.

Klein's view receives even stronger support from two genetics studies, one by Silvana Santachiara-Benerecetti, the other by Russell Thomson, Jonathan Pritchard, Peidong Shen, Peter Oefner, and Marcus Feldman. Santachiara-Benerecetti's work on mitochondrial DNA led her to the conclusion that behaviorally modern human beings arose about 60,000 years ago out of Africa and migrated eastward into Asia rather than northward into Europe, as was previously found for the more ancient anatomically modern human beings. In turn, Russell Thomson and his colleagues looked at Y chromosomes in people around the world today and computed an expected time on the order of 50,000 years to our most recent common ancestor. That dating falls within a large range of uncertainty but, in any event, moves the origin of behaviorally modern human beings closer to us by many tens of thousands of years.

Luigi Luca Cavalli-Sforza takes the final step, locating language as an invention of behaviorally modern human beings. He places it alongside the invention of boats and rafts and Aurignacian technology, which is to say, beads and pendants and other items of personal decoration used for social and ritual purposes. While Cavalli-Sforza brings the origin of language forward to about 50,000 years ago, other researchers would push the date of the invention of craft technologies such as cord-making and weaving back by several tens of thousand years. James M. Adavaso, an anthropologist specializing in textiles, estimates that weaving and cord-making probably date from 40,000 B.C., "at a minimum," and possibly much earlier.

These new findings converge to suggest the rapid cultural invention of a coordinated suite of modern human performances, dating from the same epoch, perhaps about 50,000 years ago. We have argued that all of these modern human performances, which appear as singularities in human evolution, are the common consequence of the human mind's having reached a critical level of blending capacity—namely, double-scope conceptual integration.

CHAPTER 9
ZOOM OUT

BLENDING LEADS TO SCIENTIFIC DISCOVERIES BUT ALSO TO SCIENTIFIC FALLACIES

We have seen in earlier chapters that compression through integration can yield global insight, as in the case of complex numbers. But in this chapter we have seen that a compressed global insight can also be a fallacy. Cause-Effect Isomorphism and Function-Organ Isomorphism are cases where compression can deliver misleading global insights.

Question:

- What is the merit of compression through blending if it leads to error?

Our answer:

Compression through blending often leads to deep insight into useful truth. When an earthquake causes a building to fall down, we see a tight cause-effect isomorphism between the trembling of the ground and the trembling of the building. The trembling building becomes part of the earthquake. This impulse to find similarities in cause and effect has been examined at length in anthropological studies of culture, myth, and magic—for example, by Sir James George Frazier in *The Golden Bough*.

But while compressing cause and effect so they are similar or isomorphic can be good, assuming as given that they *must* be similar or isomorphic is a fallacy. After all, long wear over centuries can cause a wall to collapse in seconds. We have seen this fallacy at work in theories of cataclysmic genetic mutation that immediately produced language capacity from scratch. In those theories, the isomorphism blends the singular nature of the effect with the cause, to give singularity to the cause. Conversely, if we assume that the cause is gradual, as in gradual natural selection, it is a fallacy to assume that the effect should also have appeared gradually. We see the fallacy at work in the theory that gradual natural selection must have produced a gradually developing protolanguage, even though there is no evidence for protolanguage. Theorists as different as Deacon, Pinker, and Calvin and Bickerton all proposed the existence of such protolanguages as stages on the evolutionary path to full language. On our view, language is a singularity and an external social manifestation of having reached the critical level of double-scope blending. There need not have been any external language-like manifestations of the earlier levels of the capacity for conceptual integration or, for that matter, any additional biological evolution after humans reached the level of double-scope blending: Cultural evolution in cultural time would have sufficed. The Cause-Effect Isomorphism Fallacy consists of taking a Cause-Effect Vital Relation and assuming that it is equally a Similarity relation: The cause and effect are identical at some level.

Another fallacy of compression appears in Klein's proposal that language was the large neural event that caused the other singularities in human evolution. In this case, because performances like language, decoration, and technology have relations of similarity at some level, we assume that they must also have cause-effect connections: One becomes not merely like the others in some ways but therefore the cause of the others. This is the flip side of Cause-Effect Isomorphism. We call it the Toblerone Fallacy, after the Toblerone ad that turns the similarity between the chocolate candy and the pyramids into the candy's being the cause of the pyramids.

All of these are fallacies of *assuming* certain kinds of compression. Yet the compressions often turn out to be right. We ourselves are proposing a compression over many singularities of human performance, seeing all of them as effects of a single cause, double-scope blending. We do not, however, use Cause-Effect Isomorphism compression: According to our proposal, the cause was gradual, continuous, and cognitive, while the effects were singular, quick, and social.

MENTAL BLENDING AND SOCIAL LANGUAGE

Capacities for conceptual integration are internal processes. Language, like art, science, and religion, is an external social process that depends on communication.

Question:

- In the blending account, it seems as if language itself plays no role in the origin of language. Can language be just a by-product?

Our answer:

Our analysis of the origin of language differs from others in proposing that an internal cognitive operation evolved over a long evolutionary period for functions independent of language, but that a certain stage of this operation—double-scope blending—was a *sine qua non* condition for the external communicative capacity of language. To have the capacity of language means to have an inventory of grammatical constructions that, although small relative to the richness of the world, are equipotential. For a relatively small inventory of grammatical constructions to be equipotential—that is, to apply generally rather than to only a small number of restricted scenes—they must be double-scope conceptual integrations. Therefore, there will be no language in the proper sense until the stage of double-scope blending is reached. But from that point, there is in principle no limit to the invention of complex constructions. In Chapter 17, we will see in detail that this double-scope capacity is indispensable for the everyday use of basic grammatical constructions and even of rudimentary words. To have any degree of command of an apparently simple word like "father" requires sophisticated double-scope blending. The same is true of all open-class words—"house," "red," "fly." Closed-class items—"of," "that," "here," "-ed" as a suffix, and so on—are themselves, as we saw, prompts for mapping schemes, typically involving integration.

To miss this point is to commit the Kanzi Fallacy, an instance of the Eliza Fallacy. Kanzi, a bonobo or pygmy chimpanzee, is to date the most adept in a tradition of ape users of signs that includes Nim Chimsky and Sarah. At the age of six, Kanzi seemed to have a command of about 150 words and could make two-word sequences. The furious debates over Kanzi and his colleagues have had to do with the question of whether Kanzi has syntax. But there is agreement that Kanzi has vocabulary, because he can manipulate some symbols that, for us, correspond to words we know. It is a fallacy, however, to assume, when Kanzi manipulates symbols that we see as words, that he is necessarily engaged in the same equipotential double-scope conceptual integration that human children use when they deploy vocabulary.

We want to make it clear that we do not dispute that Kanzi must be using powerful conceptual integration to do what he is doing in manipulating symbols and even combinations of symbols. Kanzi and human beings do indeed fall on a continuum of mental ability. But we note that none of Kanzi's attested behaviors displays the capacity for equipotentiality or reaches the level of double-scope conceptual integration.

The human child's use of words looks entirely different from Kanzi's because it *is* equipotential. There is apparently no limit to the child's rapid acquisition of new words and to their very wide application, and the child is constantly using words of everything and everybody she encounters. Kanzi, however, is stuck with few words and with limited application, and apparently has no impulse to develop them on his own or to use them except for limited purposes like making a request. We suggest that Kanzi's "vocabulary" relates to a finite number of frames of limited application and that because there is no higher-level blending capacity, those frames cannot be integrated fluidly, which is the power of blending and the *sine qua non* of language. The Eliza Fallacy here consists in taking word combinations by Kanzi and assuming that Kanzi is doing mentally what the child would be doing with those same word combinations. We have no dispute in principle with the proposal that Kanzi or Sarah might know meanings, might associate symbols with those meanings, and might put some of those symbols together in ways connected with the juxtaposition of corresponding meanings. We are making a different observation: This kind of symbol-meaning correlation need not be equipotential. For the limited frames Kanzi is using, his behavior and the child's might be quite similar, even though the underlying mental processes are different. It is a fallacy to assume that Kanzi is doing essentially the same mental work as the child. This is like assuming that because a chess-playing machine can play chess, it is doing all the fabulous double-scope blending that a human being does while playing chess. We suggest that our account is corroborated by the fact that Kanzi's vocabulary tops out at fewer than 200 words of limited application, while the six-year-old child uses 13,000 words with very wide application. The actual wide-ranging human use of even a rudimentary word turns out to be a major imaginative achievement.

THE LANGUAGE OF CHILDREN

An eighteen-month-old does not speak like an adult.

Questions:

- Is child language simple language?
- And if so, isn't that a counterexample to the theory that language flows out immediately from double-scope blending, since of course the child has that capacity?
- Isn't that also a counterexample to the claim that there can be no languages simpler than ours?

Our answers:

If we take a snapshot of a two-year-old's language, is it equipotential? Is it based on grammatical constructions that are equipotential? Absolutely yes. Is its system of grammatical constructions simpler than ours, in the sense of being smaller? Again, absolutely yes. So why would we say there are no simpler languages, properly speaking? The answer is that the child must have the capacity to do double-scope blending even at the very beginning of his use of language, and that this capacity will in the normal course of development carry him to fully complex language very quickly. From the moment the child begins to develop language, there is no way, short of death, brain damage, or nearly unthinkable absence of social support, to stop him from developing full grammar. We think the story is analogous for the origin of language: Once human beings had the capacity for double-scope blending and began to use it to develop language, they would have spooled up to a fully sophisticated system of grammatical constructions within a span of cultural time. But from the beginning, both child language, developing over a few years, and prehistoric human language, having developed over a "short" span of cultural time, depend on double-scope blending and are strongly inclined toward achieving equipotentiality.

On our explanation, it is possible for a species to use symbols—Kanzi uses symbols—before having the capacity for language. Language originated in a social community of human beings that already had symbols, elaborate social practices, and communication. It was the evolution to cognitive modernity that gave this community the ability to do double-scope blending. That change was dramatically adaptive because it brought powers for succeeding in natural and social situations. Cognitively modern human beings flourished. They then used their common ability for double-scope blending to express and evoke double-scope integration networks, not merely to represent individual frames in restricted ways. This was the birth of language.

THE BICYCLE

The bicycle was a great and singular event in recent human cultural history. Its invention had a long evolution—wheels, gears, seats, handlebars for gripping, and, earlier, mining of metal and production of steel. And let's not forget the brakes, and the air pump, and extraction of rubber from trees. None of this early evolution achieved a lower level of bicycle: The achievement of the bicycle came at one fell swoop, with little refinements thereafter. Interestingly, learning to ride a bicycle is similar to inventing the bicycle itself. It seems counterintuitive that one could move on two wheels, since with all the previous modes of transportation using wheels, it was easier to be at rest than to ride—easier to sit on the stationary horse, buggy, or chariot than to move with it. But with the bicycle, the faster you go, the easier it is to stay up, and this is counterintuitive until you've got the hang of it. One learns how to ride progressively, but success is a singularity. All at once, you can move on the two-wheeled contraption, after many fruitless and perhaps painful attempts.

No one would argue that first we must have learned to ride a one-wheeled contraption without handlebars or pedals, and, by steps, moved up to riding more and more complex machines until we finally achieved the level of the modern bicycle. No historian is looking into garages to find the intermediate steps of the bicycle, as opposed to finding earlier forms of what counts as a proper bicycle. No one proposes that children must first learn to ride a unicycle and then the more complicated bicycle. On the contrary, they start with the tricycle and then drop the third wheel. Anyone who has crashed a bicycle without being harmed knows that we can look at the damaged bicycle to see whether it can still get us home—with its handlebars twisted, or its brakes missing a pad, or its frame crumpled. Very often it can, but no one would therefore argue that the bicycle must have had as one of the intermediate stages of its invention this crumpled machine that still functions.

Riding a bicycle is a phylogenetic, ontogenetic, and technogenetic singularity that can arise only because we reach a crucial stage on very smooth, intersecting continua. We need wheels, tubes, and so on, and we also need legs (snakes and fish do not ride bicycles), eyes that look forward, refined balance, thumbs, and bipedalism. Just as bipedalism is essential for riding a bicycle although no one would suggest that bipedalism evolved to support our efforts to bicycle, so double-scope blending is essential for language although double-scope blending did not evolve to support our efforts at language. Language needs the cognitive capacity for double-scope blending and a social community that can support an effort to develop and disseminate grammatical constructions. Once these prerequisites were fulfilled, which is to say, once cognitively and socially modern human beings had evolved, we could learn language as easily as we can learn to ride a bicycle.

Ten

THINGS

Give us this day our daily bread.
—*The Lord's Prayer*

ONE OF THE MOST arresting singularities of human beings is our continual invention and deployment of, and attachment to, things. We make things, carry and consult them, teach each other how to use them, adorn ourselves with them, make gifts of them. Why?

Consider a wristwatch. A watch is a few ounces of metal and glass, with complicated interior parts one usually cannot see, strapped around an appendage. It has two or three thin rods radiating from a center that go round and round at unequal rates. As a thing in itself it is bizarre and pointless, an extra weight for the wrist, a fragile object in a position where it can easily be broken. We cannot eat it; it does not keep us warm or cool; it does not hold a soothing drink. Why make, buy, carry, and consult it?

The cognitive anthropologist Ed Hutchins has studied with great insight the way in which material culture is suffused with conceptual blending of a type that typically employs everyday objects as material anchors. Hutchins's examples include things like watches, sundials, gauges, compasses, and sliderules. The wristwatch is a material anchor for a fascinating conceptual blend.

The first step in analyzing the watch is to consider the powerful integration network in our cultural model of time for a single day. This is a mirror network, with as many inputs as there are distinct days. It is a significant achievement to be able to conceive of time as periodic. Days follow days, seasons follow seasons. The sun rises and will rise again tomorrow. We have analogy connectors running across all these days, connecting noon to noon, for example. In the generic space, there is a single abstract day. The corresponding times in distinct days are compressed in the blend into uniqueness, so that noon yesterday, noon today, and noon tomorrow in the different input spaces are felt to be the same noon in the blend. In each input space, a single day runs its course just once. In the generic space, an abstract day runs its course just once. But in the blend, *the day* perpetually runs its course and then starts again, going through the same progression of times: dawn, morning, noon, afternoon, evening, night. We will call this the Cyclic Day network. In its blended space, we "reach noon *again*." Phrases like "your *morning* coffee," "swallows disappear and bats appear in *the early evening*," and "this park closes at *dusk*" pick out structure in the blended space of the Cyclic Day network. This compression of outer-space analogy relations between the inputs into uniqueness in this blend uses the compression principle of fusing. Of course, there are many similar integration networks for different units of time, giving us blended notions of a week ("your weekly workout"), a month ("your monthly visit to your mother"), and a year ("your annual checkup"). In each of these networks, the inputs all have a similar unit of time, each of which runs its course just once. There are analogy connectors between times in the inputs, and fusing to create in the blend a natural unit of time that *repeats*, so that *the same unit of time* happens again and again. Each of these networks has linear time divided into equal segments in the input spaces but cyclical time in the blend. The *outer-space* linear ordering of the inputs—successive days that can go on to infinity—is compressed within the blend into an *inner-space* cyclical ordering of repeated motion through the *same* unique day. The Cyclic Day is a remarkable compression that conforms admirably to the governing principles and overarching goals that we discuss in Chapter 16. It compresses an infinity of time into the human scale of a single day; it compresses Many (all the mornings, all the evenings, all the "noons" . . .) into One (the morning, the evening, noon, . . .).

In this network, quite interestingly, we see inner-space time relations—the period from dawn to dusk—and outer-space time relations—the "nighttime" that bridges day to day—compressed into a single inner-space temporal cycle. The open-ended sequence of outer-space nighttimes has become a single arc in the cycle.

The watch depends on the existence of the Cyclic Day integration network. One input to the integration network for the watch is the blended space in the Cyclic Day network (and that space stays connected to the rest of its network). The other input is the rotating rods—that is, the physical appearance of the watch itself, with the thin rods in a specific position and each of them moving. That input is also inherently cyclical.

The cross-space mapping is obvious, if a little bizarre: One cycle of the Cyclic Day maps onto two cycles of the small rotating rod and 24 cycles of the larger one. When the Cyclic Day reaches noon, the positions of the Rotating Rods point at 12. After two cycles of the small rod and 24 cycles of the larger one, it is noon again in the Cyclic Day.

Our modern understanding of time as consisting of a repetition of a periodic day is emergent in the blend for the Cyclic Day network. Our more specific understanding of time as consisting of a repeating day divided into hours, minutes,

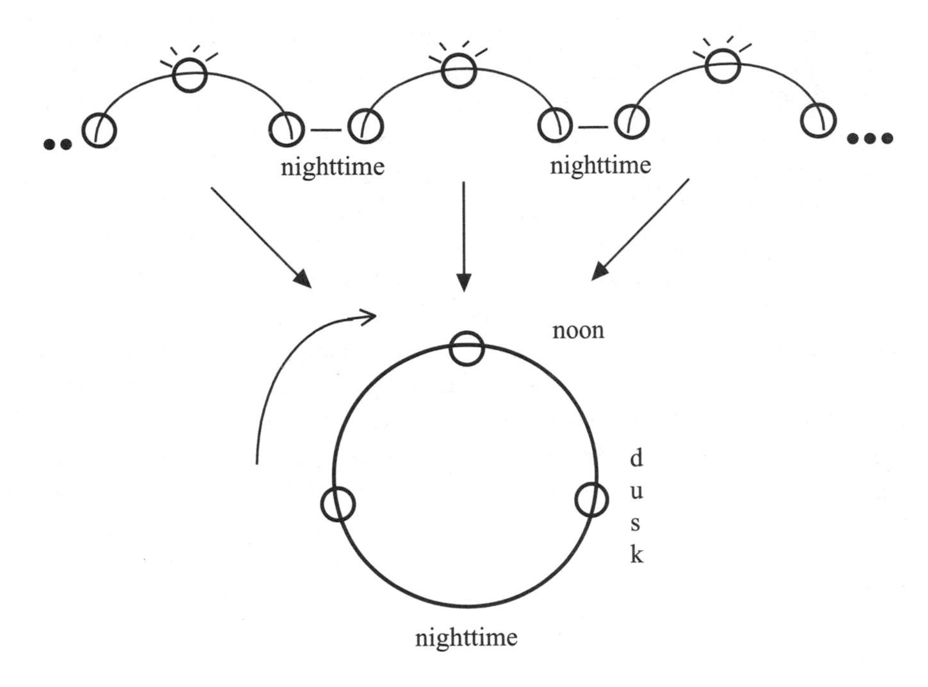

FIGURE 10.1 THE CYCLIC DAY

and seconds of equal duration is emergent in the Timepiece blend. In the cross-space mapping for the Timepiece network, each moment of the Cyclic Day is mapped onto a position of the rotating rods. In this cross-space mapping, the duration of an interval of time in the Cyclic Day corresponds to the length of an arc swept by a rod. In the blended space of the Timepiece network, the arc swept by a rod *is* an interval of time. A crucial emergent property is that equal arcs are equal intervals of time. This blended emergent notion of time is consistent because of human invention of a thing, a machine, that uses a periodic event like the rotation of the rods or the swinging of a pendulum. It is further important to be able to calibrate the machine so that it is always in the same position every noon. Once you have done that, you have automatically "divided" the Cyclic Blended Day into as many equal segments as the machine goes through from noon to noon. If we divide the watch face into k equal arcs, and there are n revolutions from noon to noon, we thereby divide the Cyclic Day into "n times k" equal intervals. If n is 2 and k is 12, we have the conventional division of the day into 24 hours. If n is 2 and k is 1, we have the conventional division of the day into A.M. and P.M. Cultural evolution since Babylonian times has arrived at the universal convention of making k equal to either 12 or 60. The same watch can have markings for both 12 and 60 by dividing each of the 12 arcs into 5 equal smaller arcs. The little hand of the typical watch corresponds to n = 2 and

k = 12. In the Timepiece blend for this hand, the motion of the little hand twice around the face of the watch, with identifiable positions at 1, 2, . . . 12, divides the Cyclic Blended Day into the 24 equal time periods we call "hours." In the different Timepiece blend for the longer hand on the watch, n = 24 and k = 60. This divides the Cyclic Day into 1,440 minutes. In the yet different Timepiece blend for the third, thin hand on the watch, n = 1440 and k = 60, which divides the Cyclic Day into 86,440 seconds. It is an extremely impressive cultural achievement with incalculable consequences for human life and knowledge to have developed these successive compressions of the Cyclic Day and the Periodic Physical Event. The deep conception of time that emerges is so compelling that we take it to be part of the fabric of time itself. We find it intuitively obvious that time is divided into equal intervals that repeat day after day, year after year.

The watch is additionally ingenious in exploiting the fortunate facts that there are the same number of seconds in a minute as there are minutes in an hour, 60, and that 12 is a factor of 60, so that the second hand ends up exactly at twelve after one minute, the minute hand ends up exactly at twelve after one hour, and the hour hand ends up exactly at twelve after one day. The watch itself is a material anchor for all three Timepiece blends—the one with n = 2 and k = 12, the one with n = 24 and k = 60, and the one with n = 1440 and k = 60.

Ed Hutchins points out that the watches we have today are historically the product of many successive blends, beginning with sundials. He notes that there is an obvious link between compasses, clocks, and dials in general. All of these blends integrate frames of time with frames of direction.

GAUGES

The pattern of blending linear scales of any magnitude with markings on a compressed circular object was extended throughout the history of gauges. We have circular dials with pointers for a host of technical measuring and navigation instruments: the radio tuner, the speedometer, the tachometer, the altimeter, the barometer, the oven temperature knob, the oven thermometer, and the thermostat. All of these gauges seem simple and transparent, but they are quite complex. Consider the oven temperature knob, for example. It indicates not the temperature of the oven but the temperature we would like the oven to be. But we also sometimes want to know when the oven has reached that temperature, as when we want to bake the salmon for a few minutes at 550 degrees but not at any lesser temperature. So ovens often have an "idiot light" that comes on when the oven is actually at the set temperature. Before the light comes on, the arrow points to a temperature in a desired mental space. Then, as soon as the light comes on, it points to a number on the knob that corresponds to the temperature.

In the case of the thermostat, the two gauges for the desired space and the reality space are superposed in the instrument, one hand for each space, and they share the same dial and its calibration.

Finally, Hutchins points to the superb selective projection involved in reading aviation dials. The dial illustrated in Figure 10.2 is giving the pilot information on when to reconfigure the wings for lift by positioning the slats and flaps on the wings. The appropriate configuration of slats and flaps depends on the speed of the plane and on its gross weight, which varies with the weight of the cargo on a particular flight and the present weight of the remaining fuel during a particular flight. After initiation of the descent from cruise altitude and before reaching a certain altitude, the pilot prepares the landing data. This involves determining the gross weight of the airplane and locating in some reference work the appropriate speeds at which the wings should be reconfigured for that weight. The pilot then moves the four clips, called "speed bugs," that are mounted on the rim of the speed gauge to mark off the speeds at which the wings are to be reconfigured. In that way, during descent, the pilot need no longer worry about the specific speed of the plane or the gross weight of the plane.

In the blend anchored by the instrument, the emergent structure is that if the hand is pointing between two particular clips, we want the slats and flaps to be in the corresponding configuration. Crucially, this structure is invariant from flight to flight and gross weight to gross weight, so that cognitively, once the clips are set, the pilot need only live in the blend. Here we see an example of modifying the real world by setting the clips in order to keep a conceptual structure constant in the blend. This blend has complicated connections to other spaces, but once he has set the clips appropriately, the pilot need not activate those connections while flying in order to succeed. In flight, all he needs to do is reset the slat-and-flap configuration when the dial is pointing at the clip.

Modern planes have linear digital displays, such as the one depicted in Figure 10.3. Ironically, as Hutchins shows, they do not reflect perceptually at human scale the differences in speed and positioning of slats and flaps, even though formally they contain all the necessary information. In the old gauge, the gross physical configuration of the display was different for different speeds, and different regions of the display corresponded to different positioning ranges; but in the new display, the gross physical configuration of the display does not change.

We take the dials for granted as simple, obvious objects, but replacing them with sophisticated and in some cases more accurate digital displays can reveal that they were supremely efficient compressions.

Modern displays also show remarkable emergent structures and compressions to human scale, most of them intended by the designers, but some discovered by users. Barbara Holder discusses the case of the "blue hockey stick" in the Airbus 320 flight control display. The blue hockey stick is a small, crooked blue

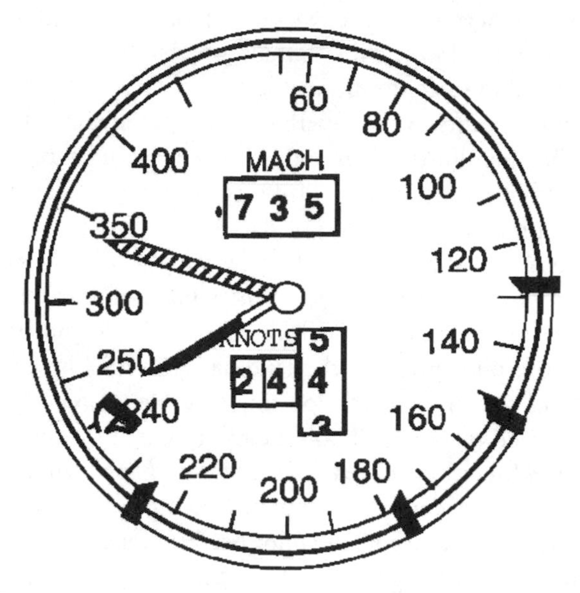

FIGURE 10.2 AN AVIATION DIAL

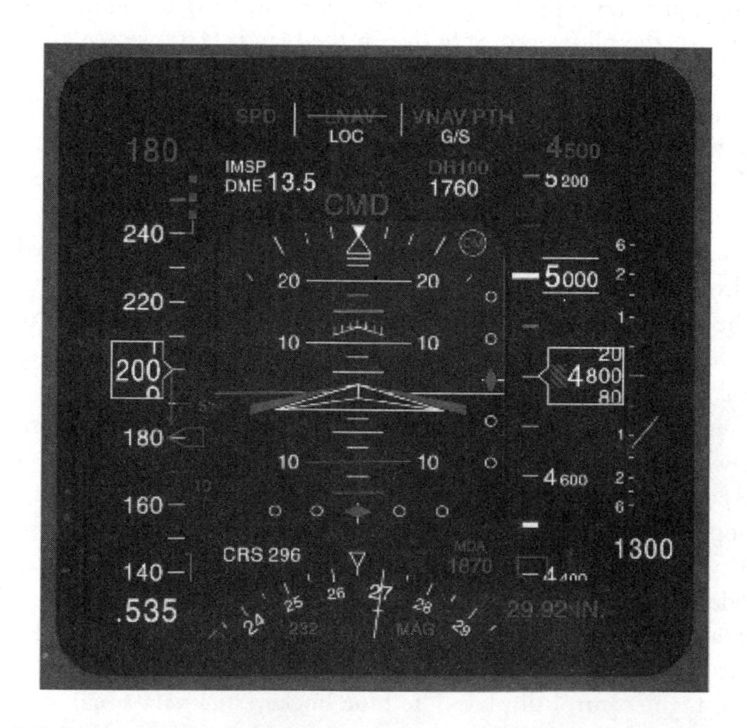

FIGURE 10.3 LINEAR DIGITAL DISPLAY IN A MODERN PLANE

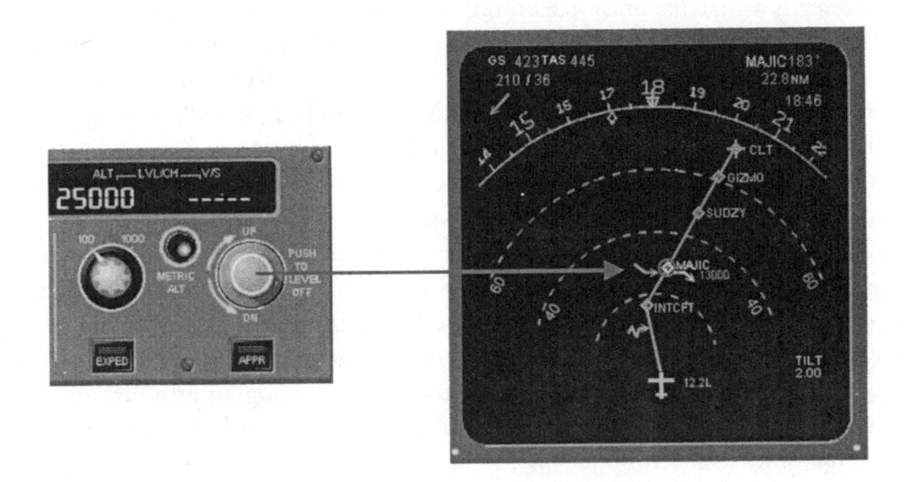

FIGURE 10.4 "BLUE HOCKEY STICK"

arrow on the screen that displays the course of the plane (see Figure 10.4). It shows the point on the future precomputed path of the plane at which the desired altitude will be reached under the current settings. During descent, the pilot will be cleared for a certain altitude at a certain geographical location. He wants to achieve that altitude by the time he reaches that location. In vertical speed mode of descent, the pilot can alter the altitude of the plane by manipulating the vertical speed knob, whose setting indicates the desired vertical speed. Manipulating the vertical speed knob has the effect, through the computation done by the plane's instruments, of changing the point on the flight path where the desired altitude will be reached. Since that point is indicated on the display by the blue hockey stick, the blue hockey stick will move when the knob is turned. This is a fabulous compression that allows the pilot to work directly in the blend; the pilot simply turns the vertical speed knob to move the blue hockey stick to the desired location on the flight path display. This compression is not taught to pilots, but some pilots discover it. Holder quotes one who did: "And what I do in a case where they [Air Traffic Control] give you a crossing restriction, I use that blue hockey stick, and just roll the vertical speed to wherever you think it should be and see where the blue hockey stick ends up; if that looks right on the map display on the nav display, away you go."

MONEY

A watch is useful because of its mechanical functioning. The conceptual network required to understand it is one we can carry in our heads. Alone on a desert island, we might be very thankful for a watch.

Money is a material anchor with different features. It, too, depends upon an elaborate conceptual network, but one that can only be supported socially. The mechanics that make money useful are in the society rather than in the money itself. Alone on a desert island, we would find a dollar bill to be practically worthless.

The history of money, like that of watches, is complex and involves many successive blends, now long forgotten. But we can look at the concept of money synchronically, just as we looked at the concept of time materially, anchored by timepieces. To get a taste of how this analysis would go, we will outline a much simplified integration network for money. Take one input, the Goods, to be the set of things that can be needed, traded, desired, inherited, given, stolen, or in some way transferred from the possession of one agency to that of another. Now take a second input, the Values, to be a metric scale using an arbitrary unit of measurement. Value scales are widely available: We can speak of something being three times as beautiful as something else or of someone being ten times as smart as someone else.

The Goods input typically has a local value structure of its own: A cow may be worth 20 chickens, and three chickens may be worth so many yards of cloth. Most goods, however, cannot be directly compared: A fisherman has no reason to exchange his net for a farmer's plough.

Goods are mapped onto the Value scale in such a way that topology of values is preserved across the inputs. So if the plough is worth 2 cows to the farmer, and the fisherman is willing to give 30 salmon to a weaver in exchange for a net or for a piece of cloth that is worth 10 chickens to the weaver, and one salmon is mapped onto position 2 on the value scale, then for consistency and preservation of local topology we must map a net onto position 60, a cow onto 120, a plow onto 240, a chicken onto 6, and the piece of cloth onto 60. The obvious consequence of projecting a full Value structure onto the Goods input is to define exchange standards implicitly for the entire domain of Goods. Clearly it would take 3 salmon to get a chicken, 4 nets to get a plough, and 2 pieces of cloth to get a cow. The mapping generalizes exchange but does not simplify it: One would have to always travel with salmon and hope that the farmer or the weaver likes fish.

The amazing step in the invention of a true Money network is to bring in a third input with objects that have no place in the original system of exchanging goods, such as identical pieces of colored paper. In the simplest case, the new objects are identical tokens. We map each token onto the same position on the scale in the Value input—say, position 1.

Next, we project all the objects from the Goods input and the tokens from the new third input to the blended space. We also project all the values from the Values input, so that every element in the blended space now has a value, including the new objects; the new objects are now collectively called *money* and individually given a name (e.g., *dollar*). The exchange system is projected from the Goods input and in the blend applies to all objects that have value.

Therefore, money can participate in the exchange system. In the input spaces, the relation of goods to values is a complex derived computation, as we saw for the chicken and salmon. The outer-space connections between the three inputs—of goods, values, and arbitrary objects—provide a complicated way to compute connections between the exchange relations of goods. But this computation requires not only doing arithmetic but also polling the entire community to discover what exchanges are acceptable. In the blended space, these outer-space connections are compressed into simple properties of the objects. Each now has an intrinsic value. In the blended space, knowing the simple value replaces doing the arithmetic and conducting the poll. In the standard buying-and-selling scenario, there is the further constraint in the emergent structure of the blend that one of the exchanged objects must be money.

The elementary social structure of buying and selling as we know it emerges in this blended space. The objects that count as money have many special properties: For example, there has to be a legally enforced public consensus on the value and nature (minting, etc.) of these special objects, and the culture must ensure that the objects are easy to carry, easy to recognize, and resistant to decomposition (see Figure 10.5). Complex financial and economic structures appear in the social practices that are part of the blended space.

Historically, the leap to the money network has always required elaborate cultural steps, especially the intermediate step of selecting as the object that will count as money something that is easily measured and transported and also easily incorporated into the exchange system of goods, such as a weight of precious metal or a quantity of spice. In retrospect, when we start to think about the culturally mature network for money, we may be amazed that anybody ever swallowed it:

> But what do we mean by worth? When we use the word, we are talking about a system of equivalencies—a transaction between the symbolic and the actual that many people believe is the most real thing there is, but which is, in fact, a sheer act of the imagination. Whose idea was it that a hard, inedible, and unprotective substance (stones, shells, metal) could be traded for food, clothing, shelter? The use of money is the purest act of faith; no anchorite who has followed a vision into the desert has acted on an idea as far-fetched as our belief that if we put a dollar in a machine we will be drinking a Diet Coke in a minute.

Although our account of the development of the money network is outrageously simplified, it illustrates our main point: Money, in the form of bills and coins, provides a key material anchor for a tight compression of the notion of goods and how to exchange them. It is amazing that the socially sanctioned production of a simple material object can result in a material anchor that radically reorganizes social practice, with dramatic consequences for human culture generally.

FIGURE 10.5 MONEY

TOMBS, GRAVES, AND ASHES

In the Homeric epics, the martial élite cremates its honored dead, while the common folk bury their dead and tend the graves. We still maintain these traditions, both of which can leave us with a material anchor for the departed person. In one case it is ashes in a mausoleum or in an urn on a shelf at home; in the other case it is a "plot" of land with a tombstone.

In the previous chapter, we presented the hypothesis that striking singularities of cognitively modern human beings arose from "double-scope" blending. That hypothesis applies equally to the invention of burial rituals and, more generally, to the invention of the concept of living with the dead. The archeological record suggests that such treatment of "the dead" also arose roughly 50,000 years ago. In the network for "the dead," one input space has the person when alive, and the other input mental space has the remains, typically looking as much as possible like the living person just before death. Some burial practices are meant to ensure that the remains have this appearance; these include embalming, death masks, sarcophagi, and cosmetic techniques used by morticians, who dress the body in its best clothes and surround it in the casket with its characteristic accessories. Many vital relations connect the input space with the person and the input space with the remains: The person and the remains are causally related; they are also related by physical change; and they can be related by disanalogy—for example, the person moves but the corpse does not. The body is a part of the person in the input with the living person, so there is a physical relation of change between the body-as-part in one input and the corpse in the other. There is also a strong literal similarity between the body-as-part and the corpse. In the input with the living person, the body and the soul—or, if you prefer, the intentional aspect of the person—are inseparable. In the blend, we have a being who has some intentionality projected from the space of the living person, and so might, for example, have some of the living person's memories, interests, and psychological characteristics. The blend has, typically, the temporal moment taken from the present mental space with the remains. The outer-space disanalogy connector between the inputs—the person was vital but the remains are not—is compressed into an *absence* inside the blend. The dead person in the blend is an absence, felt as such, but with projections from the space of the living person.

In the blended space, *person* has been projected from the input space with the living person, and the complex structure of input spaces and outer-space connections has been compressed into the inner-space property *dead. Dead* is like *safe:* It prompts for a complex network. And once the blend for *dead person* has become entrenched, it can serve as an input to a standard extension of the category *person*, so that now the category includes dead people, whose help we implore, whose wrath we avoid, and whose advice we seek.

Personal identity itself involves a diffuse network of mental spaces whose compression in the blend creates the unique person. Conceptually, a person is involved in mental spaces over many times and places, through many changes. All those spaces contribute to a blend that has the single unique person. There is a physical material anchor for this conceptual blend—the active living biological body that we can see and with which we can interact. We can hear its voice, and it can hear ours. When the person dies, the conceptual network with the unique person persists for us, if not for the person. But the material anchor is gone.

The complexity of the conceptual blends involved with dying and the dead, and of the material anchors for these blends, is immense. We point here only to a very few striking ways in which material anchors are used to develop such blends to make the dead, though absent, accessible. Cemeteries and tombs are part of the real world and have their own physical organization. But their importance lies in the role they play as material anchors for the blend of "living with the dead." The projections are relatively straightforward. Since the living person's body is mapped onto the corpse, a place where you might encounter the living person (i.e., a place where her or his body might be) is mapped onto the place where the corpse "is"—the grave. We project from the space of the living person the notion of establishing contact with the person, and from the other input the place where that contact is best established—the grave marked by the tombstone. This simple material anchoring of the blend provides the necessary implements for such widespread cultural practices as paying honors to the dead by putting flowers on graves, visiting graves at regular intervals or on publicly chosen days like All Saints Day, and, above all, communing with the dead and even talking to them on the site where "they" were buried. In the double-scope blend, the dead exist as a category, whose elements are both present and absent.

The transition of death progresses from living person near death, to the person who has just died, to the mortuary preparation, to the visitation of the body, to the funeral, to the burial, to communing with the "departed" at the grave. Right after death, the corpse itself is a very powerful material anchor for the complex integration network of the person's personal identity, and, as such, it is treated in elaborate ritual ways so as, on the one hand, to deal with the fact that it is dead organic matter and, on the other hand, not to dishonor the person for whom it is serving as temporary material anchor. During that period, the corpse is the material anchor for "making contact" or "communing" with the person. Wakes and *veillées* and other rituals designed to prevent the corpse from being left alone before burial are themselves blends of care for the living and dealing with the corpse. Once the corpse is buried, it does not serve as a material anchor. That role shifts to the tombstone and the grave plot and the cemetery, with its other graves and, possibly, a chapel or church. In establishing contact with the dead years later and communing with them over their grave, we typically call up memories of the body just before burial, and we typically think of that body—not of the actual contents of the coffin—as inhabiting the grave.

In Catullus's farewell elegy to his brother, the material anchor is the ashes in the vicinity of the death:

Multas per gentes et multa per aequora vectus
advenio has miseras, frater, ad inferias,
ut te postremo donarem munere mortis
et mutam nequiquam alloquerer cinerem.

Across many lands and seas
I have come, brother, to your sad funeral rites
to offer the final gift to the dead,
and speak in vain to your mute ashes.

In the blend, there is a unique element that corresponds to the brother in one input and to the ashes in the other. By the first projection, Catullus can call it "brother" and "you" ("frater" and "te" in Latin). By projection from the ashes, it cannot respond to Catullus. In the blend, the brother does not respond, therefore he is mute, and therefore the ashes are mute, even though in the input, absent the blending, ashes cannot be mute. The ashes in a fireplace do not speak to us, but nobody would call them "mute."

CATHEDRALS AND THE METHOD OF LOCI

Graves are not an isolated case. Enormous amounts of human time, energy, and talent go into building material anchors for spiritual and personal integration networks. Robert Scott has written a book-length study of "the idea of the

Gothic cathedral" and its anchoring in actual Gothic cathedrals, with their furnishings and rituals, in actual physical sites. Just as the grave is a material anchor for communing with the relatively inaccessible dead, so the cathedral is a material anchor for communing with the relatively inaccessible world of the divine and the departed.

In the case of a grave, the place where the corpse is buried has a natural and inevitable link to the dead person. In the blend of living with the dead, the dead person is naturally available at the grave. In the more general case of the sacred, according to Scott, the goal is "to draw sacrality to the community." There is no *a priori* reason why the sacred should be drawn to any particular place. In Scott's words, the sacred is conceived of as a "diffuse, ubiquitous, pervasive but unfocused force" that is "so to speak, hovering about in the atmosphere." One of the purposes of the cathedral is to attract the sacred and focus it in a particular location. This requires the "creation of a habitat for the sacred, a special place and an environment to which sacrality would be drawn and kept in place in such a manner that its potential to evanesce or migrate would not be realized."

The cultural and perceptual experience of the cathedral and its site must conduce the community to activate a blend that gives them a sense of the sacred. Somewhat obviously, the cathedral has features of grandeur, size, degree and quality of light, difficulty of construction and maintenance, patterns of music and prayer, and other elaborately scheduled rituals that distinguish it from any human-scale practical building and associate it with the cultural conception of the sacred. Less obviously, there are mental and cultural integrations that go far beyond the visible aspects of the cathedral. The concept of the cathedral is linked to aspects of imagination and memory that increase its effectiveness many times over as a material anchor for sacrality. Scott's hypothesis for how this happens depends upon a mental instrument that has long been recognized within Western rhetoric. This instrument is the "method of loci."

In the method of loci, someone needs to remember a complex organization of ideas, perhaps to deliver later in the form of a speech. She does this by associating the ideas with locations on some familiar path and then remembering and expressing the ideas by imagining that she is going through the locations on the path. One input space has the ideas, the other has the familiar path, and there is an Analogy mapping between two well-ordered sequences in the two input spaces.

For example, if you need to memorize an after-dinner speech, you might think of the path from the front gate of your house through the porch, the front door, the rooms, the back door, and into the yard. Then, attach each of the ideas or actions in order to the places along the path. So the thanks you give to those who have invited you is the gate to your front yard, the "opening" joke you tell is the opening of that gate, and so on through the entire dinner speech. Then, nervous at the podium, all you need to do is take a mental stroll through your house to remember what you need to say when. Hutchins provides an insightful analysis of this method as involving a layered blending template:

> The method of loci provides a well-known example of the cognitive use of material structure. In order to remember a long sequence of ideas, one associates the ideas, in order, with a set of landmarks in the physical environment in which the items will have to be remembered. The method of loci sets up a simple trajectory of attention across a set of features, let us call them landmarks, of the environment. One may establish a flow through the environment that brings attention to the landmarks in a particular order. This is a layered blend. The initial input spaces are the shape of the motion of a trajector and the set of landmarks in the environment. Together these produce a blend that is the sequential flow through the landmarks in the environment. The sequential relations of the landmarks are an emergent property of the blend. This space then becomes an input space for a more complex blend. The items to be remembered are associated with the landmarks, producing a space in which the items to be remembered are imagined to be co-located with their corresponding landmarks. In this blended space, the items to be remembered acquire the sequential relations that were created among the landmarks. These sequential relations among the items to be remembered are an emergent property of this compound blended space.

This method was part of the art of memory, developed by Cicero and others and practiced since classical antiquity. Hutchins finds examples of the method of loci in many cultures; for instance, "in the Trobriand Islands of Papua New Guinea, long narratives are structured around local geography." Since the adult islanders know the geography, the progression of a story's protagonist along a familiar path associates the order of the locations along that path with the order of the parts of the story, making the long narrative much easier to remember.

In Hutchins's examples, the material anchors used in the method of loci already exist. The Gothic cathedral, Scott argues, is a different case. The idea of the cathedral begins from an input space with theological content and an input space with a building. The method of loci is used to create a blend in which the theological organization is fused with the order of locations as one moves through the building. But then, strong emergent structure arises in this blend: The building becomes modified in imagination so as to accept topological projections from the theological space. Over generations and generations, theologians elaborate this blend, and those who present the theology use it. The result is a fabulous emergent concept—namely, the Gothic cathedral, a structure that in many complex interacting ways is fused in the blend with Christian theology. The cathedral, Scott argues, although based of course on knowledge of antecedent places of worship, exists first as a mental construct, which is only secondarily given a full material anchor by actually building one. Scott explains how extensive knowledge of sacred texts guided monks "toward an immanent experience of the Divine": "Imaginary monastic schemes" were devised for remembering those sacred texts that

> had a locational quality in the sense that they provided a place for everything and assigned everything to its place, and that the metaphors employed for doing this were architectural. Since the materials that were meant to be accessed (i.e., the sacred texts) were used for purposes of mediation, this meant that in using them, the practice of the art of memory required the practitioner to engage in imaginary movements through *imaginary spaces.*

Scott reports the examples given by Mary Carruthers in *The Book of Memory* of how the art of memory worked. He writes,

> One involved a monk, Peter of Cellar, who imagined an entire monastery in this way and invited his audience to enter and use it together with him. Another example she gives is that of Hugh of St. Victor whom she described as being ". . . careful to show exactly how each piece (of his imaginary building) is articulated in the scheme of the entire structure, and how the story and rooms are divided in them to 'place' information in the form of images within these divisions, used as mnemonic loci. . . . Hugh saw this building in his mind as he composed: he 'walked' through it and . . . he used it himself as he advised others to, as a universal cognitive machine." A third example is Augustine of Hippo, who, in his sermons, painted a literary picture of a tabernacle and then invited his fellow monks to look around and walk about it with him.
>
> Perhaps the most famous example, famous in part because it survives, is the Plan of St. Gall, an actual drawing of a plan for a wholly functional monastic community that provided the ideal space for engaging in liturgical processions and meditation in pursuit of the path to enlightenment. Significantly, though it was never actually built, it was in fact used in imagination by the monks of St. Gall as a space in which to meditate individually and as a community.

In short, the cathedral is developed as a conceptual structure in the blend before it has an accurate material anchor to support that blend. The material anchor, once constructed, supports the mental activities of monks but also enables them to communicate that conceptual structure to the lay community, and to organize their activities—much as watches and money now organize the actions and interactions of people in a society. Over centuries, the conception of the cathedral, the building of the cathedral, and the actual use and existence of the cathedral culminate in an optimal compression.

The cathedral, like places of worship in general, contains many varieties of material anchors, all coordinated: vestments, candles, special chairs and benches for special activities, confessionals, stations of the cross with their own use of the method of loci, altars, sacristies, visual images, graves, and books. To the uninitiated, this collection of material anchors can look like a bizarre and unaccountable assembly, but those raised in the tradition will have the means and competence to unpack and decompress what is actually a very powerful blend, culturally evolved through centuries of worship.

We grow so used to interacting with material anchors like money and watches that the compressions they provide seem almost as complete and obvious as the compressions provided by biology, like the perception of the blue cup. When we see banknotes exchanged for bread, the fact that one is money is as obvious as the fact that one is bread. But to the cultural outsider, such things can look entirely mysterious. The outsider who enters the cathedral lacks the elaborate conceptual integration networks that make it possible to see the material anchors for what, in the compression, they are. But to the faithful, the cathedral—with its altar, its vestments, its candles before statues—is as immediately understandable as buying bread with money or seeing a blue cup.

We began this book by discussing networks that do not seem to require material anchors: counterfactuals, metaphors, the riddle of the Buddhist Monk, the Debate with Kant. Following Hutchins's lead, we have now turned to conceptual integration networks that seem to require material anchors in order to be manipulated mentally: Indeed, it is hard to imagine a society using "conceptual money" as an effective medium of exchange without any material anchors in the form of bills, coins, numbers in account books, or electronic banking devices. Barbara Holder discusses successive blends in the development of the Automated Teller Machine.

But is there a clear difference between conceptual integration networks that require material anchors and those that ostensibly do not? Let us turn to cases where the material anchors are less obvious than money or watches. We begin with writing.

WRITING

Writing hardly seems the same kind of thing as a watch, a coin, or a cathedral. Yet when we look at it, we see physical marks on stone or paper or a computer screen, and these marks are circulated through the community. By themselves, these marks are meaningless: If we could send a sheet of writing back 10,000 years to a tribe of cognitively modern human beings, they would not have the slightest idea what to do with it, although the sheet would be a marvel. But we have elaborate conceptual and linguistic mental systems that can use these marks in culturally supported ways. Just as we look at the watch to see what time it is, we look at a sentence in a letter to see what someone is saying to us.

The blend seems natural to us even if it is immensely rich in its projections and elaborations. Suppose a woman is reading a letter from her fiancé, a soldier at the front. What is she doing? From one perspective, she is looking at and distinguishing marks on a sheet of paper. But a lab rat or a pigeon can probably distinguish marks on a sheet of paper, too—and the woman is clearly doing something the lab rat or the pigeon cannot. There is one input in which the woman is alone, looking at a material object. There is another with the fiancé and his general capacity to speak to her. In the blend, her fiancé is speaking to her. The projections are selective and imaginative. There is emergent structure in the blend: They cannot answer each other in all the usual conversational ways, and there is no audible sound from the fiancé. The fact that the writing consists of words comes from the space of speaking. And the specificity of those words/marks comes from the space with the specific marks on the paper, combined with a general mapping, evolved by the culture, for connecting equivalence classes of sounds to equivalence classes of marks—that is, connecting the mark "boy" to the sound "boy" (or, more precisely, a category of marks like boy, **boy**, *boy*, boy, Boy, BOY, *boy* . . . to a category of sounds that consists of all the ways of pronouncing the word "boy"—with high or low pitch, with a British or Australian accent, in a whisper, . . .).

A proficient reader ends up with a general competence for constructing integration networks for writing and reading. One input has someone talking, the other has some medium with marks, and in the blend, the marks and the speech are fused in impressive ways. The emergent integrated activities of "expressing oneself through writing" and "understanding others through reading" are strikingly different from speech in nearly all aspects.

The writing and reading blend is of immense cultural importance to us. It cannot exist without the material anchors of distinctive marks on material substances. But the use of these material anchors depends on a very powerful prior conceptual blend that compresses a certain infinity of marks (boy, **boy**, *boy*, boy, Boy, BOY, *boy*, . . .) into a single entity, the written word "boy," and that entity itself is construed as identical to another compressed infinity, the spoken word "boy."

Once we have learned it, the writing and reading blending network seems simple and inevitable. But it includes complex projections and social conventions that we take for granted. For example, to read a book in English, we must map speech in time onto linearly ordered locations from left to right horizontally on the page, and understand that at the end of the line, the speech jumps back to the beginning of the next line, and that turning the page (the commonest action we take with a book) has no counterpart in the speech space.

SPEECH

Speech may seem immaterial, hardly like a watch or a cathedral or even an inscription on a tombstone. But in fact it is a material anchor. Consider the scene in which the woman is actually listening to the speech of her fiancé. He has returned unscathed, and they are having coffee in the kitchen. From one perspective, what is happening is that longitudinal waves in the air are striking her eardrums, and she is aware of this. But from that same perspective, a lab rat or a pigeon would be doing the same thing, and again, she is clearly doing something they are not. For her, the longitudinal waves give rise to "sounds" that are like physical objects. Our ability to categorize sounds in such a way that two sounds count as the same for the purpose of communication accounts for the permanence that gives sounds the status of material anchors. She knows a complex mapping that connects particular equivalence classes of sounds to particular linguistic structures like words and clauses that are publicly shared and mentally represented.

The complexities of these conceptual integration networks are much greater than we have portrayed. One would need to bring in phonetics, phonology, and morphology to describe aspects of them. Our superficial description nevertheless hints at the type of blending that is going on in writing and speech and at the ways in which material anchors of various kinds are indispensable to the mental and social activity.

SIGN LANGUAGES

It is now universally recognized that language can have modalities other than voice. Spoken languages use oral-auditory modality, but sign languages use visual-gestural modality. The structural and conceptual complexity of sign languages is of the same magnitude as that of spoken languages. Voice is the indispensable material anchor for the public sharing and learning of integration networks in spoken languages; gesture is the equivalent material anchor for sign languages. But because the modalities are different, their material anchors show some interesting differences as well.

A number of distinguished scholars—such as Scott Liddell, Karen Van Hoek, and Christine Poulin—have studied the ways in which connections between mental spaces are reflected and prompted for in the modality of sign. Liddell, who has explicitly studied blended spaces in American Sign Language (ASL), shows that mental representations of one's immediate surroundings constitute a special type of mental space, which he calls a "grounded mental space." The immediate physical surroundings are the material anchor for that mental space. Elements in this grounded mental space have corresponding physical locations in the immediate surroundings that can be pointed to as part of communication. Sign languages use blending and pointing in interesting ways to allow speakers to refer again and again in complex ways to the same referent. When the things being talked about are not physically present, signers can make them conceptually present by creating grounded blends.

Liddell presents the following analysis of a revealing case:

> [A] native speaker of American Sign Language is narrating, describing an interaction between the cartoon character Garfield and his owner in which Garfield is looking up at his owner. This immediately follows a section of the narrative in which the owner has just told Garfield that he has removed the batteries from the remote control for the television. The signer produces the two-sign clause CAT LOOK-AT to describe Garfield's initial response to the owner. The subject of the clause is the sign CAT. The predicate LOOK-AT is illustrated in Figure [10.6], Space B.
>
> The meaning being expressed is that the cat is looking up and to the right (toward its owner). During the narration the real cartoon characters are not present. In order to explain both the direction of the sign LOOK-AT, the turning of the head, and the direction of the eye gaze in Figure [10.6], I propose that the signer's head position and eye gaze are no longer his own. Instead, they are demonstrations of Garfield's turning his head to the right and looking up at his owner. The signer has also conceptualized Garfield's owner as standing to his right. This conceptualization involves blending elements from spaces A and B to produce the grounded conceptual blend shown as Space C.
>
> Within the blend Garfield's owner is now standing to the right of the signer-as-Garfield and the signer looks up to the right in order to show that the blended character is making eye contact with the owner. This is no longer Real Space, because Real Space is a mental representation of only one's immediate physical environment. Here the signer's head position and eye gaze are to be interpreted as Garfield's head position and eye gaze in the blend. That is, the signer's head and eye gaze provide a demonstration of what Garfield did. So the signer has become Garfield, at least partially, since his head and eye gaze are conceived of as Garfield's. In Figure [10.6] lines connect Garfield in A and the signer in B to the blended Garfield in C.

In other words, the living body of the signer has become the material anchor for an absent character, and a location where another character is absent is brought into play as a material anchor by pointing to that location. This is a blend of a present scene and an absent but reported scene. We take what we see as prompting for a blended space that we are to unpack so as to construct the input spaces, the connections between them, and the selective projections to the blend. A given element in the blend, such as the direction of gaze, is in one input the direction in which the signer is looking and in the other input the direction in which Garfield is looking.

Liddell points out that similar blends are also found in systems of gesture that accompany spoken languages, and indeed we find such blends involving gestural material anchors so natural that we might have to think twice to see

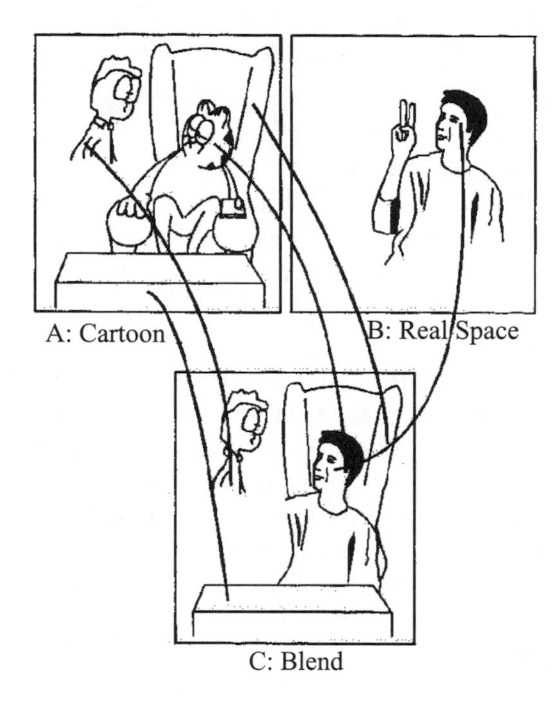

FIGURE 10.6 A BLEND WITH SIGNER AS GARFIELD. IN C (THE BLEND), THE CONCEPTUAL SETTING IS THE CARTOON AND THE PHYSICAL SETTING IS THE HERE AND NOW. (SOURCE: ADAPTED FROM LIDDELL, 1998, FIGURE 4.)

how complex these performances really are. They bring in the power of double-scope blending, which is an astonishing capacity that human beings take for granted because every one of us can use it easily, beginning from early childhood. We just can't help thinking that way.

CHAPTER 10 ZOOM OUT

Over time, cultures have developed a range of objects that prompt for elaborate conceptual integration networks.

Questions:

- How does the young child learn to use these objects?

- What capacities does the child have, and what line of development does the child go through in becoming able to manipulate all these integration networks?

Our answers:

In Chapter 9, we discussed the evolution of the capacity for double-scope integration around the time of the Upper Paleolithic. But having a capacity is not the same as having its products. Human beings, equipped with the capacity for double-scope integration, had to go through the arduous cultural work of producing integration networks, using those networks as inputs to further networks. In retrospect, looking back at the cultural invention of a repertoire of blends, we may be able to pick out the developmental sequence, as in the development of speaking tens of thousands of years before the invention of writing, with speaking as an input to writing. We also saw in the case of the development of numbers a long sequence in which each newly achieved number blend became an input to a later number blend. In this history, we need always to keep in mind the distinction between the *operation* of conceptual blending and the *cultural products* of conceptual blending.

Clearly, the child learns very quickly what often took the culture centuries or thousands of years to develop—how to write or use money or use clocks. It might seem strange that what cultures struggled so hard with is so straightforward to most of the children who must learn it. It might seem as if it should take longer for the child, or as if the culture should have moved much faster.

But, in fact, recent work in psychology—in particular, Jean Mandler's "How to Build a Baby"—has shown that infants have developed complex conceptual systems long before they start talking. We know that the child comes into the world with double-scope capacity. We see from work like Mandler's that the infant, as early as seven months old and perhaps earlier, is already deploying powerful integration networks. At least as early as the stage of pretend play using props, and certainly no later than eighteen months, the child has clearly constructed robust double-scope networks.

With these capacities, the child comes into a world already populated with material anchors for the culture's conceptual blends. What for the adult is a blended space in an elaborate integration network comes to the child at first as a single integrated space. The child plays with money, toy watches, and books long before having the concepts of buying, telling time, and reading. Through imitation, the child can develop some of the routines for manipulating those material anchors before developing the integration networks they are meant to evoke. What for the culture is the blended space of an elaborate integration network can be for the child a starting point for acquiring that network. That space, by virtue of the constraints on integration, is at human scale, involving direct perception and action. This makes it a good place for the baby to begin.

The child is on the path of both acquiring networks that lie behind what its culture offers and developing new ones through blending of inputs.

The case is no different for language. In our view, the culture is using material anchors at human scale as well as all the directly perceptible features of the environment—from mother's voice to the baby blanket to the trees to the baby's own body and actions—to entrain the child toward the development of its repertoire of integration networks. There will be external evidence for the acquisition of these networks—in particular, the culturally appropriate manipulation of material anchors such as intonation patterns, words, and phrases, simultaneously with and inseparably from gestures, facial expressions, eye movements, dress, handy physical objects, social interactions, and anything else at human scale that can be manipulated as a prompt for constructing meaning.

This view sets up a vast research program of investigating these particular networks and anchors and the development of the mastery of such networks and anchors that counts as cultural competence.

Eleven

THE CONSTRUCTION OF THE UNREAL

"I see nobody on the road," said Alice.
"I only wish I had such eyes," the King remarked, in a fretful tone. "To be able to see Nobody! And at that distance, too!"

—*Lewis Carroll, Through the Looking Glass*

PEOPLE PRETEND, IMITATE, LIE, fantasize, deceive, delude, consider alternatives, simulate, make models, and propose hypotheses. Our species has an extraordinary ability to operate mentally on the unreal, and this ability depends on our capacity for advanced conceptual integration.

Evolution has a way to consider alternatives and make choices: Organisms that differ from each other are already, from a God's-eye view, alternative experiments that evolution runs in the natural world. Which alternatives are fitter is gradually made clear, over generations, by the calculus of differential reproduction. The great evolutionary change that produced cognitively modern human beings was a matter of evolving an organism that could run off-line cognitive simulations so that evolution did not have to undertake the tedious process of natural selection every time a choice was to be made. Human beings can run several scenarios, mentally check the outcomes, and make choices, all in minutes rather than generations. Conceiving complicated new scenarios in nearly any domain while making complicated new inferences and choices is now something that can be run as part of mental and cultural life. The cognitive capacities of modern human beings not only allow individuals a far greater power of conception and choice, they also allow cultures to transmit choices that have been made and tested by entire communities.

VARIETIES OF THE UNREAL

In Chapter 2, we quoted Nelson Goodman's classic observation: "The analysis of counterfactual conditionals is no fussy little grammatical exercise. Indeed, if we lack the means for interpreting counterfactual conditionals, we can hardly claim to have any adequate philosophy of science." By now, this observation is not in dispute in any of the branches of learning that discuss counterfactuals. Goodman's point—that science depends in essential ways on counterfactual reasoning and, therefore, on the availability of counterfactual constructions in language—is equally valid in the social sciences, where in fact the importance and the problems of counterfactual reasoning are explicitly recognized. Gary King, Robert O. Keohane, and Sidney Verba, in *Designing Social Inquiry: Scientific Inference in Qualitative Research,* argue that there is no form of causal inference in the social sciences that does not depend upon counterfactual reasoning. Analyzing causality for social events is a matter of contrasting what in fact happened with counterfactual scenarios of what might have happened under different conditions.

> [The] *counterfactual* condition is the essence behind this definition of causality. . . . Thus, this simple definition of causality demonstrates that we can never hope to know a causal effect for certain. Holland (1986) refers to this as *the fundamental problem of causal inference,* and it is indeed a *fundamental* problem since no matter how perfect the research design, no matter how much data we collect, no matter how perceptive the observers, no matter how diligent the research assistants, and no matter how much experimental control we have, we will never know a causal inference for certain.

Although the key role of counterfactuals is widely recognized, it is often thought that conceiving of a counterfactual scenario is a simple matter of making a change in the actual world and observing the consequences of that change. King, Keohane, and Verba, for example, despite their experience with the complexities of local politics, ask us to consider the real situation of an election in a specific congressional district in which the Democratic incumbent runs against one Republican (nonincumbent) challenger and receives fraction x of the vote. To do the counterfactual reasoning, we are to "imagine that we go back in time to the start of the election campaign and *everything remains the same, except that* the Democratic incumbent decides not to run for reelection and the Democratic Party nominates another candidate," who receives fraction y of the vote. King, Keohane, and Verba define the causal effect (in this case, of the incumbency of the Democratic nominee) as the quantity x-y.

But as Goodman first recognized, changing any one element opens up complicated questions of what else would need to be changed in order for that one element to differ. Counterfactual scenarios are assembled mentally not by taking full representations of the world and making discrete, finite, known changes to deliver full possible worlds but, instead, by conceptual integration, which can compose schematic blends that suit the conceptual purposes at hand.

It is also commonly argued that counterfactual thought is always directed at causal analysis. Neal Roese and James Olson, in introducing *The Social Psychology of Counterfactual Thinking*, state as an established discovery of social psychology that "all counterfactual conditionals are causal assertions." But on the contrary, as we will see, great ranges of counterfactual thought are directed at important aspects of understanding, reason, judgment, and decision that are not concerned principally with causality.

Let us now turn to some striking examples of this process of assembling counterfactual blends.

Our first example is a counterfactual but not a causal assertion. A woman who had already been in a coma for ten years was raped by a hospital employee and gave birth to a child. A debate ensued concerning whether the pregnancy should have been terminated. Counterfactual scenarios arose such as "It is right to figure out *what she would want* [emphasis added]. It is wrong to try to figure out what we want." The *Los Angeles Times* article reporting the case ended with a comment by a law professor, who said, "Even if everyone agrees she [the comatose woman] was pro-life at 19, she is now 29 and has lived in PVS [persistent vegetative state] for 10 years. Do we ask: 'Was she pro-life?' Or do we ask more appropriately: 'Would she be pro-life as a rape victim in a persistent vegetative state at 29 years of life?'"

In the blend, the woman is in a persistent vegetative state, but has the reasoning capacities and general information that she would have had at age 29 under ordinary circumstances. The purpose of this blend is not to construct a plausible situation in which a woman is reasoning about her inability to reason. Nor, obviously, is the purpose to establish the causes of the woman's coma, her pregnancy, or anything else about the situation. The counterfactual blend is instead offered with the purpose of casting light on the element of "choice" in the input space in which the woman is indeed in a coma, so that we can come to a considered judgment about what action is legitimate. The issue is not causality but propriety. The law professor is committed to framing this woman as having the right to choose, but what does it mean for a woman in a coma to choose? Her abstract opinion, voiced ten years before her specific dilemma, does not meet our frame for "choice," so the professor is offering an alternative: In the blend, the pregnant woman can make an informed choice about the specific dilemma, and this choice should be projected back to the input to guide our actions.

This example makes it crystal clear that counterfactual reasoning is not a matter of imagining what we would have to change in the real world for the counterfactual scenario to be possible. We cannot fabricate a possible scenario in which someone who has been in a coma for ten years has also been fully aware during that time and can now reason about her inability to reason. The blend is as impossible as the Buddhist Monk's meeting himself or *Great American II*'s racing against *Northern Light*, but possibility or impossibility is entirely beside the point in such cases.

Although they may attract the attention of the philosopher who is disconcerted that truth can be sought by imagining impossible situations, such cases go mostly without notice in everyday discourse. This counterfactual, for example, was offered in legal discussion and was meant to guide our actual reasoning and judgment. It was reported by the newspaper as a perfectly rational comment needing no explication or apology. The question for the judge, the lawyers, and the readers of the newspaper was only whether the law professor was right or wrong. Nobody thought she had said anything bizarre. In fact, her comment could have been followed quite naturally with remarks like "That's absolutely true" by people who supported her position or "That's false" by people who did not. Both sets of respondents could continue without hesitation to reason in the blend. The supporters could say "She was so young at 19 that her opinions were not yet her own." The detractors could say "She was steadfast and decent and would not have been swayed by mere fashionable amorality."

It is worth noting that there is no levity involved here. The law professor's counterfactual is at the center of one of the most crucial and prickly social negotiations in current American society, and her blend is a genuine cognitive effort to offer a real solution to a deep and troubling question. There is no reason to expect blends to be pleasant, as we can see in this countervailing blend offered by the pro-life side: "Every Third Baby Dies by Choice." The billboard carrying this slogan depicts nine cute babies, sitting in a line, but the third, sixth, and ninth babies are shaded grey. The ninth looks to the side, its mouth open in a cry that, while not particularly dramatic, is nonetheless unmistakably a cry of unhappiness or distress. In a standard counterfactual blend ("if it had lived . . . "), the specific baby would be alive, doing what babies do. The grey baby in the blend is simultaneously an embryo or fetus that has been aborted and a baby some months old. The Change relation of development from embryo to baby has been compressed in the blend. The Cause-Effect relation between embryo and child has been compressed. The Time span has been compressed. This is the packet of vital relations that we compress in our usual template of personal identity, so the ad gets them for free. Consequently, there is in the blend a counterpart for what happens to the embryo in one of the inputs. In the blend, what happens to the embryo and the baby is the same since the embryo and the baby are the same. And we are told that the baby is dead. The ninth baby's distress can be projected back to the reality space of the fetus. It is only in the blend that the cry can be interpreted as a normal baby's reaction to a distressing situation that happens to be the impending death of a fetus.

Like any important subject of political or social debate, abortion elicits numerous powerful and competing blends that become part of the conventional cultural discourse, available for journalists, interviewers, political candidates, or anyone else to use directly. The two examples we have given are very strongly double-scope: The woman in a coma has a frame incorporating absolutely contradictory elements from the frames of the two inputs (e.g., coma versus consciousness), and the billboard about every third baby takes the termination of the organism from one space but, from the other, features of the organism that arise only later. Seana Coulson provides a detailed and insightful analysis of several attested abortion debates and interviews. Her cognitive rhetorical analyses show the intricate mapping schemes, cultural models, and conceptual blends deployed in these discussions.

While these examples are acrobatic, counterfactual reasoning is an everyday event that usually goes unremarked: "If I had milk, I would make muffins," "If I were you, I would quit." Grammar uses the same conventional forms to prompt for both everyday and acrobatic blends. Just as we can use the same form in "Ann is the boss of the daughter of Max" and "Prayer is the echo of the darkness of the soul," so we can say both "If I had the money, I would buy a house" and "If she could think about herself being pregnant in a coma, she would not choose to give birth." Even though the blends feel very different, the same mapping schemes are at work, prompted by the same language forms. In English, a typical form for explicitly setting up counterfactual blends uses two clauses, an antecedent clause with "if" and a consequent clause; these may appear in either order ("I would buy a house if I had the money"). A combination of tenses, moods, and time reference in the two clauses either suggests or forces counterfactuality. "If I had measles, I would have spots" only suggests counterfactuality; it could be followed by "And I have spots, so I'm going to the doctor to check." But "If you had come to my party today, you would have had a lot of fun" forces counterfactuality.

"If Clinton were the *Titanic*, the iceberg would sink" is a striking counterfactual that circulated inside Washington, D.C., during February 1998, when the movie *Titanic* was popular and President Clinton, already famous for sexual scandals, had just been accused of another escapade, this time with a young intern in the Oval Office. Yet he seemed to be surviving the rumor without damage. This counterfactual turned out to be prophetic: Because of the scandal that ensued, Clinton would become only the second president in history to be impeached. He would survive impeachment, and six months later almost everyone would have essentially forgotten the entire incident. But at the time the counterfactual was coined, the history had only gotten started.

The counterfactual blend has two input mental spaces—one with the *Titanic* and the other with President Clinton. There is a partial cross-space mapping between these inputs: Clinton is the counterpart of the *Titanic* and the scandal is the counterpart of the iceberg. There is a blended space in which Clinton is the *Titanic* and the scandal is the iceberg. This blend is double-scope. It takes much of its organizing frame structure from the *Titanic* input space—it has a voyage by a ship toward a destination and it has the ship's running into something enormous in the water—but it takes crucial causal structure and event shape structure from the Clinton scenario: Clinton is not ruined but instead survives. In the Clinton input, the events were largely speculative. It was not clear how many people were involved in arranging these events or what their motivations were. But the *Titanic* space supplies to the blend a tight compression at human scale: A ship runs into an iceberg, and the clear and dramatic consequence of that single cause follows quickly. The blend is counterfactual to the space of the historical *Titanic*, which sank. The historical *Titanic* is ranked as the supreme vessel on a scale of unsinkability, and therefore the iceberg as the supreme obstacle on a scale of immovability. The blend uses the compression from that space and retains the scales. But it reverses the causality of the sinking so that the Clinton-*Titanic* is now even more unsinkable than the real one. The blend is deliberately hyperbolic: Icebergs can be submerged but cannot sink. In the blend, Clinton is stronger than even the laws of physics.

There is a generic space whose structure is taken as applying to both inputs: One entity involved in an activity motivated by some purpose encounters another entity that poses a threat to that activity. In the generic space, the outcome of that encounter is not specified.

As we discussed in Chapter 8, the products of conceptual integration fall along a number of gradients and, at first blush, do not look like the same kind of thing. Superficially, metaphors, counterfactuals, literal framing, analogy, and hyperbole feel like different species, and it is easy to fall into the Cause-Effect Isomorphism Fallacy by assuming that the differences in the products must be caused by differences in the basic mental operations that produce them. That fallacy has conditioned most of the study of these products in philosophy, linguistics, psychology, and the humanities, which have sharply separated the phenomena on the basis of surface differences and so assumed an equally sharp partitioning of the underlying mental operations. The Unsinkable *Titanic* is interesting because it provides us with a blend that exhibits many of these supposedly quite different phenomena all at once. It has an explicitly counterfactual grammatical form and an explicit counterfactual blend—Clinton is not a ship and the *Titanic* did sink—so the blend is counterfactual with respect to both its inputs. Its cross-space mapping is metaphoric and recruits basic metaphors like PURPOSIVE BEHAVIOR IS JOURNEYING, FAILING IS BEING STOPPED, FAILURE IS DOWN, and ADVERSARIAL OPPOSITION IS PHYSICAL COLLISION. In this cross-space mapping, the ship on the journey is metaphorically mapped onto Clinton, the person with the purpose. Metaphors typically have source domains that provide tight compressions at human scale. That is exactly the case here. The compression is projected into the blend from the *Titanic* input. The conceptual integration network develops a disanalogy between the two input spaces, and, as we saw, the final rhetorical effect is hyperbole. In this single conceptual integration network we see many features traditionally thought to be distinct and even incompatible.

Most of the counterfactual blends we have considered thus far, such as the Buddhist Monk blend we analyzed in Chapter 3, are noticeably counterfactual. But counterfactual blends are more often unremarkable, their intricacy hidden from conscious sight. "Kant disagrees with me on this point," the unremarkable but intricate statement we analyzed in Chapter 4, depends upon a counterfactual blend, as do all of our "safe" examples ("This beach is safe," "The child is safe").

To see how easy it is to construct intricate counterfactual blends, let us consider an unremarkable blend used in the middle of a serious argument. Roger Penrose, in a book about consciousness titled *Shadows of the Mind*, presents a diagram summarizing his view of the connections between mental, physical, and platonic worlds. He writes: "[S]ome might argue for a reversal of the directions of some of my arrows. Perhaps Bishop Berkeley would have preferred my second arrow to point from the mental world to the physical one. . . . I am somewhat uncomfortable about directing the third arrow in the seemingly 'Kantian' orientation that is depicted [in the diagram]." Here, the past counterfactual ("would have preferred") prompts for a counterfactual blend in which Berkeley can see the diagram and express opinions about where the arrows should point. "Would prefer" would also work, prompting us to construct the sort of timeless space we noticed in the Debate with Kant. In the blend, we find Bishop Berkeley, Penrose, ourselves, and other thinkers evoked by the phrase "some might argue."

It is very easy to construct and work from a blended space in which we, Berkeley, Kant, Penrose, and innumerable others are all together, able to converse. The conceptual frame for this conversation is based on the debate frame, but it has special intricacies because of selective projection, topology, and other principles of blending. In the blend, for example, Berkeley can disagree with Penrose without being his contemporary—which is to say, without knowing who Penrose is or knowing any modern neuroscience. The reader works with an intricate and surprising conceptual integration network and its counterfactual blend without being aware of the network, its intricacy, the blend, or its counterfactuality.

The Berkeley counterfactual is actually a multiple blend, of the sort discussed in Chapter 14. Many things are going on in such a blend, motivated by many inputs. The arrows printed in Penrose's book can now move; Bishop Berkeley can see them and object to their orientation. The verb "prefer" sets up mental spaces for Berkeley's conceptions. Compression over these many inputs is manifest. There is compression of Space and Time in placing Berkeley, Penrose, and the moving arrows together. There is compression of cause and effect: In the blend, preference alone moves the arrows. There is also the compression of a difficult philosophical discussion, carried on over centuries, into the simple action of pointing an arrow in one of two possible directions.

THE COUNTERFACTUAL ZOO

Explicit counterfactuals, with their surprising variety of forms, provide one of the most recognizable laboratories for the investigation of blending. To give a flavor of this variety, we list here a few counterfactuals with pointers toward the problems they raise.

"If all circles were large, and this small triangle 'D' were a circle, would it be large?" This wonderful and absurd counterfactual, invented by David Moser, has the surprising property that people very consistently and unhesitatingly answer "yes." The reasoning seems unproblematic: Now that D is a circle, it must obey the laws applying to members of that category, one of which is, by stipulation, that circles are large. How can people be willing to build this absurd "possible world," in which all circles are large, despite the law of plane figures that they can be of any size, and in which a triangle of our world shows up as a circle? The answer is that people are not considering the causal connection between our world and this counterfactual. They are instead setting up a straightforward integration network in which identity and projection have been stipulated. One way to construe this counterfactual statement is to make successive blends. The first has one input in which there are plane figures of any size, including small triangles and small circles, and another input in which it is a criterial property of circles that they are large. This blend is a world that can have small triangles but only large circles. Next, we blend the output of the first blend, containing D, with a mental space containing a counterpart circle of unspecified size for D. In this new blend, it is still the case that circles must be large, but now D is also a circle, and hence large. Notice that the same schematic counterfactual form elicits quite a different answer in "If all the coins in your pocket were quarters, and this dime were placed in your pocket, would it be a quarter?" Here, the response is "no," because the projection principles are construed differently. We apparently find it harder to view as criterial "is a quarter" for the set of coins in our pocket than to view as criterial "large" for the set of circles. We surmise that logical and mathematical systems are relatively poor laboratories for thinking about everyday counterfactual reasoning because the everyday world is very rich in possibilities for selective projection, adjustment, and transformation. "Being a circle" is a stark and rigid category, but "being in your pocket" has available to it all the possibilities of human life.

"If you'd only put yourself in my shoes, you'd have some sympathy, and if I could put myself in your shoes, I'd walk right back to me." This simple line from a country-western song points to a very complex, but common, use of counterfactual blending to manipulate identity, dispositions, and situations. We will return to this topic in the next chapter. For X to be in Y's shoes usually means that X keeps X's identity and dispositions but imagines being in Y's situation. The first counterfactual in this song lyric uses that conventional meaning, but the second does not. In the first counterfactual, the jilted lover asks the departing lover to imagine herself in his situation. In the second counterfactual, the departing lover does not change situations but inherits new dispositions from the jilted man. In the first case, a change of situation for the woman generates in the blend a change of disposition toward the man. In the second case, a change of dispositions for the woman generates a change of action toward the man.

"In France, Watergate would not have hurt Nixon." This counterfactual can have many readings, but a typical one contrasts the American and French cultural and political systems (see Figure 11.1). It brings in aspects of the French system from one input and the Watergate scandal and President Nixon from the other. In the blend, we have a Watergate-like situation in France, but running the blend delivers attitudes quite different from those in the American input, and so the president is not harmed. Again, the speaker is not making a causal assertion. The point is not to imagine what changes would need to be made in our world to have Watergate take place in France or to have Nixon be president of the Republic.

"If the Earth were as close to the sun as Venus, life as we know it would never have evolved on our planet." The point of this counterfactual is to emphasize the special conditions on Earth that made the origin of life possible. The point is not at all to imagine the nearly unthinkable changes that would have to be made in physical laws to allow the Earth to evolve in the orbit of Venus or to evolve in its own orbit and somehow be magically moved to the orbit of Venus.

"If cars were men, you'd want your daughter to marry this one" (said of a Volvo in an advertisement). This counterfactual underscores the arbitrariness of some of the cross-space mappings at work in integration networks. The world in which cars are men does not strike us as at all possible; and, even if conceivable, its "distance" from us on a scale of possibility or similarity is in no way measurable. Nor is the counterfactual causal in the sense of the social psychologists: The ad is not inviting us to make a causal assertion about the world in which cars are men so that we can deduce a causal assertion about our world, in which men are men. Rather, what the blend is exploiting is an analogical mapping that links *wanting to buy* and *wanting to marry.* This analogy is built on the assumption that you want to buy and marry what is best and most reliable. When buying and marrying are blended, so are optimal cars and optimal sons-in-law. Notice once again that no preexisting objective similarity between cars and men, buying and marrying has to exist for this blend to be intelligible.

"Coming home, I drove into the wrong house and collided with a tree I don't have." This statement is counterfactual because it depends upon the evoked but counterfactual scenario of driving into the right house and therefore not colliding

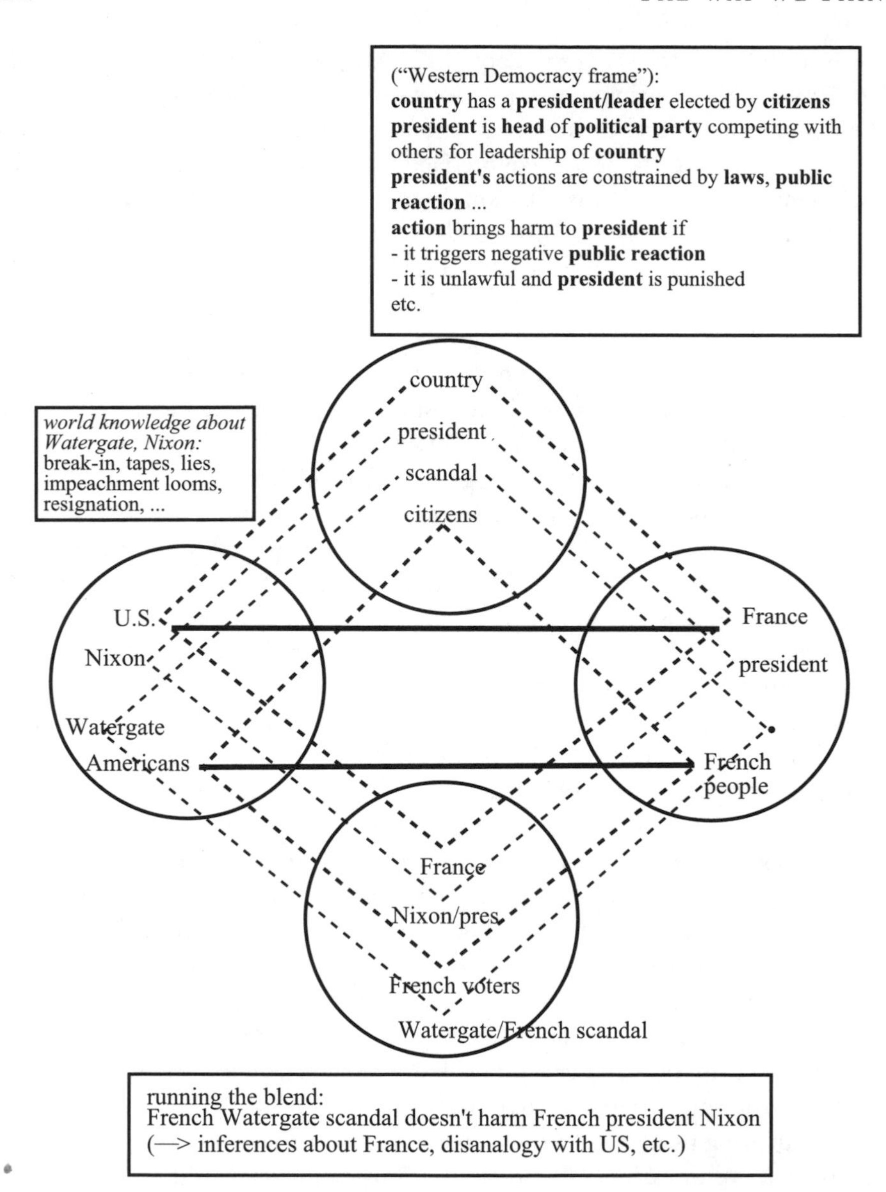

FIGURE 11.1 NIXON-IN-FRANCE NETWORK

with a tree. The grammatical trigger here is not "if . . . then" but rather the adjective "wrong." In one input, the driver drives into the parking place at his home. In the other, he drives onto the property of some different house and collides with a tree. These inputs share the frame of parking a car at a house, and there are identity connectors between the cars and the drivers, but there are disanalogies between the two inputs, having to do with the value of the role *house* and the existence of a tree at a particular location. In the blend, we have, from the space of what actually happened, the house where the driver did drive, and the tree, and the collision. The disanalogy between the houses is compressed in the blend into a property of the house: It is now the "wrong" house. And the disanalogy having to do with the tree is compressed into a property of the tree: "a tree I don't have." It is tempting to think that this is a property of the tree independent of the blend, but note what happens if our companion on a walk through some public woods says, pointing to a tree, "That is a tree I don't have." We are likely to interpret the speaker as meaning that he does not own a tree of that type. It would be quite strange if he actually meant to point out that he did not own that particular tree. In the statement we are looking at, "a tree I don't have" is not interpreted to mean that the driver does not own that particular tree but, rather, that there is a counterfactual relation between the blend and the input with the driver parking his car at his home: There is no tree in the corresponding spot at his home, and no collision when he drives through that spot, either. Very generally, when disanalogy operates on the existence of a value for a role, that disanalogy is a good candidate for compression into nonpossession, as in "That car does not have air conditioning," "Arkansas has no coastline," "Africa does not have bears," and "My house doesn't have that porch."

"Caffeine headache," "money problem," "nicotine fit": These straightforward phrases—referring to a headache that comes from lack of coffee, a problem that involves a lack of money, a fit brought on by lack of nicotine (presumably from not smoking enough)—all set up an integration involving a counterfactual link between spaces. "Caffeine headache" brings up two situations, one in which you have your coffee and one in which you have a headache. There is evident identity, analogy, and disanalogy between these two situations: In both, it's late morning, and you are at work. But there is the coffee only in the first and the headache only in the second. A blended network is constructed in the following way: There are input spaces corresponding to the two contrasting situations, links of analogy, disanalogy, and identity between them, and projection of the frame of morning activities from both inputs to the blend. From the input with the headache, we project the headache. And from the desired input, we project the causal relation and the causal element. In the blend, the headache is now the effect of something (see Figure 11.2).

The blend is the new construal of the situation (see Figure 11.3). The input with coffee is counterfactual with respect to the blend. In the blend, there is a counterpart for coffee that causes the headache. It is what we refer to by means of the expression "absence of coffee." The expression "caffeine headache" brings in the label "caffeine" from the coffee element in the counterfactual input and applies it to its counterpart in the blended space.

In the linguistic construction shared by "caffeine headache," "money problem," and "nicotine fit," the first noun picks out the element in the desired

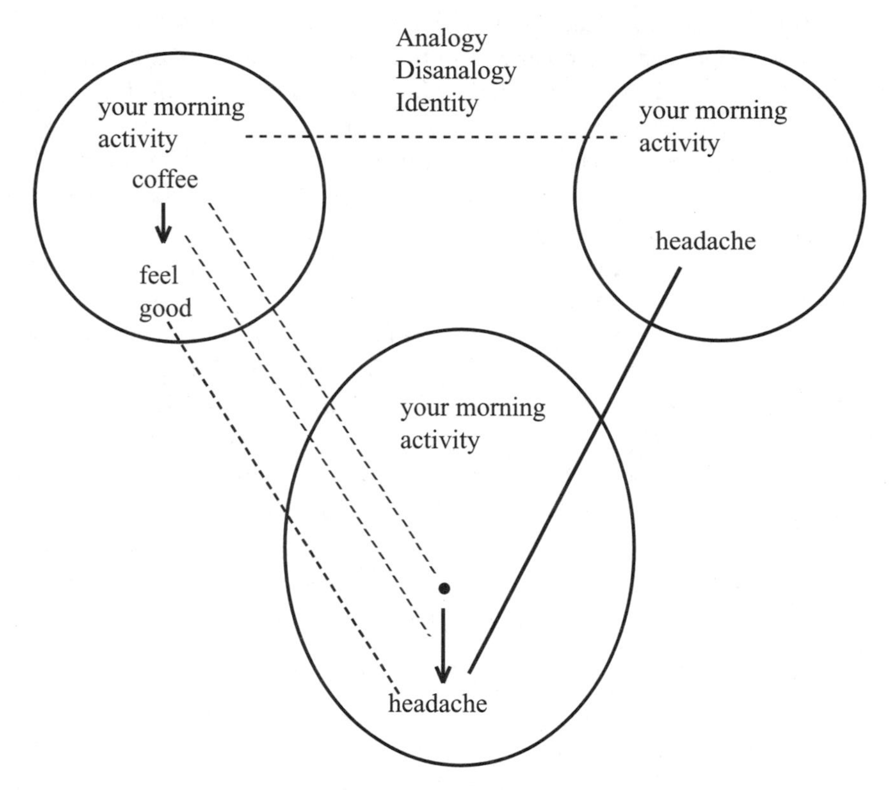

FIGURE 11.2 THE CAUSED HEADACHE NETWORK

input whose absence in the blend is causal for the unwanted state, and the second noun picks out the bad state that obtains in one of the inputs and in the blend. Thus we also have "security problem," "arousal problem," "insulin coma" and "insulin death" (in the case of hyperglycemia, which results from absence of adequate insulin), "food emergency," "honesty crisis," and "rice famine."

These examples demonstrate the way in which blending has multiple possibilities. For example, we could read "caffeine headache" as referring to a headache *caused* by the caffeine. For both networks, there is a cause-effect relationship in the blend—in the first case, between *absence of caffeine* and *headache*, and in the second case, between *presence of caffeine* and *headache.* In both, the Cause-Effect vital relation is further compressed into Property. There can now be "caffeine headaches," "whisky headaches," and "sex headaches."

The notion of *absence* is not explicitly indicated by any part of the expression "caffeine headache." It emerges from the entire network, as prompted by this grammatical construction. But there are linguistic expressions for indicating this compression explicitly: "absence of," "lack of," "want of," even "no," as in "I have a no-caffeine headache."

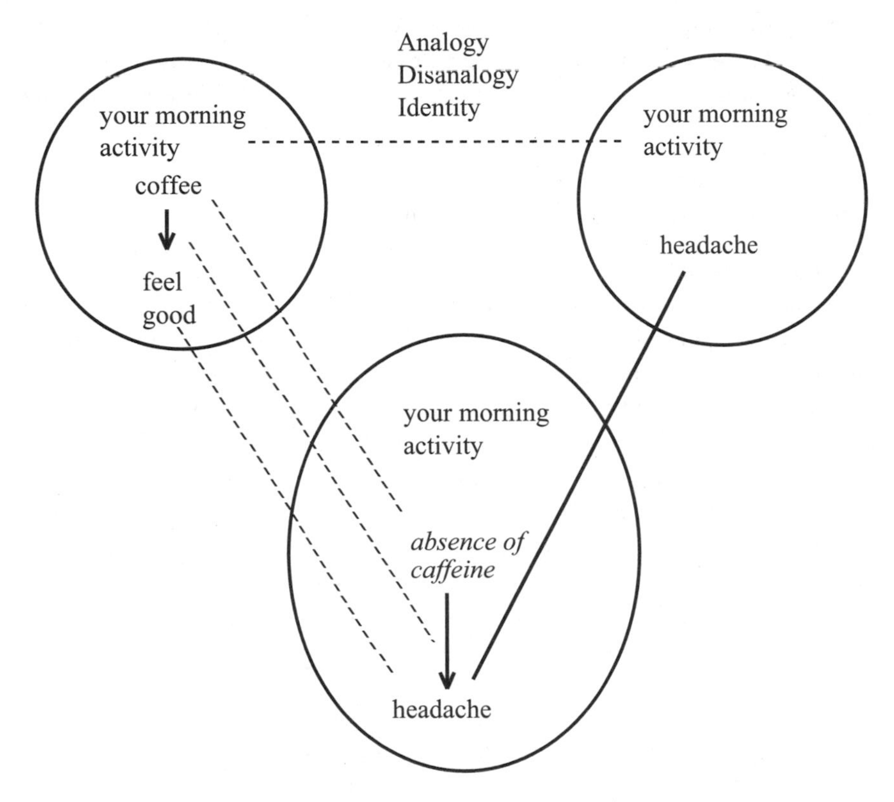

FIGURE 11.3 THE CAFFEINE HEADACHE NETWORK

These counterfactual networks are frequently very hard to notice, since we construct them so effortlessly as part of backstage cognition. Consider, for example, the January 31, 2000, edition of *USA Today* on that year's Super Bowl. In the last play of the game, the ball carrier for the Titans was tackled one yard from the goal line. We unavoidably construct the contrasting space in which the runner advances one more yard and scores. Blending these two spaces gives us a blended space in which there is now an element that is *absence of one more yard of progress*. We might express this blend by making the negative element explicit: "The Rams won by stopping the Titans from advancing one more yard." But the front-page headline actually read "Rams win by a yard." It then becomes possible to refer to the Rams' "one-yard win"—a phrase that, in its integration patterns, is identical to "caffeine headache." This example reveals something else as well. The more conventional pattern for expressions like "winning by a yard" and "winning by a nose" is, of course, that of a race where the winner crosses the finish line a yard ahead of the runner-up. This doesn't feel intuitively like a counterfactual expression; it seems as if we can "see" that fateful yard right there on the photograph of the finish. But when you think about it, you realize that this more standard notion of a "one-yard win" is indeed counterfactual. The crucial yard is the one that the loser failed to cover, just as the crucial yard in the Super Bowl win was the yard that separated the ball carrier from the goal line.

BLENDS, INPUTS, AND IMPLICIT COUNTERFACTUAL SPACES

It is common to imagine that reality is simply opposed to unreality, and that counterfactual spaces are those that do not refer to reality. Thus it is an easy but false step to imagine that when we are thinking about reality, about events such as what we did this morning or what we might do this afternoon, we are not doing counterfactual thinking. We are doing counterfactual thinking, the logic goes, only when we imagine what we might have done this morning if we had won the lottery, or what we would do this afternoon if we were a millionaire. On this view, studying the history of World War II does not involve counterfactual thinking, but wondering what would have happened if Churchill had become prime minister in 1938 does.

There is every kind of evidence against this logic. Thinking of a caffeine headache that we in fact have is certainly thinking about reality in the most practical terms, but it involves fundamentally a space in which we had the coffee, and that space is counterfactual with respect to the blend in which we have a special kind of headache, a caffeine headache. Counterfactuality is forced incompatibility between spaces, and when one is thinking about reality, counterfactuality is often a vital relation between spaces that involve some of the same people and the same events.

Counterfactuality is not an absolute property. A space will be counterfactual depending on the point of view one takes—that is, on the space that serves as the viewpoint. Consider, for example, "The suitors don't know that the beggar is really Odysseus in disguise, so they abuse Penelope's hospitality. They are thinking that if Odysseus were here, they would all be in great danger." The grammatical cues present as counterfactual the space in which Odysseus is watching and they are in danger, even though readers know it is real.

In this book, we use "counterfactual" to mean that one space has forced incompatibility with respect to another. But there is a narrower and more common use of the term to mean that one space has forced incompatibility with respect to a space we take to be "actual." This narrower meaning is also useful. But the important point is that there is no difference in the cognitive operations of mapping and blending between the two cases. "Counterfactual thought," with all of its mechanisms, is happening in any case; it just happens that we sometimes mark one of the spaces as "actual." As we saw in Chapter 5, typically any integration network has implicit counterfactual spaces attached to various of the "actual" spaces we are focusing on. Expressions like "safe beach," "fake gun," and "If Clinton were the *Titanic*," pointedly require us to build counterfactual spaces and to deploy a precise mapping scheme. But these spaces are also often involved in networks without such expressions. In the Impotent Smoking Cowboy integration network, the spaces with virility and potency are crucial but are not prompted for grammatically.

TRUE SCIENCE AND EMOTIONS FROM COMPLEX FALSEHOOD

We began this chapter by saying that the capacity to juggle counterfactual spaces is a consequence of the evolution of cognitively modern human beings and their remarkable capacity for double-scope blending. Counterfactuals are a good exemplar of double-scope blending because the oppositions between the spaces are so manifest. One cannot overstate the importance of counterfactuals in human life. Many thinkers, from many different fields, have realized that counterfactuals have an enormous and basic indispensability in human life. We will look at two striking cases that illustrate this indispensability in very different ways. One has dramatic consequences that stem from invisible and unstated counterfactual blending. The other has dramatic consequences, too—but they stem from profiled and explicit counterfactual blending. In both cases, reality is profoundly affected by cognitive work in the unreal.

An especially strong form of depression was studied in Britain in the 1980s. Sufferers had purchased a lottery ticket a few weeks before the drawing, knowing full well that the odds against winning were enormous. They expressed no hope of winning and rationally declared that they were buying the ticket for fun. Yet once the drawing was held and they had lost, they slumped into debilitating depression. The symptoms were significantly different from "gambling depression," which is an effect of addiction to gambling. The victims of "lottery depression" had symptoms like those of people who have suffered severe losses, such as destruction of a house or loss of a parent. The interpretation given by therapists was that in the two weeks or so between the purchase of the ticket and the drawing for the winner, these victims had fantasized, consciously or unconsciously, wittingly or not, about what they would do upon winning the lottery. The actual drawing made them lose everything they had acquired in the fantasy world. In that world, they did indeed suffer a severe loss. The amazing thing is that the fantasy world seems to have had profound effects on the psychological reality of the real world, given that the patients had no delusions about the odds of winning, and said so clearly.

The lottery players constructed a hypothetical blend that became counterfactual upon the drawing. We have talked earlier about living in the blend and about the all-powerful role of the unconscious in building up the most mundane everyday meanings. Here, in lottery depression, the unconscious has the same great power and the players have lived in the blend for several weeks. This is considered a psychiatric condition, but the effects of "living in the blend" also show up in neurobiological syndromes. V. S. Ramachandran reports on patients who have a paralyzed arm as the result of cerebral hemispheric damage and who also have a syndrome known as anosognosia. Such patients are obviously sane, but the anosognosia causes them to be convinced they can move their paralyzed limbs. When Ramachandran asked such a patient whose left arm was paralyzed whether she could use her left hand, she said that of course she could and that her two hands were equally strong. When Ramachandran asked her to touch his nose, her hand lay motionless in front of her, but she "confabulated" the belief and the perception that she was following his instruction. When asked "Are you touching my nose?" she replied "Yes, of course." Other patients with the same syndrome commonly concoct rationalizations such as "I'm sick of it; I don't want to move my arm," or "I have severe arthritis."

The victims of lottery depression are running multiple conceptions simultaneously, some of them conflicting with each other, and it seems that the brain is very well designed to run such multiple and potentially conflicting conceptions. Consider another of Ramachandran's patients with anosognosia, Mrs. Macken. Using a technique devised by Eduardo Bisiach, an Italian neurologist, Ramachandran squirted ice-cold water into her ear. This sets up a convection in the ear canal that fools the brain into thinking the head is moving. Amazingly, the cold-water irrigation of the left ear brought about a complete, though temporary, remission of the anosognosia. Mrs. Macken was suddenly fully able to acknowledge and admit her paralysis. Moreover, she correctly stated that she had been paralyzed for three weeks. "This was an extraordinary remark," Ramachandran reports,

> for it implies that even though she had been denying her paralysis each time I had seen her over these last few weeks, the memories of her failed attempts had been registering somewhere in her brain, yet access to them had been blocked. The cold water acted as a "truth serum" that brought her repressed memories about her paralysis to the surface.

Once the truth serum wore off, Mrs. Macken not only returned to her denial but insisted that under the cold-water treatment she had answered that her arm was fine. Another patient with anosognosia overcame her denial of her paralysis permanently. When Ramachandran asked this patient what she had told him about her arm each time he had visited her, she answered that she had always reported that her left arm was paralyzed, even though on every previous visit she had told him her arm was fine. Both of these patients were running conflicting conceptions and continually revising their memories according to whichever conception was in control.

There are many spectacular scenarios—from hypnotism to delusion—where we can see clearly that human beings are living in a blend, and sometimes in more than one blend at once. In cases of lottery depression, the lottery player, before the drawing, is living in parallel blends—one in which he has won and one in which the ticket is incidental, for fun only—but it is only when depression sets in and has to be explained that it becomes clear that the player had been living in the blend in which he won. Lottery depression is just one of many ways of "living in the blend" that have been studied by Daniel Kahneman, Paul Slovic, and Amos Tversky. One of their fundamental results is that the same objective facts are much more painful for subjects when framed as a loss than when framed as a gain. In lottery depression, the player has a long time between the buying and the drawing in which to accumulate fantasies of important gains. Those "gains" are lost in one fell swoop when the winning number is announced. The patient had only a dim consciousness, if any, of how solid those fantasy gains had become for him. The real event of drawing the winning ticket is part of both blends—the one in which the player wins, and the one in which the ticket is only incidental. This is a devastating event in the winning space, since it throws out the fundamental presupposition of that space; it changes a settled fact in a way that is completely unjust and outrageous.

A contrasting case, in which the counterfactuality is not only explicit but the driving goal, is the fundamental deduction technique of *reductio ad absurdum* in mathematics (and, for that matter, logic and natural sciences). In *reductio ad absurdum,* we seek to prove the falsity of a proposition *P* within the system. The network is explicitly constructed by blending two inputs. The first is built up from our settled knowledge of a consistent mathematical system. In that input we activate some facts and deduction procedures from the mathematical system and the certainty that it is noncontradictory. In the other input, we have the same facts, but also the proposition *P*, considered as true. In the blend, we have *P* from one space, the currently activated facts from both spaces, and the deduction procedures from the input built up from the settled mathematical system. We can now run the blend with its facts, deduction procedures, and P, and recruit as we go along any potentially helpful new facts that we choose to activate from the settled system of mathematics linked to the first input. We seek to develop in the blend a contradiction as emergent structure. If a contradiction emerges, we know that the blend is counterfactual with respect to the settled input, which is independently known not to be self-contradictory. Since the only difference between the blend and the settled input is the truth-status of P, we conclude that *P* is not true in the settled input.

For example, suppose a mathematician is working with number theory and wants to prove that there is no highest prime number. She assumes that there is such a highest prime number, *h*, and then operates on it with some of what is already known about arithmetic. She constructs the number $h! + 1$, where $h!$

(pronounced "h factorial") is the product of all the integers from 1 to *h*. Now, *h!* + 1 is much bigger than *h*. Therefore, by assumption, it cannot be prime, and so, by definition, must have a prime factor other than 1 and itself. But if that factor is any integer between 1 and *h*, then it also must be a factor of 1, and so can only be 1. So *h!* + 1 must have a prime factor bigger than *h*. But this implies that there is a prime number bigger than *h*, so in the blend, *h* is the largest prime and there is also a prime larger than *h*. This is a contradiction, so the assumption that there is a highest prime must be false.

The method of *reductio ad absurdum* is not a special procedure in mathematics. It is among the bedrock procedures of mathematics and has been used frequently from Euclid to Bourbaki, often in the development of the most important theorems. The canonical view of falsifiability in science, such as Popper's, incorporates the notion of *reductio* in all experimental procedures. Given a hypothesis A, the way to falsify A is to construct a blended space that includes A, some of what we take to be unquestionable, and some experiments and their results. We then run the blend to show that A has consequences that conflict with the experimental results. Therefore, the blend is self-contradictory, and this is taken as evidence that A must be false. In everyday life, we do much the same thing. Hypothesis: Billy does not love Jane. Combine this with what we think we know of human nature: Therefore, he will not sacrifice himself for her and, in particular, will not type her doctoral dissertation for her. But he did. The blend is contradictory; therefore, he must love her. QED. Without *reductio*, there would be no detective novels, and fewer love stories.

Some of the blends we have seen in previous chapters are everyday *reductio* arguments. In the Bypass, you die from being operated upon by well-meaning but incompetent children-surgeons. This is what comes of neglecting education, and so is a *reductio* of present policies. You must therefore "act now" to improve education. In the Impotent Smoking Cowboy, the impotence in the blend is a consequence of continued smoking, and so is a *reductio* of continuing to smoke. The general goal of *reductio* is to show a catastrophe in the blend, which is therefore to be avoided. In the special case of mathematics, the catastrophe is mathematical self-contradiction, which we avoid by rejecting the assumption. In other realms the catastrophe can simply be an undesirable consequence.

It may seem as if *reductio* is principally a negative technique for keeping foolishness out of our systems of truth, but it is often a technique of great mathematical discovery.

Consider hyperbolic geometry. As both Morris Kline and Roberto Bonola describe it, the laborious birth of non-Euclidean geometry took fifteen hundred years. Euclid had defined parallel lines as straight lines in a plane that, when extended indefinitely in both directions, never meet. He had presented a sequence of proofs independent of the parallel axiom showing that two straight lines are parallel when they form with one of their transversals equal interior alternate angles, or equal corresponding angles, or interior angles on the same side that are supplementary. But proving the converses of these propositions appeared to require what is known as the *parallel axiom:* "If a straight line falling on two straight lines makes the interior angles on the same side less than two right angles, the two straight lines, if produced indefinitely, meet on that side on which the angles are less than two right angles." This axiom seemed to many geometers, probably including Euclid, to lack the desirable feature of self-evident truth. Rather than assume it as an axiom, they sought to derive it from the other axioms and from Euclid's first twenty-eight theorems, none of which uses the parallel axiom.

It was Gerolamo Saccheri (1667–1733) who made the crucial attempt, focusing on a quadrilateral ABCD where angles DAB and ABC are right angles, and where line segments AD and BC are equal (see Figure 11.4). Without using the parallel axiom, it is easy to prove that angles BCD and CDA must be equal. Saccheri did this. If we assume the parallel axiom, BCD and CDA must be right angles. Therefore, if we deny that BCD and CDA are right angles, we thereby deny the parallel axiom. Saccheri did just this, in the hope of deriving a contradiction from the denial, which would prove the parallel axiom by *reductio ad absurdum*.

But if BCD and CDA are not right angles, they are still equal, and so must be either obtuse or acute. Saccheri sought to show that, in either case, a contradiction follows. He assumed that they are acute, with the result that in the blend, DAB and ABC are right while BCD and CDA are acute and equal (see Figure 11.5). The blend is impossible in Euclidean geometry, but Saccheri never found a clear-cut contradiction. He carefully drew many conclusions about this blend that he regarded as repugnant elaborations of the blend's inherent falsity but that today count as foundational theorems of hyperbolic geometry.

Saccheri's reasoning, far from being exotic, employs the uniform strategy of all *reductio* arguments in logic and mathematics: A system of inferential principles that is taken to be consistent is applied to a structure that may not be consistent. Saccheri imagined that he was conducting what could only be a *reductio ad absurdum* argument, and he hoped that the expected contradiction would inevitably emerge. Those who came after him reinterpreted the same proofs not as *reductio* arguments but as steps in the development of a new and consistent branch of geometry.

It happens that there are many equivalent ways to produce a blend that delivers hyperbolic geometry. All that is needed is a blend that requires the interior angles of a triangle to sum to fewer than 180 degrees.

Saccheri is not credited with the invention of non-Euclidean geometry. Kline summarizes and simplifies the history as follows: "If non-Euclidean geometry means the technical development of the consequences of a system of axioms containing an alternative to Euclid's parallel axiom, then most credit must be

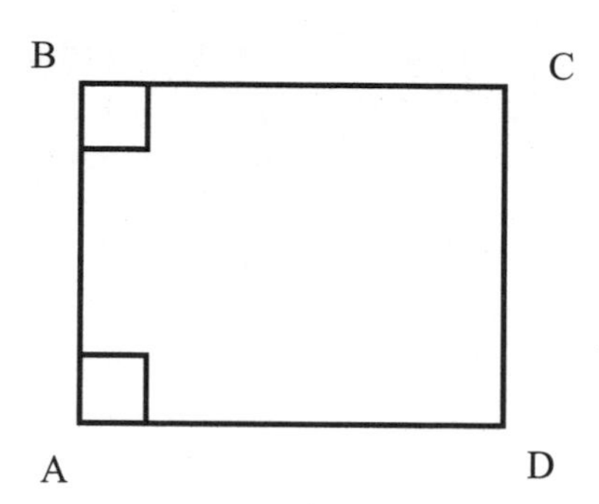

FIGURE 11.4 QUADRILATERAL ABCD WHERE ANGLES DAB AND ABC ARE RIGHT ANGLES AND LINE SEGMENTS AD AND BC ARE EQUAL

accorded to Saccheri, and even he benefited by the work of many men who tried to find a more acceptable substitute axiom for Euclid's." Credit is given instead to Carl Friedrich Gauss, Janos Bolyai, and Nikolas Lobatchevsky for recognizing (but not proving) that hyperbolic non-Euclidean geometry is mathematically consistent, and to Gauss for recognizing that physical space might be non-Euclidean.

The essence of the *reductio* strategy is to blend two spaces, one of them containing already accepted truths in the mathematical system, the other containing the assumption to be proved false. We then run the blend to develop inner-space compressed structure: The blend contradicts itself. It is therefore counterfactual with respect to the settled mathematical input. Since the only thing that differed in the blend before we began running it was the assumption, that assumption cannot be in the mathematical system.

REPRISE

Many of the integration networks we considered in earlier chapters include explicit or implicit spaces that are counterfactual with respect to our "actual" world—in other words, that are "false." The "false" blends are used in powerful ways to operate on the rest of the network, ultimately yielding inferences that are relevant for reality. They can yield both new structure or inferences for a particular space and new outer-space connections between the inputs. The "literal falsity" of these spaces is irrelevant to reason. The Iron Lady example has a "false" space in which Margaret Thatcher runs for U.S. president but is opposed by the labor unions. Inferences developed inside that "false" space are meant to cast light on aspects in the input space with the United States. The Buddhist Monk has an even more exotic "false" blend in which someone exists in two locations and ultimately meets himself, and this inner-space structure of multiple location and self-meeting is decompressed to give us the "true" answer to a question about

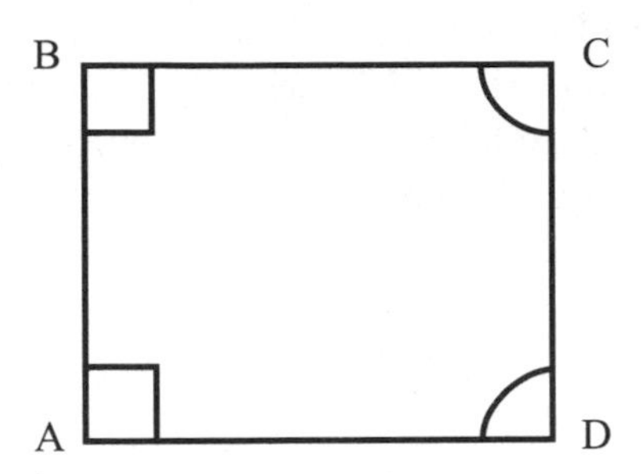

FIGURE 11.5 QUADRILATERAL ABCD WHERE ANGLES BCD AND CDA ARE ACUTE, ANGLES DAB AND ABC ARE RIGHT, AND LINE SEGMENTS AD AND BC ARE EQUAL

real-world events. The Pronghorn has a "false" blended space in which the pronghorn is being chased by nasty predators and has memories of having learned to run fast because of them. This "false" inner-space structure is again decompressed into the factual outer-space connections of adaptation and inheritance. By now we see a general pattern for many of these "false" blended spaces: When we want to establish, understand, or manipulate actual outer-space connections over various spaces, it is good to compress them into a blend, which often means compressing them into powerful inner-space relations in the blend. The "literal falsity" of those relations is irrelevant to the reasoning process. We see exactly this pattern in Regatta: The blend has the "literally false" inner-space structure of two boats from different centuries racing from San Francisco to Boston. This compressed "literally false" inner-space structure corresponds to the actual outer-space connections between the two input spaces, each with one boat in its own time. This way of constructing integration networks is particularly useful for casting light on the input spaces by finding a counterfactual compression that, through proper alignments between the blend and inputs, preserves the structure we need in the inputs but operates over that structure efficiently and memorably, giving a global insight at human scale. The Bypass, the Impotent Smoking Cowboy, the Mythic Race, and the Debate with Kant all have the same pattern of "literally false" inner-space structure in the blend that decompresses into outer-space "true" relations between the inputs.

There are also integration networks in which the blended space contains inner-space structure that is "false" but the point of the network is not to emphasize the relationship between that structure in the blended space and the outer-space relation between the inputs. For example, in the blended space of the Image Club, there is a man having sex with a high school student, and in the Micronesian navigation blends, there is an island, the ETAK, that serves the purposes of the navigator; but from a perspective outside the blend there is no real high school student involved and no real island at that location. In these networks, the principal purpose of the blended space is to provide integrated guidance to satisfactory action, and, while for various purposes it might be necessary to have access to the inputs, decompression might interfere with action. Sometimes, it is preferable to stay in the blended space.

CHAPTER 11
ZOOM OUT

COUNTERFACTUALS ARE EVERYWHERE

It is common to think that counterfactual thought is rare and marked in language by a handful of very specific grammatical constructions, such as "if, then" or "What if?" with verbs of the right form.

Question:

- Is counterfactuality a restricted grammatical phenomenon?

Our answer:

Far from it. Language is rife with forms that prompt for the construction of a counterfactual relation between one space and another. "Paul believes he'll get his daughter admitted to Berkeley because he thinks Mary is the dean of admissions" invites us, in complicated ways, to imagine that although there is a Paul, his daughter, Berkeley, Mary, and a dean of admissions, nonetheless Mary is not dean of admissions, and therefore that Paul's belief space is "false." If Bill is calling Paul and is annoyed that he can't fax something to Paul, he might ask, "Why don't you have a fax machine? You could be reading my proposal right now!" We must construct a "false" but preferred blend in which Paul has a fax machine and is reading the proposal. In fact, "in fact" automatically asks us to build a counterfactual relation, and in fact the only sense in which the space in which Paul has a fax machine is "false" is that we have already set up an imaginary base space in which there is an imaginary Paul, an imaginary Bill, an imaginary phone call and proposal, but no fax machine in Paul's imaginary house! Relative to that space, the blend with Paul's fax machine is "false." But from a "real" point of view, they are both "false." "Pont-Ivy was meant by Napoleon to have become the capital of France" requires a "false" space in which Pont-Ivy is the capital of France. "I didn't buy a car. It would take too much room in the garage" illustrates the fact that negation routinely sets up counterfactual blended spaces, which can be elaborated. The pronoun "it" picks out the car in the "false" space in which "I" did buy a car and it does not fit in the garage. There is no limit to how much we might elaborate this "false" blend: We could go on with "it would have cost me a fortune in gas, I couldn't have gotten insurance for it," and so on. As we noted in Chapter 5, expressions like "too bad," "missed," and "otherwise" also routinely prompt counterfactual spaces. Some verbs, like "prevent," and nouns like "dent," also prompt for counterfactual spaces.

COUNTERFACTUAL NETWORKS

We have defined counterfactuality as a forced incompatibility between spaces in a network.

Question:

- Is this the only place in thought that counterfactuality can be found?

Our answer:

Interestingly, an entire integration network can be counterfactual with respect to some other meaning. In Chapter 7, we presented the Toblerone network, in which Toblerone chocolate is framed as the motivation for monuments, and in the blend, the pyramids of Giza have the shape they do in imitation of the shape of Toblerone chocolate. This entire network is counterfactual: We think it is false to bring the "monument" frame and Toblerone chocolate together in the first place, much less to construct ancient Toblerone worship in the blend as the context for the building of the pyramids.

MODUS TOLLENS

We have given a straightforward explanation of *reductio ad absurdum* as blending and have pointed out that it is pervasive in everyday reason.

Question:

- Isn't all this just the simple logical operation of *modus tollens*, in which, given that p implies q and given that not-q, then it follows that not-p?

Our answer:

It is true that if an assumption p implies a contradiction q, as in our example with prime numbers, it follows that not-q is true tautologically and therefore not-p is true, by *modus tollens*. Clearly, in that sense, the logical law of *modus tollens* captures something true and important. But *modus tollens* has nothing to say about how q could possibly be found. This is where the construction of the counterfactual blend is crucial, and it takes considerable running of the blend before q pops out. In the case of hyperbolic geometry, for example, q never popped out even after centuries of effort, and yet the blend was not run in vain, since what did emerge was the much greater discovery of multiple geometries. The long work of the mathematician or logician who finally finds the contradiction q can, in retrospect, be compressed into a *modus tollens* statement, but the logical law of *modus tollens* is not in itself a way to discover contradictions. In fact, even more simply, inference, as in "p implies q," is after-the-fact shorthand for what may be a long, difficult, imaginative cognitive process. Constructing the contradiction is a cognitive achievement. We often construe truths of mathematics or logic Platonically, as having eternal, independent existence, and so we naturally distinguish the cognitive accomplishment from the eternal truth. This creates the illusion after the fact that "p implies q" was always there. What is true of mathematics is true *a fortiori* of everyday life: Proving absence of love by inference from an untyped thesis recruits considerable cultural and cognitive resources, and the conclusion is a product of all that imaginative work in a counterfactual blend. The *modus tollens* formula is only a superficial report that such a process has taken place (somehow) and delivered a certain result.

TRUTH FROM FALSEHOOD

We have seen many counterfactual blends that persuade us.

Questions:

- How could falsehood ever persuade us of anything?
- How can truth come from falsehood?

Our answers:

We saw that blends can, on the one hand, compress long causal chains into inner-space structures that give us global insight or, on the other hand, set up inner-space structures to do cognitive work that can then be decompressed into outer-space relations between the inputs. That inner-space blended structure bears precise relations to the structures in and between the inputs. What is important in such blends is the compression and the decompression. Whether the inner-space structure in the blend is true or actual is not the point. For example, the coherent (but "false") world of the pronghorn antelope remembering fleet predators is systematically decompressed into the "true" outer-space relations of Change and Identity specifically involving adaptation and inheritance. The false mathematical blended space in which there is a largest prime number finds its inner-space contradiction nicely correlated with a "true" outer-space counterfactuality: A mathematical space with a largest prime is counterfactual with respect to our existing system of mathematics.

NONTHINGS

Here are some things that we say: "Put the vegetables on the plate in front of the missing chair," "I have a tooth missing," "The absence of fog made landing easy," "Nobody offered a proposal; *it* would have been shot down," "I had no luck," "Do you take your coffee with milk or with no milk?"

Question:

- Expressions like "the missing chair," "a tooth," "a proposal," and so on, seem to be pointing to something, but what they point to looks like a gap in the real world. Is this phenomenon just a convenience of language, or does it say something about the way we think?

Our answer:

The compressions we saw in examples like "caffeine headache" routinely create elements that are actual in the blend but are nonetheless *nonthings*. In the blend, *absence of caffeine* is a thing that is real and causal. It is constructed by projection from the thing in one space (caffeine) and the fact that the caffeine in one input has no counterpart in the other input. In the blend, we have by compression a thing that, from the outside, we view as a nonthing: absence of caffeine. Inside the blend, this new element can be manipulated as an ordinary thing, and the usual routines of language for referring to things can be deployed. In the case of "the missing chair," the missing chair is a thing in the blend that, viewed from the outside, is a nonthing. It can be pointed to and takes up physical space. It inherits thing-hood from the counterfactual space in which there is a chair. It inherits its physical characteristics of being a gap from the "actual" input, in which there is not a chair in the corresponding position. We suggest that it is no accident that expressions like "nobody," "nothing," and "no luck" are ordinary noun phrases for picking out things in a space. That is why it is easy to get them in all the normal places in grammatical constructions: "He was seen by no one," "I had no money," "No brains is your problem," "I expect no one to understand me," "He has a no-nonsense attitude."

NONTHINGS WE CAN'T DO WITHOUT

We've just seen that nonthings do indeed play a large role in everyday thinking.

Question:

- Do they have any role in serious thought, such as science and mathematics?

Our answer:

Nonthings have been a great engine of mathematical and scientific discovery. The invention of zero is a signal case. In general, the history of the development of numbers is a history of reconstruing the number system so that we see "gaps" in it that are themselves reconstrued as numbers on their own. The numbers 0, 1, 2, 3, . . . were reconstrued as having "gaps" between them, and these gaps were reconstrued as fractions like one-half. The fractions were in turn reconstrued as having gaps between them, and these gaps were reconstrued as irrational and transcendental numbers. The same pattern holds for the invention of negative and complex numbers.

The history of mathematics shows that the concept of number has been repeatedly revised by creating blends in which we have two (or more) inputs—one with numbers of some kind, the other with elements of some kind. There is a partial counterpart mapping between them in which every number has a counterpart in the other input space but not conversely. All the numbers and the category *number* are projected to the blend. From the other input, all the elements and much of their organization is projected in the blend. Everything in the blend comes to be viewed as belonging to the category *number,* and as a result of this reconstrual, some elements in the blend are now viewed as filling in gaps in the antecedent number system, the one in the input space.

In such a network, those elements that have a number counterpart in the cross-space mapping are fused with that number in the blend. The rest of the elements in the blend are new numbers. Topology of mathematical orderings and operations is preserved.

For example, if in one space we have whole numbers and in the other space we have proportions of objects, then in the blend we have all the proportions, all categorized as numbers. Those proportions that had whole-number counterparts are fused with those counterparts, so that, for example, 6:3, 12:6, and 500:250 are fused in the blend with 2. But now 3:4, 256:711, and 5:9, which had no whole-number counterparts, are now also numbers in the blend. They now count as filling in gaps between whole numbers.

Some topology from both inputs is preserved. The number input, of course, has ordering, addition, and multiplication. The proportion input also has topology. In that input, every proportion a:b has two parts, a and b. Given any two proportions that have the same second term, the one with the larger first term is larger. And, taking one more step, given *any* two proportions a:b and c:d, a:b is equal to ad:bd and c:d is equal to cb:db, so we can compare a:b and c:d by the previous principle: If ad is larger than cb, the proportion a:b is larger than the proportion c:d. These topologies and operations are preserved under projection to the blend. Equivalence of proportions like 3:9 and 1:3 in the proportion space is compressed into identity of number in the blend: 1/3 = 3/9 = 5/15 . . . . In this way infinities of elements in the proportion input are compressed into single numbers in the blend.

The blend has considerable emergent structure. It turns out that there is an "addition" operation in the blend that will correctly preserve addition from the number input and ordering from the proportion input. It can be defined as a/b + c/d = (ad + cb)/bd. Similarly, topology preserving multiplication in the blend can be defined as a/b × c/d = ac/bd. Quite interestingly, some topology that is useful in the proportion space is not projected to the blend. We can compose two proportions a:b and c:d into a+c:b+d, and this is a useful operation (distinct of course from addition of rational numbers): If 2 warriors oppose 3, and another 3 oppose 2, and they compose, suddenly the battle is equal, because 2+3:3+2 is balanced. But the topology of this kind of composition is not preserved under projection to the blend, because in the blend 3/2 is identical to 6/4, but 2+6/3+4 does not equal 1. In the blend, adding 6/4 to 2/3 is the same as adding 3/2 to 2/3, but in the space of proportions, composing 6 warriors against 4 with 2 warriors against 3 is not the same as composing 3 warriors against 2 with 2 warriors against 3.

The blend of numbers and proportions is elaborate and highly imaginative, with selective projections and striking emergent structure. This blend is of course none other than the domain of "positive rational numbers including zero," or what children in second grade learn as whole numbers and fractions.

For a child to learn how to add fractions, she must give up the very useful operation of composing proportions (a:b and c:d = a+c:b+d), and also give the status of number to many proportions that previously had no such status. She must additionally see as identical proportions that, in the number of elements they involve, are quite different: In the blend, the proportions 2:4 and 53:106 become identical even though the first looks as if it involves six things and the other looks as if it involves 159.

Going from the input of whole numbers to the blend of rational numbers has a dramatic consequence: Previously, 2 was the next number after 1, and 3 the next number after 2. There was no other number between 1 and 2. In fact, there could not be any other number between 1 and 2. Whether another number might exist between 1 and 2 was an issue that could not even arise. But in the blend of rational numbers, we have the staggering emergent property that there is now an infinity of numbers between 1 and 2. Once we have the blend of rational numbers it becomes intelligible to ask whether, given two consecutive whole numbers, there is another number between them, and we can now figure out, as a theorem, that the answer is yes. We can then ask whether, given any two rational numbers, there is another number between them, and again prove, a little surprisingly to the second-graders, that the answer is yes. Still more surprisingly, between any two numbers, no matter how close, there is always an infinity of others!

Once we have the blend and reify it, we can adopt the view that the previous conception of number was "missing" several numbers that were "there" but not yet "discovered."

Another spectacular blend arises when rational numbers are mapped onto segments of different lengths on the same line. As before, the geometric topology of the line and its segments is mapped exactly to the blend. Segments that have rational-number counterparts are fused with those counterparts. If one segment of the line is given the rational number 1 as its counterpart, then it becomes the unit segment and all the other counterparts are determined by the proportion of the segment to the unit segment. A segment half the length of the unit segment is, in the blend, 1/2. Are there any segments without number counterparts? As the Pythagorean geometers discovered to their horror, the answer is yes. It is well known the hypotenuse of an isosceles right triangle whose legs are unit segments has no rational-number counterpart.

Yet another blending network in response to this "gap" produces another dramatic emergent structure, the irrational numbers. In hindsight, we can now "see" that another infinity of numbers was "missing" from the rationals. Again, things that could not exist have become elements in the blend, and we see "gaps" in the original input. Another round of extremely imaginative blending will "reveal" further gaps: the transcendental numbers. Still other blends create zero, negative numbers, imaginary numbers, ... In all cases, the blend has many elements with no counterpart in the preceding number input, and in hindsight, those numbers are regarded as "missing" from the previous concept of number. The symbol 0 was used by the Alexandrian Greeks to denote the absence of a number, but "the absence of a number" became a full number, able to participate in addition, subtraction, and multiplication, in the blends developed by Hindu mathematicians in the seventh century A.D. The history of the concept of number is one "caffeine headache" after another. The recent work of George Lakoff and Rafael Núñez on the embodied nature of mathematics provides considerable additional evidence for the crucial role of blending in the development of mathematics and mathematical thought.

Not only number theory but the entire set-theoretic foundation of mathematics proceeds by making blends that have elements where before there was nothing. A standard set-theoretic way of deriving numbers is to begin with nothing and call it the "null set." This null set is now something—it is a set, commonly denoted Φ. Now consider the set that has as its only element Φ. It is commonly denoted {Φ}. There is now also automatically a set of two elements {Φ,{Φ}}. And therefore a set of three elements {Φ,{Φ},{Φ,{Φ}}. And so on: In the set-theoretic foundations, these sets *are* the whole numbers (through an implicit blend with the everyday notion of number). The blend of positive or negative rationals can be "defined" using equivalence classes of pairs of such sets, and additional frills such as "cuts" will provide a set-theoretical foundation for the real numbers.

King Lear's great mistake is embodied in the line "Nothing will come of nothing." Not only Cordelia's love but also the history of mathematics proves him wrong.

NONEVENTS

We have seen nothing become a thing. Our mental universe is populated by a dark matter of nonthings.

Questions:

- Can nothing become an event?
- Can nothing become an action?

Our answers:

Nonevents and nonactions are nearly everywhere in our cognition. Physical reality is a material anchor for conceptual blends that typically carry many projections from counterfactual spaces. "The jar-lid won't come off," "The stack of books has not fallen," "The stack of books will fall," "The jar-lid refuses to come off," and "The stack of books wants to fall over" all present networks in which one input has nothing happening and the other input has something happening. In the blend, the nothing happening becomes an event that is contrasted with the other event: The stack of books stays upright versus the stack of books falls. "Missing a shot" evokes a blend that contains a nonevent: Both inputs have the shot, one input has the ball going somewhere other than into the goal, the counterfactual input has the ball going into the goal. In the blend, the ball's not going into the goal becomes a "missed shot," a nonevent.

Missed shots can even be blended with nonshots, as in the wisdom printed on a sports T-shirt: "You miss 100 percent of the shots you don't take." This observation prompts for an elaborate integration network with many kinds of nonevents. Two of them are already conventional. First, any "missed shot" is already a blend; both inputs have a ball thrown and going somewhere, but only one of the inputs has the ball going into the basket. Both of these spaces are full and complete events. The contrast between them is compressed in the blend: The blend now has a new category of event—namely, missed shots, which are not in either of the inputs. Second, any "shot not taken" is already a blend: In this case, both inputs have a player doing something with the ball, but only one of those spaces has her throwing the ball toward the basket. Both of these spaces are full and complete events. The contrast between them is compressed in the blend: The blend now has a new category of event, shots not taken, which is not in either of the inputs. Each of these conventional blends creates a new category of shots—missed shots and shots not taken. They play a crucial role in the game of basketball. Statistical measures are routinely applied to shots in basketball—successful three-pointers, unsuccessful two-pointers, and so on. Those measures can now be applied to *nonshots*. Interestingly, for all other categories of shot, we have to play the game to get the statistical measures, but the shots-not-taken blend guarantees in advance that no nonshot goes into the basket, and the missed-shot blend guarantees that all of these are "missed shots." Therefore, 100 percent of shots not taken are missed shots.

This blending extravaganza takes all the statistical machinery from the *missed-shots* blend. Since one wants to avoid missing shots, the implication is that one should avoid shots not taken; that is, one should shoot more. But there is topology in the *shots-not-taken* blend: If you don't take the shot, you retain control of the ball, which is useful. That topology is different from the topology in the *missed-shots* network, where you take a shot and so give up control of the ball. In the *shots-not-taken = missed-shots* blend, the topology of shots-not-taken is lost at the expense of the topology of taking a shot.

CONVENTIONAL CULTURAL COUNTERFACTUAL COMPRESSIONS

We have seen that language provides counterfactual compressions, as in "caffeine headache," "missing chair," and "gap."

Question:

- Is this a purely linguistic artifice?

Our answer:

It seems that cultures find it efficient to evolve compressions that can be easily transmitted. On the one hand, cultures want to channel thought so as to block off large ranges of possible double-scope integrations. There are, for instance, many ways to deal with the possibility of collision at an intersection, but culture does not want you to try them all out, so it delivers one governing solution: the traffic light. The traffic light is a material anchor for a complex compression. The decompressed form has a different mental space for each of the possibilities of cars going through intersections: no cars, one in this direction, one in that, one in each direction, one that has just gone through as another arrives, and so on, including values for different speeds, different kinds of drivers, and so on. All of these spaces have the same frame, but there are very many ways to specify that frame. In some of these spaces, there would be a collision. Those are the ones we want to be counterfactual with respect to the actual world. One way to arrange for that is to disallow all possibility of movement along crossing paths simultaneously through the intersection, and one way to do that is to make certain that when it is possible to move along one path through the intersection, it is impossible to move along the crossing path. This solution is absolute but overpowerful, since it means that often a driver will sit at an intersection watching no cars go through the intersection. Still, the overpowerful solution avoids the catastrophe, so it is adopted.

The driver with the green light has confidence that the social system has arranged for one particular blended space to be actual. In that blended space, we have projected from the no-collision space the fact that no cars on the cross street are entering the intersection, and we have projected from the collision space that there may be many cars on the cross-street, including some that are, at present, obviously on course to collide with the driver. The emergent structure of the blend guarantees that there will be no collision in the blend, but not for the same reason that there is no collision in the no-collision space.

A RIDDLE

What is greater than God?
More evil than the devil?
The poor have it.
The rich need it.
And if you eat it, you die.

Hint: It is what you should stop at if you are looking for the answer.

Twelve

IDENTITY AND CHARACTER

> These names recur incessantly; within the great uncertainty of the race, they are anchors that link the episodic and tumultuous passage of time to the stable essences of great characters, as if man were above all a name that becomes the master of events.
>
> —*Roland Barthes, on the Tour de France*

IN THE *ODYSSEY,* ODYSSEUS works through nearly every situation conceivable—fighting, sailing, disputing, womanizing, hiding, pleading, persuading—and remains Odysseus throughout. It is a central aspect of human understanding to think that people have characters that manifest themselves as circumstances change. When someone acts in a certain way in a novel situation, we might say "That's just like him. I would never have done that." Character transports over frames and remains recognizable in all of them, to the extent that we can ask "What would Odysseus do in these circumstances?" despite the fact that those circumstances are unknown in Odysseus's world. Children's literature gives us contrasting characters in the same situations—Toad, Ratty, Mole, and Badger in *The Wind in the Willows*. But adult literature does the same—Achilles, Diomedes, Hector, Ajax, Menelaus, Odysseus. Roland Barthes, in *Mythologies,* analyzes the Tour de France as an epic in which the characters manifest their essences through their performances in the Tour:

> Il y a une onomastique du Tour de France qui nous dit à elle seule que le Tour est une grande épopée. Les noms des coureurs semblent pour la plupart venir d'un âge ethnique très ancien, d'un temps où la race sonnait à travers un petit nombre de phonèmes exemplaires (Brankart le Franc, Bobet le Francien, Robic le Celte, Ruiz l'Ibère, Darrigade le Gascon). Et puis, ces noms reviennent sans cesse; ils forment dans le grand hasard de l'épreuve des points fixes, dont la tâche est de raccrocher une durée épisodique, tumultueuse, aux essences stables des grands caractères, comme si l'homme était avant tout un nom qui se rend maître des

> événements: Brankart, Geminiani, Lauredi, Antonin Rolland, ces patronymes se lisent comme les signes algébriques de la valeur, de la loyauté, de la traîtrise ou du stoïcisme. . . . Gaul incarne l'Arbitraire, le Divin, le Merveilleux, l'Élection, la complicité avec les dieux; Bobet incarne le Juste, l'Humain, Bobet nie les dieux, Bobet illustre une morale de l'homme seul. Gaul est un archange, Bobet est prométhéen, c'est un Sisyphe qui réussirait à faire basculer la pierre sur ces mêmes dieux qui l'ont condamné à n'être magnifiquement qu'un homme.
>
> There is an onomastics to the Tour de France that tells us all by itself that the Tour is a great epic. The cyclists' names seem to come from an ethnic age long ago when a few superb phonemes trumpeted the valor of one's blood (Brankart the Frank, Bobet the Francian, Robic the Celt, Ruiz the Iberian, Darrigade the Gascon). These names recur incessantly; within the great uncertainty of the race, they are anchors that link the episodic and tumultuous passage of time to the stable essences of great characters, as if man were above all a name that becomes the master of events: Brankart, Geminiani, Lauredi, Antonin, Rolland—these patronymics read like algebraic signs of valor, loyalty, treachery, or stoicism. . . . Gaul incarnates the Arbitrary, the Divine, the Marvelous, the Chosen, complicity with the gods; Bobet stands for the Just, the Human, Bobet denies the gods, Bobet symbolizes man on his own; Gaul is an archangel, Bobet is promethean, a Sisyphus who succeeds in toppling the rock onto the very gods who condemned him to be, magnificently, only a man.

Family lore paints a similar picture—Uncle Joseph had this character, Aunt Emily that, and we *can tell* what they would have done in any given situation. In the way we think, a character can stay essentially the same over widely different frames, and a frame can stay essentially the same when populated by widely different characters.

In some of the blends we have analyzed, it is crucial that specific characters from one input get projected with their essences intact into the blend. In the Debate with Kant, for instance, Kant must come in with his characteristic philosophical and logical dispositions. We do not just import to the blend what he actually wrote; rather, we construct what he might have said given his character. A counterfactual blend like "If Churchill had been prime minister in 1956, there would have been no Suez disaster" implicitly assumes an essential character for Churchill that can be projected to a time and circumstance when he was no longer active. For people—whether bartenders or political scientists—to use such expressions persuasively, there must be an uncanny consensus regarding the existence of such an essential character. Characters, like frames, are basic cognitive cultural instruments. We may dispute every aspect of their accuracy or legitimacy or invariance, or even their very existence, but cognitively we cannot do without them.

IDENTITIES AND FRAMES

To this point, our taxonomies of integration networks have emphasized the role of frames. Simplex, mirror, single-scope, and double-scope networks were all defined, as main types, by the relations of the organizing *frames* of the inputs and their relation to the *frames* in the generic space and the blended space. But identity and character are an equally important aspect of the way we think. We can think of frames as transporting across different characters (the *buy-sell* frame stays the same regardless of who is buying and selling), or we can think of character as transporting across different frames: Odysseus remains who he is regardless of his situation.

Consider, for example, a corporate employee who describes his situation to his wife, who has no role in that profession. She says, "If I were you, I would quit." Here, we must construct a blend with the organizing frame from the company but a blended person in the role of employee. That person has the identity of the husband in the general public sense, as well as his experience and many of his characteristics, but now has the disposition, judgment, and will of the wife. Importantly, in this case we are not blending frames.

Recall also our example from the previous chapter, "In France, Watergate would not have hurt Nixon." Our analysis there focused on the blended U.S.-France frame and took it for granted that a president "like Nixon" was available for that blended frame. But a more perspicuous treatment would say that Nixon is like Odysseus, a character who can make a place for himself in any frame. In the blend, we have not only a U.S.-France blended political frame but also the appearance of a character.

This transportability of character is manifest from a campaign slogan used against Nixon: "Would you buy a used car from this man?" To clarify Nixon's character, we insert him into a frame where it is cast into high profile. It makes no difference that selling used cars has no place in Richard Nixon's actual life. What matters is that the new frame gives a great compression and a global insight into the character. This campaign slogan is just one instance of a very general cognitive strategy for prompting people to understand character. If we want to describe a co-worker, we might say "He's the kind of guy who, if he bought a lottery ticket, would immediately dine out on his winnings." In the used-car and lottery-ticket examples, the focus is on discovering the essential character that transports across frames.

This subject—the stability of character across different activities—is immensely complicated and infinitely explored in the world's literatures. And as we have just pointed out, conceptual blending plays a central role in this conception of character. We are not in a position here to go into further detailed analysis on this point, but some general patterns are clear. First, we are able to extract regularities over different behaviors by the same person to build up a generic space for that person—a personal character. Second, we are able to extract regularities over different behaviors by many people to build up a generic space for a kind of behavior. They interact. The phrase "He's the kind of guy who does X" asks us to do both: to establish a generic for him and a generic for the kind of behavior exemplified by X. There is a further kind of extraction, the kind Theophrastus and his successor La Bruyère do in their works on character, where we create a generic space for a "kind of person"—the Vain Man, the Liar, the Social Climber—who is a blend of several of these kinds of generic behavior. A great deal of popular psychology—finding out "who you are," discovering your real self—consists in finding and constructing these kinds of generic spaces for character. The same is true of the science of personality. The works of Sigmund Freud, Abraham Maslow, and Mihaly Csikszentmihalyi are directed at locating generics for character and how it got that way. This achievement of blending is a mainstay of humor: A recent humorous blend reports that God's mysterious and otherwise inscrutable acts can be accounted for once we realize that he has bipolar disorder—a condition whose cyclic nature results, for example, in seasons on Earth.

In these cases, character is clarified by transporting it across frames to locate the shared generic. But in other cases, the frames are clarified by transporting across them the same essential character. So, in "In France, Watergate would not have hurt Nixon," the focus is on the political frame as instantiated in the United States and France, respectively. The purpose is to draw a disanalogy between those two instantiations of "Western democracy," and this is helped by putting Nixon into both of them.

What we see are general principles: To clarify a single frame, fill it with different essential characters; to clarify the relationship between frames, fill them with the same essential character; and to clarify essential character, transport it across different frames.

Naturally, frames and characters are not always distinct, since some characters seem to be attached to their frames. Odysseus builds a place for himself inside many different kinds of frames, but Sherlock Holmes brings his detective frame with him and causes a frame-blend just by appearing: If Sherlock Holmes comes to your dinner party, someone is going to get killed.

Again, we are not debating the merit of assessing character by shifting frames or assessing frames by shifting character. Rather, our point is that making such assessments involves general imaginative routines. What is remarkable, given an identity like Odysseus or Churchill, is that we would feel confident in running a blend of that identity and some frame, regardless of whether the person ever occupied such a frame. If anything, we seem to have especially great global insight if the frame is alien to the individual: "He'd give you the shirt off his back."

The overall folk theory that we see here is that we use character to converge on an understanding of frames and we use frames to converge on an understanding of character. In the folk theory, frames and characters are interlocking aspects of human reality. You can't have one without the other, although in some cases the emphasis falls more on character and in other cases it falls more on the frame. What makes all this slippery is that character is in principle all the behaviors in all possible frames, but frames themselves may be just as substantially linked to characters. *Saint, diplomat, hooker, mediator,* and *conqueror*, for instance, can work in both ways. Consider *prostitute*. It can be construed as a frame that anybody can fall into, or as a frame with character implications: Someone who can fit into it aptly must have a certain character. Construing *prostitute* as just a general frame, we can investigate character by asking how such a character would perform in that frame: How would someone with the character of Mother Teresa, Margaret Thatcher, Cleopatra, or Bill Clinton operate within the prostitute frame? Construing the frame as having implications for character, we can investigate whether someone has the character befitting a prostitute by imagining whether that person could operate within that frame. Under the first construal, Mother Teresa reveals her sainthood by accepting the sacrifice with fortitude, by never complaining, by trusting God. The frame cannot impinge upon her character, for "To the pure, all things are pure." Under the second construal, Mother Teresa's character prevents her from ever becoming a prostitute. One might take the view that to convert from prostitute to saint, as is said to have happened in the case of Mary Magdalene, requires a change of character.

CONCEPTUAL INTEGRATION, MENTAL SPACES, FRAMES, AND CHARACTER

Just as conceptual integration networks arise that emphasize frames—simplex, mirror, single-scope, double-scope—so integration networks arise that emphasize either the blending of a character and a frame or the blending of a character and a character. Here we will consider some examples of these kinds of integration network.

Recall the Debate with Kant, a mirror network with a blended space in which Kant and a modern philosopher were brought in as distinct elements. Using character blending, we could readily create a different kind of integration network in which Kant and the modern philosopher are fused in the blend instead of projected to distinct elements. Perhaps the modern philosopher has taken up a classic philosophical problem of the sort Kant might have considered, or that he pursued only briefly. The philosopher, stymied, asks herself, "If I were Kant, how would I attack this problem?" This scenario launches a blend—one that can then be run creatively and indefinitely—in which there is a single philosopher. The modern philosopher uses her knowledge of Kant's character and identity, and her own intellectual character, tastes, and interests, in order to run the blend—that is, to "become" Kant in some respects as she approaches the problem. Whether her imitation of Kant is objectively accurate is beside the point; what matters is whether the blended character she imagines and "becomes" is intellectually productive. Remarkably, she can get new insight, have a different cast of mind, acquire new capacities to an extent as she inhabits this blend, and thus make discoveries that were otherwise unavailable to her.

But wait a minute. Who is this "Kant" anyway? When we were focusing on frames, we could see structure coming in from them. In the case of character, identity, or ego, what is projected into the blend? The philosopher has a rich acquaintance with the life and works of Kant in the way others might have a rich acquaintance with Freud, Odysseus, Jesus, or Socrates. Far from being just a set of features, "Kant" for her is a rich knowledge of his behavior over many different scenarios, including his personal styles and manner of action and the feelings he had in situations that he later reported in letters. Cognitively, this amounts to a very rich set of neural patterns that may be activated in the running of the blend. From a frame perspective, this is still a mirror network. But from an identity perspective, it is double-scope, since it brings in sharply different aspects of the identity of Kant and the modern philosopher.

Now imagine that the modern philosopher is a computer scientist and neuroscientist—Patricia Churchland, perhaps—and is tackling a problem involving connectionist computation and neuronal group selection. Her activity still fits the schematic frame of a musing philosopher, but the frames in the inputs are nevertheless significantly different. Connectionist theories and neural darwinism did not exist in Kant's day; he could not have tackled them. Yet this fact is in no way an obstacle to constructing and running the blend. It may turn out to be efficient for Patricia Churchland to blend herself with Kant in order to approach the relevant issues in the deepest and most insightful way.

Not only the identities but also the frames themselves may be strikingly different and yet lead to a creative blend. So Dawn Riley, a sailor lost at sea with a failed electronic guidance system but with charts and instruments and a sextant she never learned how to use, might draw upon her lifelong interest in Kant when asking herself: "If I were Immanuel, what would I do now?" Like the previous ones, this blend is double-scope for Identity: Kant never left his house, let alone sailed the Pacific, and Dawn has previously always kept her philosophy and her sailing distinct. But the blend is now also double-scope for framing since a new blended frame of sailor/philosopher emerges from it. The person in the blend now has the public identity of Dawn Riley but a new character that delivers, she hopes, some way of proceeding that is not available from the character of Dawn Riley alone. She does not know when she asks this question how the blend will be put together or what structure will emerge. Blending is neither deterministic nor compositional, which is good in this situation: If it were, it would be far less likely to deliver new insight.

Conceptual integration is a basic mental operation over mental spaces, and it satisfies the same structural and dynamic principles in all cases. Frames give one basic way to organize a mental space. Character, Identity, and Ego give another. Inevitably, conceptual integration works over both frames and character when it operates over mental spaces.

IF I WERE YOU . . .

Once we see this system of canonical integration networks and the principles that underlie them, it becomes clear how to approach a great range of data that logic and semantics previously found puzzling. One kind of data consists of statements like "If I were you, I would be with Ursula today," said by Marianne to Robert on a Saturday when both are in the office. One input has Robert, who has chosen to go to the office; the other has Marianne. In the blend, Marianne and Robert are fused to give a new character influenced by both Robert's relationship with Ursula and Marianne's judgment. The blended character has chosen to be with Ursula, so in the blend we have a different frame: companionship. Conventionally and pragmatically, bringing up an alternative can be evaluative. Marianne's comment can thus be construed as critical of Robert (he should be with his wife) or as laudatory (he is truly dedicated to his work).

"If I were you, I would hire me," said by the prospective employee to the boss, asks for a similar blend of identities. In one input we have "you" considering whether to hire "me," and in the other input we have "me" with an essential character, judgment, and disposition. The expression makes "you" and "me" counterparts by means of a Disanalogy connector, and makes "me" and "me" counterparts by means of an Identity connector.

The "you"-to-"me" outer-space Disanalogy connector is compressed in the blend into a unique person, with the public identity and powers and interests of the employer, but the speaker's good judgment and the speaker's familiarity with himself. But the blend also has the "me" who is the prospective employee as viewed by the employer. This is a straightforward projection from the interview input to the blend. That the two "me"s are projected from different inputs correlates with the grammatical preference for not marking the pronoun as a reflexive: "If I were you, I would hire me" is preferred to "If I were you, I would hire myself" (which is likely to be understood as saying that the boss should hire himself for the job).

But if the job applicant is grammatically in the third person, we do not say "If he were you, he would hire him" but rather "If he were you, he would hire himself" (with the intended interpretation that if "he" is the applicant Bill and "you" the employer Mary, then the employer in Mary's position and with Bill's knowledge and judgment would hire Bill). There is a stronger constraint on the third-person pronouns in "*he* would hire *him*" that prevents them from pointing to connected elements when there is no reflexive marking. In our example, the elements *are* connected and therefore only the reflexive option is available (*he* would hire *himself*).

Using an expression like "I would hire me" to pick out structure in a blend may look exotic, but such examples are quite common and usually go unnoticed, as in the following comment by right-handed pitcher Bret Saberhagen: "If I was a writer [who votes for the Cy Young Award] and looking at the numbers and how the pitchers have performed throughout the year, I would say I would be the leader." Similarly, Michel Charolles points out the following line from Samuel Beckett: "S'il avait été sa mère, il se serait détesté" ("If he had been his mother, he would have hated himself," with the reading that in the counterfactual blend, the mother hates the son); and Eve Sweetser reports the following attested example: "Quand j'étais étudiante, j'aurais voulu me rencontrer" ("When I was a student, I would have liked to have met me"), meaning that the speaker wishes her own professors had given her the personal attention and concern that she tries to give to her students.

Jeff Pelletier reports that he was once invited to interview for a job in Ancient Philosophy, in which he had some training but no real specialization. His Ancient Philosophy professor advised him, "If I were you, I would go; but if I were in your position, I probably wouldn't go." She meant that Pelletier, given his confidence and facility, could succeed in the interview despite his lack of expertise but that she herself was too timid to try such a thing. Yet, unlike Pelletier, she actually was an expert in Ancient Philosophy. Such examples show the power but also the great flexibility of counterfactual blending of identities. "If I were you, I would go" prompts for a blended job applicant who inherits from his professor her wisdom about obtaining appointments in the academic world and about the student Pelletier; so the blended applicant, who has the public identity of Pelletier, now possesses more knowledge about Pelletier than the real Pelletier does about himself. But this blended applicant has Pelletier's confidence and courage, at least as perceived by the professor. Contrastingly, "If I were in your position, I probably wouldn't go" prompts for a counterfactual blend in which the professor, perhaps at a younger stage of life, is in the frame of being interviewed for this particular job in Ancient Philosophy. The blend of identities here is much less double-scope: The professor has all of her character, but Pelletier's relative ignorance about Ancient Philosophy. In the complete advice given by the Ancient Philosophy professor, the first part is advice to her student, but the second part is only a comment on herself, perhaps meant to clarify her reasoning in suggesting that he, with his different character, could succeed.

In addition to permitting the blending of two different identities, counterfactual blends evoked by similar grammar may emphasize the blending of a character with a frame. "If I had been his wife, I would have been his widow long ago," an example reported by Nili Mandelblit, emphasizes the persistence of character:

The speaker dislikes the man so much that marriage would not have altered her feelings; the frame is overcome by the character and must be deformed in the blend, where the wife quickly kills the husband. It should be noted that a reading that blends identities is also available: If the man already has a wife, and it is the wife's experiences that prompt the outrage, then that outrage, blended with the steely determination of the speaker, gives a character and a murder that are not available from either input.

These points are also apparent in the following example, spoken by President Bill Clinton at a fund-raising event for his wife Hillary's campaign for senator: "I would be here for my wife if she were not my wife, because we have got to have people with a lifetime commitment to the future, to the children." There is leeway in constructing the counterfactual blend: By stipulation it has Bill Clinton and Hillary Rodham, but not married, and she has her lifetime commitment to the future. Beyond that, it is a matter of pragmatics whether, in the blend, she and Bill have been lifelong friends, went to Yale Law School together, and so on, or she is simply anybody with such a commitment running for office. Technically, the frame-role *my wife* in one input is used to access its value, Hillary, and what we find in the blend is a counterpart of that value, not a counterpart of the role *my wife.* The counterfactual expression would be appropriate with a different interpretation if Bill Clinton had refused to participate in the fund-raiser on the grounds that it was bad form for him to support his wife so publicly.

The fact that Hillary Rodham is Bill Clinton's wife is clearly not a side issue. Frame structure can play an essential role in behavior apparently linked to identity. Consider a dialogue from the movie *The Naked Lie*, in which a prostitute has been found murdered. Webster, an unpleasant, self-centered character, shows no sympathy. Victoria challenges him:

> Victoria: What if it were your sister?
> Webster: I don't have a sister, but if I did, she wouldn't be a hooker.

Later in the movie, Victoria, talking to someone else, says "You know that sister Webster doesn't have? Well, she doesn't know how lucky she is."

At first, Webster rejects the counterfactual blend on the logic that there is no value of the role *Webster's sister* to be projected into the blend. But then he is forced to accept a blend in which only the role *sister* is projected from a schematic kinship frame, and the value of that role is emergent in the blend. So in the blend, we have a specific *Webster's sister*, with an identity, but we know nothing about her except that she is Webster's sister, and therefore her "character" is entirely defined by this property. According to Webster, the property of being his sister confers upon her character an essence that prevents her from being a prostitute. He is viewing his own kinship frame as having implications for character, so that a sister of his could no more be a prostitute than could Mother Teresa. Victoria's later comment about "that sister Webster doesn't have" takes the previous counterfactual blend in which Webster does have a specific sister, call her Ann, as an input for a further blend with reality. It projects Ann from the counterfactual input to this new blend and projects from reality the fact that Webster has no sister. So in the new blend, Ann exists but is not Webster's sister, which makes her very fortunate. She cannot know of the fate she was spared, since she is not aware of her counterpart in the counterfactual input. As we have seen earlier, many words call for alternative counterfactual spaces: "dent," "drop," "miss," and so on. "Lucky" is another one: We can understand it only by constructing another space, if only a very schematic one, in which the person in question has less success. In the dialogue in *The Naked Lie*, the needed counterfactual space has already been built up during the discourse: It is the previous counterfactual blend in which Ann is Webster's sister. Indeed, that is the only space that contains any information about Ann, so it must be activated to understand anything about her.

REDEMPTION, VENGEANCE, AND HONOR

Let's revisit this bit of dialogue from Chapter 7:

> "Do you remember how when you were little you were so intent upon hiding your treasures that you hid them so well even you could not find them again? Do you remember that you hid your new penny when you were four and we never found it? That's just what you have done with Angela. You've been talking for two hours about all your troubles, but what they boil down to is that you have hidden away your love for her so deeply that you can't see it. Once again, you've hidden your penny, even from yourself."

Earlier, we pointed out that in this blend, hiding treasures is projected to the blend from the input about childhood and provides the blend's organizing frame. The frame of hiding treasures provides a way of construing the love situation with Angela. But there is more going on. First, constructing the integration network develops a generic space in which there is a deep characterological psychology, an essence, that is shared by all the spaces in the network and is the speaker's rhetorical point. Second, this remark is addressed as a revelation to the man in love, and to understand it he must perform double-scope blending on his own identities: The adult man in love must be blended with the child hoarding objects. This blending is a way to bring out this essence, which is itself abstract and invisible except in its concrete manifestations. Hiding love and hiding a penny are both behaviors. In the blend, they are the same and so take on the double concreteness that comes from the two vivid input spaces.

In Chapter 8, we discussed identity blending in XYZ mirror networks, as when Paul, who had been traumatized by the loss of his daughter Elizabeth, and so behaved coldly to his much younger daughter Sally, finally comes to his senses, and, when asked by Sally why he is treating her in this new way, replies, "Because you are my long-lost daughter." If we emphasize frames, this is a mirror network: Each space has the roles *father* and *daughter* and Paul as the value of the role *father*. But if we emphasize character, this is a double-scope network. Although Paul is blending himself with Paul, he is psychologically a much different man than he once was. He recovers some of his former self by blending his present and former selves. The emergent structure here is evident: He undergoes a considerable psychological transformation. Sally is also blended with Elizabeth, and it is this radical double-scope identity blend that makes Paul's realization psychologically possible.

These examples point to a general dynamic principle of human psychological and emotional life: Outer-space vital relations, often connecting a person in one space to himself in another, can be compressed into inner-space character traits understood to be part of the essence of the person. The "hiding your penny" example gives an idiosyncratic blend and an idiosyncratic personal essence. But culture sets up general models of just this type and makes them into cultural categories like *redemption, restoring honor, vengeance, vendetta, curse.*

Redemption is a matter of entering or in fact creating a later situation that counts as equivalent to a prior situation in which one failed. From a frame point of view, these are mirror networks. One succeeds in the later situation. In literary, dramatic, and cinematic representations of redemption, the protagonist frequently hesitates at just the instant when he failed before, but this time does the right thing. We do not take such a plot as the story of a person who failed once and succeeded once, with equal weight given to the two events. Instead, we take the second event as the one that reveals the essence of the protagonist and proves that the first was a fluke. The success does not simply neutralize the failure, setting the scale back to zero. It restores the protagonist's identity, making him "whole" "once again." Objectively, it is odd that any later performance should have an effect on the evaluation of an earlier performance: The failure cannot be changed, and none of the terrible consequences that provide the guilt or shame and so fuel the need for redemption can be changed in the slightest detail. In the input spaces, no redemption is possible. But in the blend, the two situations become one, and the character (if not the behavior) of the protagonist comes from the later input, thus providing in the blend and in the generic space a stable and good character from which the earlier input space is merely an unfortunate deviation.

Art and conversation often point very explicitly to these blend structures. In the movie *The Sixth Sense*, a child psychologist who catastrophically failed one of his patients many years before goes to supernatural lengths to locate a boy with the same symptoms, and to treat him. Why? He says, "I thought that helping this new one would be like helping the old one." There is no way, objectively, that the former patient can be helped, because he is dead, arguably as a result of the psychologist's failure. An elaborate folk notion of character is at work here. One failure can suggest a generic space in which the protagonist has essential bad character. Does he? The later success in the same kind of situation denies that essential bad character in the generic space, and so denies that it could have been the cause of the earlier failure. The blend goes even further: In the blend, times are fused and the situations are blended, so success in the new situation counts as retrospective success in the old. No one is deluded: The old failure stands as unchangeable history. But in the integration network, the psychological context and weight of that failure are completely changed. In *The Sixth Sense*, the child psychologist feels that he is "making it up to" the first boy.

There are also genres of redemption story in which the failed protagonist does not seek redemption, perhaps thinking it is not to be attained, and goes on with life, only to be informed by a human sage or divine messenger that he has in fact redeemed himself. Alternatively, after the fact, it dawns on him and on those around him that an act of redemption has just taken place. The original failure and the later success happen independently but then get blended psychologically, producing a global insight about character running over time. The cultural category of *redemption* thus depends on blending, but its reality for the culture can be so strong as to seem a fundamental defining property. Vice-President Al Gore, when asked during his 2000 presidential campaign what should be done to a baseball pitcher who had made racist comments, proposed that, instead of punishment, he should be given a chance to redeem himself. He added, as indisputable warrant for his view: "America is all about redemption." With this comment he transfers redemption from the character of the individual to the character of the nation.

Redemption's absolute opposite is *curse*. In that case, the later event also has a failure, and in both the blend and the generic space, and in fact in the two inputs, there is an essential property that is the cause of these failures. In one version of *curse*, this essential property attaches to the character of the actor and it is not caused by any outside agent. In other versions, the essential property is attached to the actor by some outside agent, such as a god or even a human being with spiritual connections. There are intermediate cases, too, as in a curse on a family, imposed by an outside agent but then carried by the descendants as an inherited property of character. In social life, anybody's curse counts, but has greater weight the closer the person cursing is to the person cursed.

Revenge is a similar cultural category. A later situation has meaning only with respect to an earlier one, and again the result is not one loss versus one win but instead a true integration in which the elements of the revenge situation are elements of the original situation. Enacting the integration network does not change the facts of the original situation as an input, but it does change the overall defining context in which it sits, and it also eliminates the possibility that losing is a permanent or fundamental feature of the avenger.

In *redemption, curse,* and *revenge,* we see cross-space vital relations compressed into inner-space cultural elements in the blend. All these cases do essential compressions on Identity and Time. *Curse* also does compression on Identity of the outcomes (loss – loss), while *redemption* and *revenge* do compression on Disanalogy on the outcomes (loss – win). Further, overarching compressions create cultural elements like *honor.* Like *absence* or *gap,* honor is a word that calls up an implicit counterfactual, specifically a mental space in which something dishonorable has happened, such as cowardice, an unanswered insult, or an attack on friends or family. But once honor has been "lost" in such a mental space, it can be "regained" or "restored" by later action. Once again, we do not view this simply as a dishonorable event followed by an honorable event, a mere sequence of one loss versus one win. Rather, the later scenario has meaning by relation to the earlier one, and although the integration does not eliminate the input space in which honor was lost, it does eliminate the condition of being in a state of dishonor.

There are many similar notions, all depending on the general availability of blending for the creation of cultural meaning: vendetta, punishment, restitution, *lex talionis,* and many others. Crucially, they interact, so that, for example, an action may be simultaneously a redemption, a revenge, and a punishment. The coherence of the system shows up especially vividly in overarching concepts like *honor:* Under the general warrant of honor, one can put together an especially rich integration network that all at one swoop provides meanings of redemption, vengeance, and punishment, as when the hunter, who must leave his family to provide for them, returns to the evidence of a dishonoring criminal act committed in his absence, tracks down the criminals, and, in punishing them, redeems himself and regains honor.

CHAPTER 12 ZOOM OUT

SCENARIOS

In discussing the blending of frames, blending of characters, and blending of frames and characters, we have noticed that notions like double-scope and mirror apply to both frames and characters.

Question:

- Do we have a neat picture then in which we have frames on one hand and characters on the other, so that all we have to do is bring them together by composition?

Our answer:

Alas, or perhaps thank God, no. In our exposition, we have often spoken as if frame and character are separable. In a sense, they are, and the linguistic system presents them as such: We can think of the frame of *air travel* without any essential attachment to the character of the traveler, and the language gives us words like "passenger" to pick out the frame with no reference to the character. In the other direction, we can think of a character like Bob Hope (who traveled a great deal) without any essential attachment to the frame *air travel*, and of course the language gives us the name "Bob Hope" to pick out that identity with no reference to any frames at all. If we refer to Bob Hope as "passenger," we are abstracting away in the direction of the frame; if we call him "Bob Hope," we are abstracting away in the direction of the identity. But in fact, one can never abstract all the way in one direction or the other. Any identity comes with considerable attachment to frames, and any frame comes with considerable attachment to identities: *Father* is a very abstract frame, but fully attached to our own and our friends' fathers. By the same token, if your father is "John Smith," you will have a difficult time trying to think of John Smith without activating the frame of *father*. It is accurate to say that the blend evoked by "Paul is the father of Sally" uses as one input a schematic frame of *father* and as the other input the identities Paul and Sally, but the frame is embedded in a myriad of potentially active connections to characters, and the identities Paul and Sally are embedded in a myriad of potentially active connections to frames. All of this feltwork of knowledge is available for recruitment to the inputs and projection thence to the blend. Knowing someone means knowing what that person will do in the most diverse situations, including novel or impossible ones, and knowing *that* depends on knowing what the person has done in the past, and being able to apply frames to the old and new situations. Similarly, knowing a frame is knowing specific instances and how various characters operate inside it. There is no limit to the amount of detail in frames or identities, and at the neurocognitive level of activations, frames and characters are always intertwined. Naive metaphysics and folk theories supported by language and conscious apprehension keep frames and identities neatly apart, masking their more intricate backstage involvements.

THE EFFLORESCENCE OF THE BLEND

In many of our examples of blends, and in particular the identity and counterfactual blends, a clear meaning emerges, and there is substantial agreement as to what that meaning is.

Question:

- Doesn't this show that blends are, after all, in some sense compositional, algorithmic, and deterministic?

Our answer:

Again: Alas, or perhaps thank God, no! For each of the language forms associated with a particular blend, there were a large number of different blends that would have been equally possible in a different context. The meaning that is prompted by a language form in particular circumstances seems subjectively like the only meaning associated with that form. This is natural, since the backstage cognition that leads to the emergence of that meaning is largely unconscious, and once a meaning has emerged, we usually have no reason to explore an alternative construction. The Eliza effect comforts us in the belief that the meaning we got was entirely contained in the language form and did not require any special cognitive choices. Special circumstances—the punchline of a joke, for instance—may bring some of the choices to conscious awareness. By contrast, "If I were you, I would quit" might seem to express a unique blend, but in fact it can prompt for many blends, as shown by the following continuations:

. . . but I am independently wealthy; you shouldn't quit by any means.
. . . but I am a hothead and would regret it later and would have to go on my knees begging for my job back.
. . . and so should you.
. . . but you shouldn't.
. . . but that's only because the boss needs me so much he would offer me a raise to get me back.
. . . since I couldn't live with myself knowing how badly I had treated me.
. . . because being you would make me so utterly miserable I couldn't possibly get any work done.
. . . since I would have a wealthy father.
. . . since you have another job offer.
. . . since I have another job offer.
. . . since your beloved boss has another job offer and will be leaving soon.

THE IDENTITY OF THE MISSING

In Chapter 11, we saw nonthings, nonevents, and nonactions. In Chapter 10, we saw blends that permit us to make contact with the dead at tombs.

Question:

- Can nothing be a person, too?

Our answer:

Our worlds are full of nonpeople. We discussed how tombs and ashes serve as material anchors for making contact with the departed. Using the same kinds of networks with material anchors, we can address a photograph of someone dead or alive, a wedding ring of a dead spouse, even a letter from someone not here. A song, a place, a memory of someone is enough to induce us to create, in the blend, a person who is absent but who nonetheless has intentional features: We ask our former mentor for advice, one of our grandparents pops up to scold us, or we have a conversation with a departed spouse. These are not merely static memories: We can hear our dead grandmother talking about our daughter even though the living grandmother died before the daughter was born. Psychologically, there is nothing mysterious about this once we recognize how blending works. Selective projection from long-term memory on the one hand and from present circumstances on the other is available for constructing such blends, which like all blends can have their own elaborate emergent structure and material anchors.

Using the identical patterns of blending, teachers coming into a classroom notice the "absent" students. Those students exist; they are not dead; they are present somewhere in the world. But they are also "absent" in the classroom by virtue of a blend that brings in the counterfactual space in which they are in class as expected. These blends, which compress the counterfactual outer-space vital relation into a property (absence) in the blend, have the same general structure as blending networks that let us contact the dead or that let us conceive of caffeine headaches that could have been avoided by having a cup of coffee. In other words, just as we have the nonthings, nonevents, and nonactions described in Chapter 11, we have nonpeople.

The pointing techniques of gesture and signed languages that we mentioned in Chapter 10 are equally available for these networks of nonpeople. We can designate the nonperson by pointing to where the actual person sat last time we met. Empty chairs and phones that keep ringing, like graves, photos, and locks of hair, provide material anchors for missing people. We can feel anger at the absent student in the empty chair, or curse the nonperson on the other end of the phone. In signed languages, through referential shift (see Chapter 10), the speaker can become a material anchor for an absent person being talked about.

The example, earlier in this chapter, of Webster and his sister is a more noticeable case of nonpeople playing some important role in assessing reality: The sister doesn't exist, she was never Webster's sister, and she is extremely lucky. And she is important to our understanding of Webster and Victoria. Similarly, in the Debate with Kant, the modern philosopher implicitly invites us to assess his intellect according to his performance in debating the missing Kant.

The Debate with Kant is an integration network in which Kant, who is not in the space with the modern philosopher, is present in the blend where the debate takes place. But the language for evoking this blend is so conventional that it does not point explicitly to the construction of the blend. By contrast, there are linguistic instruments whose purpose is to draw attention to the blend. We have seen one such instrument, the word "ghost," which is used in saying that *Great American II* is "barely maintaining its lead over the ghost of *Northern Light,*" and in saying that the American Pronghorn runs as fast as it does because it is being chased by the "ghosts of predators past." Todd Oakley discusses an elaborate example of this usage of "ghost" to point to the blend, taken from *Maus II, a Survivor's Tale,* by Art Spiegelman.

[Art is talking to his wife Françoise about his brother Richieu, whom he never met.]

> Art: I wonder if Richieu and I would get along if he was still alive.
> Françoise: Your brother?
> Art: My *Ghost-Brother,* since he got killed before I was born. He was only five or six. I didn't think much of him when I was growing up. He was mainly a large, blurry photograph hanging in my parents' bedroom. The photo never threw tantrums or got in any kind of trouble. It was an ideal kid, and I was a pain in the ass. I couldn't compete. They didn't talk about Richieu, but that photo was a kind of reproach. He'd have become a doctor and married a wealthy Jewish girl . . . the creep. But at least we could've made him go deal with Vladek. It's spooky having sibling rivalry with a snapshot.

The missing brother is the most important person in Art's life, and perhaps in his parents' lives. The present reality and the people who inhabit it are assessed in relation to that missing person.

Missing persons come in all types. Most of our examples stress the sorrow caused by the absence, but these people can bring honor, will, and satisfaction as well. Consider just one case, in which a skipper is trying to win the America's Cup for his nation. America's Cup racing now has an officially designated spot for "the seventeenth sailor," a passenger who rides in the boat but is forbidden to help sail it. Suppose the competing nation has a legendary skipper, Horatio, now deceased, who several times came very close to winning the America's Cup. This time, the skipper wins, and when he is asked at the press conference how he did it, he says, "My seventeenth man today was Horatio." In this case, there is a mirror network, where in each input there is a skipper from the same nation trying to win the America's Cup. In the blend, these skippers are not competitors but allies. The modern skipper becomes the successor and redeemer of his mentor. The Time and Space outer-space vital relations are compressed by superposition, and the outer-space vital relation of analogy is strengthened by the successor relation, and that successor relation between the input spaces is compressed in the blend into the inner-space role *collaborator.* The blend has impressive emergent structure of many sorts, not the least in now having a seventeenth sailor who, instead of being a mere passenger, is the most effective member of the crew. The diffuse chain of events over many years in which the country attempted repeatedly to win with different boats and different crews and different skippers, racing in different places, and finally did win, is now compressed into a single human-scale race and a glorious victory brought about by the collaboration between a living skipper and a Ghost Skipper.

DRAMA CONNECTORS

Dramatic performances are deliberate blends of a living person with an identity. They give us a living person in one input and a different living person, an actor, in another. The person on stage is a blend of these two. The character portrayed may of course be entirely fictional, but there is still a space, a fictional one, in which that person is alive. In the blend, the person sounds and moves like the actor and is where the actor is, but the actor in her performance tries to accept projections from the character portrayed, and so modifies her language, appearance, dress, attitudes, and gestures. For the spectator, the perceived living, moving, and speaking body is a supreme material anchor. The outer-space relation is one of Representation. Typically, Representation is supported by outer-space Analogy, so that, for example, a middle-aged female character will be played by a middle-aged female actress. In the blend, these outer-space relations are compressed into uniqueness.

In principle, actors are linked to characters by virtue of performing in the real world actions that share physical properties with actions performed by the characters in a represented world. This allows us, as Erving Goffman observes, to be aware, when we see a play, of more than one framing. While we perceive a single scene, we are simultaneously aware of the actor moving and talking on a stage in front of an audience, and of the corresponding character moving and talking within the represented story world. Common to the two frames are some language and action patterns. Movies are technologically more complex, but again we can be aware simultaneously of the character and the actor. In *Gone with the Wind*, we see Rhett Butler and also know that we are seeing Clark Gable. We interpret Rhett's actions as part of the story, and the very same actions, attributed to Clark Gable, as part of movie making. In this sense, in each particular theatrical representation, there will be rich shared generic structure between "reality" and "fiction."

The spectator can decompress the blend to recognize outer-space relations between these input spaces, as when we notice that the actor has not quite got the accent right or Hamlet trips over the stage lights. But the power of drama does not come from these outer-space connections: We do not go to a performance of *Hamlet* in order to measure the similarity between the actor and a historical prince of Denmark. The power comes from the integration in the blend. The spectator is able to live in the blend, looking directly on its reality. Goffman points out that in extreme cases, spectators have been known to lose the framing of themselves as spectators and of the actors as actors, and have rushed onstage to stop the murder or had a heart attack when the heroine is hacked to bits.

Dramatic performances can proliferate spaces and connections in the network. For example, there may be an additional space of historical reality that has relations to the story of the drama, as when the eponymous character in Shakespeare's *King Henry V* is connected to a historical King Henry V, and is also connected to the actor Laurence Olivier. A man can play his great-grandfather in a reenactment of a Civil War battle. In a cameo appearance, someone can play himself. The possibilities are many.

Experiencing a dramatic performance requires further complex blends. We will not look at these in detail, but they have remarkable properties. Perhaps most notably, the spectator will live in the blend only by selective projection: Many aspects of her existence (such as sitting in a seat, next to other people, in the dark), although independently available to her, are not to be projected to the blend. Her normal animacy and agency, her motor powers and her power of speech, her responsibility to act in response to what she sees, must all be inhibited. The actor, meanwhile, is engaged in a different kind of blend, one in which his motor patterns and power of speech come directly into play, but not his free will or his foreknowledge of the outcome. In the blend, he says just what the character says and is surprised night after night by the same events. The toddler engaged in pretend play and the parent watching have yet different projections and experiences.

The importance and power of living in the blend would be hard to overestimate. In some cases, biology has arranged for us to live in the blend, as when we perceive a blue cup as obviously blue, and its blueness as the cause of our seeing it as blue. We also live in the blend when we use watches and gauges and complex numbers, but in these cases the blend is a product of cultural evolution, and the inputs and their outer-space relations are much more accessible. In drama, the ability to live in the blend provides the motive for the entire activity. These are advanced double-scope blends, perfectly natural for human beings but apparently unavailable to other species. In principle, there is no reason that bonobos and even jackals could not put on plays in their groups. They can perform actions, they have memories, they have organized goal-driven behaviors, and they have elaborate social structures and social needs that could be served by dramatic performances. But they do not have double-scope blending, so the possibility never arises.

Thirteen

CATEGORY METAMORPHOSIS

> The imaginary number is a fine and wonderful recourse of the divine spirit, almost an amphibian between being and not being.
>
> —*Gottfried Leibniz*

WE FREQUENTLY ORGANIZE NEW material by extending a conventional category. Usually, these category extensions are provisional. A handout for an academic talk had one column with elements listed 1 through 7, and another column with elements listed A through F. During the question period, people began referring unselfconsciously to "Number E." The inputs to this blend are the counting numbers and the alphabet, each ordered in its customary fashion. The generic space has only a well-ordered ordinal sequence. It defines the counterparts in the two inputs. The blend has the well-ordered ordinal sequence, but also has, linked to it and thus to each other, two paired sets of counting numbers, one of which is the "real" counting numbers and the other of which is the letters of the alphabet. But the blend does not have, for example, arithmetic properties from the input space with counting numbers, or spelling from the alphabet.

SAME-SEX MARRIAGE

In other cases, blending may lead to a permanent category change. Consider the phrase "same-sex marriage," which we discussed in Chapter 7. Expressions with this syntactic form—a noun from one conceptual space and a modifier from a different conceptual space—can be systematically used to trigger blends. For "same-sex marriage," the inputs are the traditional scenario of marriage, on the one hand, and an alternative domestic scenario involving two people of the same sex, on the other. The cross-space mapping may link typical elements such as partners, common dwellings, commitment, love, sex. Selective projection then recruits additional structure from each input. For example, social recognition, wedding ceremonies, and mode of taxation are projected from the input of "traditional marriage," while same sex, absence of biologically common children, and culturally defined roles of the partners are projected from the other input. As illustrated in Figure 13.1, properties of this new social structure emerge in the blend.

Such a category metamorphosis can change fundamentally the structure of the category. Traditional marriage may, for many or most, have had criterial attributes such as heterosexual union for the sake of progeneration. That structure would be lacking in a new category of marriage that includes same-sex union. Other criterial attributes would structure that new category.

COMPLEX NUMBERS

The history of science, and of mathematics and physics in particular, is rich in conceptual shifts. It is customary to speak of models either replacing or extending previous models, but the pervasiveness and importance of blending has been underestimated.

We have already seen in Chapter 11 how the category *number* changed to include zero and fractions, and, in the case of fractions, how complicated was the blending that produced the new version of that category. After the fact, it looks as if new elements have simply been added to the old ones, because we still use the same words for them. But in fact, in the metamorphosis of the category, the entire structure and organizing principles have been dramatically altered. It is an illusion that the old input is simply transferred wholesale as a subset of the new category. As we saw for same-sex marriage, the metamorphosis of the category can change it fundamentally.

Consider again the development of complex numbers. Historically, the mathematical conceptual development in which complex numbers became endowed with angles and magnitudes was slow and fraught with difficulty. Square roots of negative numbers had shown up in the formulas of sixteenth-century mathematicians who correctly formulated operations on these numbers. But the very mathematicians who did this, Cardan and especially Bombelli, also felt that such numbers were "useless," "sophistic," "impossible," and "imaginary." Descartes held the same opinion a century later. Leibniz said no harm came of using them, and Euler thought them impossible but nevertheless useful. The square roots of negative numbers had the strange property of lending themselves to formal manipulations without fitting into a mathematical conceptual system. A genuine concept of complex numbers took time to develop.

This development began from a preexisting blend of numbers and one-dimensional geometry: The points on a line had been blended with whole numbers and fractions so that each number was a point and each point was a number.

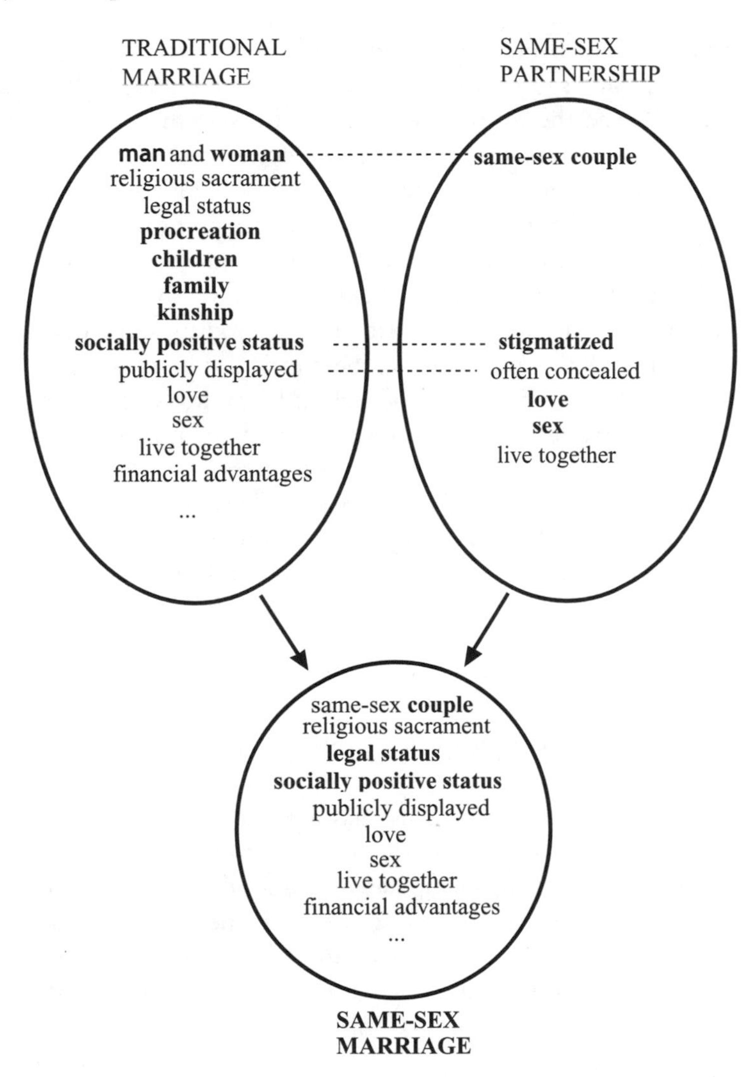

FIGURE 13.1 SAME-SEX MARRIAGE

Today, this blend is so entrenched in our culture as to seem a brute fact of physical reality, but the intellectual struggle to achieve it shows how imaginative it was.

The blend of numbers and one-dimensional space was extended by Descartes to create the coordinate plane, in which each point was defined by a pair of numbers, including negative numbers. The next step was taken by the seventeenth-century mathematician John Wallis, who observed in his 1685 book *Algebra* that if negative numbers could be mapped onto a directed line, complex numbers could be mapped onto points in a two-dimensional plane. Wallis provided geometric constructions for the counterparts of the real or complex roots of $ax^2 + bx + c = 0$. In effect, he provided a consistent model for the mysterious numbers, giving some substance to their formal manipulation. Although Wallis's mapping showed the formal consistency of a system including complex numbers, this was not enough to extend the concept of number. As Morris Kline reports, Wallis's work was ignored: It did not make mathematicians receptive to the use of such numbers. This is an interesting point in itself. Mapping a coherent space onto a conceptually incoherent one is not enough to give the incoherent space new conceptual structure. It also follows that coherent abstract structure is not enough, even in mathematics, to produce satisfactory conceptual structure: In Wallis's representation, the metric geometry provided abstract schemas for a unified interpretation of real and imaginary numbers, but this failed to persuade mathematicians to revise their domain of numbers accordingly. Only after the new conceptual structure of *complex number* develops in the blended space is the domain of numbers actually extended.

In this blend, but not in the original inputs, it is possible for an element to be simultaneously a number and a geometric point, with Cartesian coordinates (a, b) and polar coordinates (ρ, θ). In the blend, numbers have interesting general formal properties, such as

$$(a, b) + (a', b') = (a+a', b+b')$$
$$(\rho, \theta) \times (\rho', \theta') = (\rho\rho', \theta + \theta')$$

Every number in this extended sense has a real part, an imaginary part, an angle, and a magnitude. By virtue of the link of the blend to the geometric input space, the numbers can be manipulated geometrically; by virtue of the link of the blend to the input space of real numbers, the new numbers in the blend are immediately conceptualized as an extension of the old numbers.

The entire conceptual integration network has two inputs: two-dimensional geometric space and real numbers. Complex numbers and their properties emerge in the blended space. As in Wallis's scheme, the mapping from points on a line to numbers has been extended to a mapping from points in a plane to numbers. This mapping is partial from one input to the other—only one line of the plane is mapped onto the real numbers in the other input—but it is total from the geometric input to blend: All the points of the plane have counterpart complex numbers. And this in turn allows the blend to incorporate the full structure of the geometric input space (see Figure 13.2).

When a rich blended space of this sort is built, an abstract generic space comes along with it. Having the three spaces respectively containing points (input 1), numbers (input 2), and complex points/numbers (blend) entails a fourth space with abstract elements having the properties "common" to points and numbers. The abstract notions in this case are "operations" on elements. For numbers, the operations are addition and multiplication. For points in the

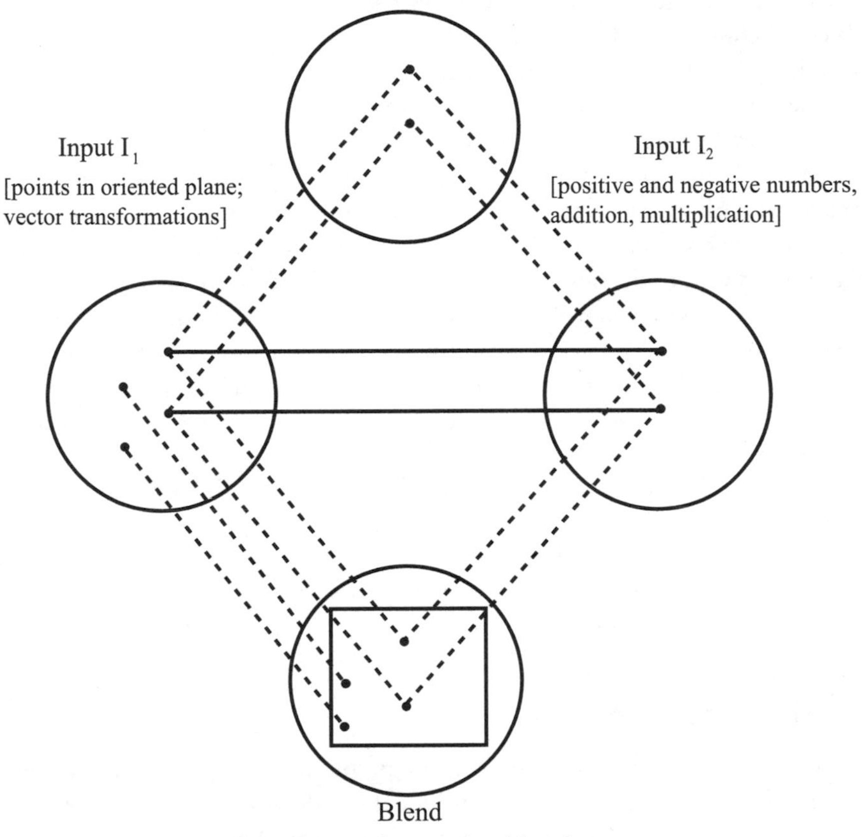

FIGURE 13.2 COMPLEX NUMBERS

plane, the operations can be viewed as geometric transformations such as rotation and stretching. In the blended space of complex numbers, numbers and vectors are the same thing; so addition of numbers is just vector addition, and multiplication of numbers is just rotation and stretching of vectors.

In the generic space of the fully completed integration network, specific geometric or number properties are absent. All that is left is the more abstract notion of two operations on pairs of elements, such that each operation is associative and commutative and has an identity element; each element has under each operation an inverse element; and one of the two operations is distributive with respect to the other. Something with this structure is called by mathematicians a "commutative ring." It is typically manipulated unconsciously by mathematicians who study geometry, arithmetic, or trigonometry until it becomes itself an object of conscious study in mathematics. In the development of complex numbers, it took roughly three centuries for mathematicians to reach that point.

The emergence of the concept of complex numbers with arguments and magnitudes displays all the properties of blending. There is an initial cross-space mapping of numbers to points in planar geometric space; a generic space; a projection of both inputs to the blend, with numbers fused with geometric points; and emergent structure by completion (arguments and magnitudes) and by elaboration (multiplication and addition reconstrued as operations on vectors).

The blend takes on a realist interpretation within mathematics. It constitutes a new and richer way to understand numbers and space. Yet it also retains its connections to the earlier conceptions provided by the input spaces. Conceptual change of this sort is not just replacement. It is the creation of more elaborate and richly connected networks of spaces.

The evolution of the concept of complex numbers highlights the deep difference between naming and conceptualizing; adding expressions like $\sqrt{-1}$ to the domain of numbers, and calling them numbers, is not enough to make them numbers conceptually, even when they fit a consistent model. This is true of category extension in general.

COMPUTER VIRUS

We can see an effort in the direction of permanent category extension in the current study of "artificial life." Consider the following remark: "Several scientists insisted that the term *virus* is more than a nice metaphor. . . . Although computer viruses are not fully alive, they embody many of the characteristics of life, and it is not hard to imagine computer viruses of the future that will be just as alive as biological viruses." This is a dramatic double-scope blend, with one input organized by the frame for a manufactured product, *computer,* and the other organized by the frame *biological virus.*

Over the last fifteen years of the twentieth century, this invention of a new category arose from two unconnected situations. In the first, hackers caused mischief by writing software code that interfered with the operation of computers. In the second, as we discussed in Chapter 6, biologists and complexity theorists wrote programs like "The Blind Watchmaker" to simulate the evolution of organisms on computers. The hacker scenario led to the initial, relatively thin blended concept of "computer virus." The cross-space mapping for this original blend was based at first on some vague shared properties of viruses and the hackers' nefarious programs:

- The element is present, but unwanted; it comes in, or is put in, from the outside; it does not naturally belong.
- The element is able to replicate; new copies of it appear that have the same undesirable properties as the original.
- The element disrupts the intended functioning of the system.
- The element is harmful to the system and, hence, to its users.

That integration network was rapidly developed to create a much richer category of *computer virus,* with associated categories like *disinfectant, vaccine, safe interface,* and *computer health maintenance providers.*

In the meantime, the modeling of biological evolution on computers was itself evolving: It went from modeling life to the more general study of the algorithms and processes that allow this modeling, collectively called "artificial life." Artificial life was a new computational concept, built upon the analogy with biological processes. Artificial life and biological life were still viewed as sharply distinct, but also as sharing some interesting and fundamental dynamics. Just as in the evolution of complex numbers the intriguing but noncategorial blend of geometric space and numbers advanced to the stage in which a richer and more integrated blend counted at last as an extension of the concept *number* itself, so in the development of the notion of artificial life the noncategorial blend of computational processes and biological life advanced to the stage in which the richer and more integrated blend was dramatically nominated as an insight into the nature of the category *life.* The blend was now offered as a discovery that *life* was other than what we had thought.

Computer virus and *artificial life* both evolved into legitimate categories of their own. The final step, envisioned by the scientists quoted above, would consist of establishing a conventional cultural and scientific megablend of *biological life, computer virus,* and *artificial life.* The generic space would characterize this blend as constituting a crucial scientific category, and the blend would give us details unimagined by Darwin or Lucretius.

WORDS AND THEIR EXTENSIONS

As the neuroscientist Antonio Damasio has pointed out, words are not different from other neurobiological elements: They are attached to networks, become activated, are connected to other networks, and so on. This means that in blending (which of course must be fully instantiated in the brain), words, like other elements, will attach to activation patterns for mental spaces and will be selectively projected to blends. We explicitly analyzed this kind of projection for "caffeine headache," "money problem," and "nicotine fit" in Chapter 11. The word "caffeine" is projected from the counterfactual input in which there was caffeine and no headache, and so it ends up being part of the linguistic label for a kind of headache that occurs in situations where there is no caffeine at all.

We can see this projection easily in "caffeine headache" because the contrast between caffeine and no caffeine jumps out at us. But in fact it is the normal course of events for the use of any word. This is why words end up "having" multiple "meanings," and why we can usually use existing words to talk about new categories. The operation of selective projection in blending, when applied to words as elements attached to input spaces, yields four principles for extending the uses of words:

1. Through selective projection, expressions applied to an input can be projected to apply to counterparts in the blend. In this way, blends harness existing words in order to express the new meanings that arise in the blend. For example, "virus" refers to something in the original input space of health, and it can be projected to refer to the relevant counterpart in the blend of computers; but crucially, what "virus" picks out in the blend is not at all what it picks out in the input from which it was projected. We can now say "I got a virus from your floppy" to pick out the structure in the blend.
2. Combinations of expressions from the inputs may be appropriate for picking out structure in the blend even if those combinations are inappropriate for the inputs. Grammatical but meaningless phrases can thus become grammatical and meaningful for the blend. For example, once we have the complex-number blend, we can say meaningfully, using preexisting words and grammatical patterns, "the square root of negative one." And once we have the same-sex marriage blend, we can say meaningfully "The brides married each other at noon." We can do this even though neither of these expressions is meaningful for the preexisting inputs.
3. We often have terms for emergent structure in the blend and can use them even if those terms cannot be applied to the inputs themselves. For example, in the Debate with Kant, we can say that "Kant has no answer," and this does indeed tell us something about the input space with Kant, even though "answer" has no application in that space.
4. Blending routinely and inevitably extends the uses of words, but we rarely notice these extensions. "Safe," for example, as we discussed in Chapters 2 and 11, has many more "surface" "meanings" than we realize.

It is important to realize that these principles, far from being sloppy, often make precise and consistent reference possible. Mathematics loses none of its rigor by using words like "number" in various ways. In some contexts, "number" picks out elements that do not have angles; in others, "number" picks out elements that do have angles. "Number" retains all of its old meanings but acquires a new one to pick out elements in the complex-number blend.

We have often seen these principles at work. For example, the discussion of "father" XYZ blends in Chapter 8 provided many extensions of the use of "father." Some of these uses pick out meanings that we see as metaphoric, or creative, or provisional, or permanent. This gradient for the word "father" falls out naturally because "father" is in each case attached to one of the inputs; blending as a conceptual operation applies to those inputs; and by principles 1 and 2, "father" comes to participate in phrases that pick out structure in the blend but not the inputs. Extending or modifying the use of a word is not a property of words *per se* but a by-product of the operation of conceptual integration and the fact that words are projected from inputs to the blend. The cognitive operation of conceptual blending is not restricted to language. But a mind that can blend and that also knows language will inevitably develop multiple meanings for words through blending. If words show up in inputs, they can be projected like any other elements of that input. This will change their domain of application, unnoticeably in most cases, but noticeably when the emergent meaning in the blend seems remarkably distant from the domain of the input from which they came. When we notice this distance, we call it by one of many names: extension, bleaching, analogy, metaphor, displacement. On our view, polysemy—that is, multiple meanings for a single word—is very common, a standard by-product of conceptual blending, but noticed only rarely.

Human beings face a fundamental problem: Conceptual systems are vast and rich and open-ended, but linguistic systems, however impressive, are relatively quite thin. How can a linguistic system be used to convey the products of conceptual systems, and how can these products find expression in language, given the stark mismatch in their respective infinities? If forms of language had to represent complete and invariant meanings, language could communicate very little. The evolutionary solution to this problem is to have systems of forms prompt for the construction of meanings that go far beyond anything like the form itself. The "of" found in a range of examples like "Paul is the father of Sally," "father of cruelty," "father of the Catholic church," "Vanity is the quicksand of reason," and "Wit is the salt of conversation" does not single out any particular blend or even any particular projection; it only prompts for finding a way to construct a conceptual network that will have a relevant meaning. What we have to do to construct that network is nowhere represented in the linguistic structure. The single word "of" is thus associated with an open infinity of mappings. Yet this infinity of mappings is anything but arbitrary. It is constrained by the requirements on conceptual integration networks. Different grammatical forms prompt different infinities of conceptual mappings.

Because linguistic expressions prompt for meanings rather than represent meanings, linguistic systems do not have to be, and in fact cannot be, analogues of conceptual systems. Prompting for meaning construction is a job they can do; representing meanings is not.

Fourteen

MULTIPLE BLENDS

Thou antic death, which laugh'st us here to scorn,
Anon, from thy insulting tyranny,
Coupled in bonds of perpetuity,
Two Talbots, winged through the lither sky,
In thy despite shall 'scape mortality.
O, thou, whose wounds become hard-favour'd death,
Speak to thy father ere thou yield thy breath!
Brave death by speaking, whether he will or no;
Imagine him a Frenchman and thy foe.
Poor boy! he smiles, methinks, as who should say,
Had death been French, then death had died today.

—*William Shakespeare*

CONCEPTUAL INTEGRATION ALWAYS INVOLVES at least four spaces: two input spaces, a generic space, and a blended space. Until now, we have mainly used this minimal template. Now, however, we turn to a more general account of conceptual integration as a dynamic operation over any number of mental spaces that moreover can apply repeatedly, its outputs becoming inputs for further blending.

In this more general scheme, we still find the defining features of conceptual integration: cross-space mappings between inputs, selective projection, generic spaces. But there need not be a single generic space for a multiple blend network. As we will see, there are two main ways in which networks can be multiple blends: Either several inputs are projected in parallel, or they are projected successively into intermediate blends, which themselves serve as inputs to further blends.

DRACULA AND HIS PATIENTS

Here is an excerpt from a newspaper editorial about former U.S. President Clinton's health care reform.

> What President Clinton did, bravely and brilliantly, I think, was to gamble that the repertory actors of the health care industry have frightened Americans so badly that we are willing to accept anything, including higher taxes, rather than to continue being extras in a medical melodrama that resembles nothing so much as an endless "Dracula" movie where the count always wins, right up to the last drop. . . . The Dracula crowd will scream "socialized medicine" and moan that you won't be able to pick your own doctor.

In this passage we find two blending networks, each of whose input spaces is connected by a metaphoric mapping. In the first network, one input space is the health care industry. It has professionals (doctors, hospital administration, health insurance agents) and the public (who are patients, pay for services, and pay for insurance). The other input space gets its structure from movie making and contrasts privileged repertory actors (who have stable employment, good pay, and arrogance perhaps) with "extras" (who are at the mercy of producers' whims, have no protection, and are exploited). The repertory actors (movie professionals) are the counterparts in this cross-space mapping of the health care professionals, and the lowly extras are the counterparts of the weary public.

The second blending network *also* has the input space of the health care industry. Its other input space has a conventional horror story, that of Count Dracula the Vampire, who sucks the blood of his victims. In the blended space, the health professionals are vampires and the patients are their victims. These health care vampires extract money/blood from their victims/patients.

Both of these blending networks feel metaphoric and have metaphoric cross-space mappings, and clearly the two metaphors are different and could be used independently: In one the doctors and HMO's are vampires, and in the other they are repertory actors.

But now we see the way that conceptual integration can operate over any conceptual array. These two metaphoric integration networks *share* an input space: the health care industry. Additionally, the other two input spaces (repertory acting and Dracula the Vampire, one of them in each of the networks) have a natural and conventional cross-space mapping: The story of Count Dracula is often told in movies, where there are actors and extras who play the parts of Dracula and his victims. This natural mapping links actors to characters, according to the Drama connectors discussed in Chapter 12. This Drama cross-space mapping is not metaphoric.

The phrase "the repertory actors of the health care industry" is a "Y of" prompt to begin to build an integration network whose inputs are the spaces of movie making and of health care. When we learn almost instantly that these actors have "frightened Americans," we include that act in the blend, but its relation to the other elements is in many ways indeterminate until we are offered

"Dracula." Then one way of accounting for the fright is clear: There is another space, with the horror story of Dracula that has a metaphoric cross-space mapping to the health care industry and a natural Drama mapping to the movie-making input. A multiple blend is now off and running. The Dracula space is now also an input to the blend, in which we have doctors/actors/vampires extracting money/privileges/blood from the patients/extras/victims.

With three mutually connected spaces for movie making, movie content, and health care, the counterpart structure is as follows:

MOVIE MAKING	MOVIE CONTENT	HEALTH CARE
repertory actors	vampires/Dracula	health care professionals
extras	victims	public

Notice that this structure maps extras onto victims even though in real movies, victims may well be played by leading actors.

All three spaces are selectively projected to a blended space in which event-participants inherit features of all three inputs. One group gets frightened, they are extras, they are the public (patients), and they get their blood sucked in more ways than one. The other group (the "Dracula" crowd) consists of repertory actors/vampires/exploiters.

How do we know this is a multiple blend rather than two intertwined metaphors? First, the mapping of actors/extras to exploiters/victims would be a very poor metaphor. Repertory actors are hardly the stereotype of bloodthirsty exploiters. But in the blend, where the force of the victimization relation is projected from the vampire space, the compatible actors/extras structure serves to add useful features and subtract others. For instance, the repertory actors are the ones you see all the time, who come back in different guises, the only ones who matter as actors. In the blend, there is a movie "establishment" just as there is a health care establishment, and the "extras" are not part of it or part of its preoccupations. In the vampire-movie space, on the other hand, victims are at the forefront of the action and of the spectator's attention, and they are important. But those features are not projected into the blend, where they would clash with the more relevant "extras" property of being unimportant, which corresponds with the intended message for the health care input that the public is being ignored.

The linguistic form employed in the passage is also characteristic of blending: "The repertory actors of the health care industry have frightened Americans so badly." Here we find vocabulary from the three inputs in a single sentence. Repertory actors, whatever their faults, do not typically frighten extras or Americans. "Frightened" is projected from the vampire space in accordance with the principles for projecting words that we presented at the end of the previous chapter. And we find in the blend the kind of opportunistic recruitment often noted for other examples:

> "The count always wins, right up to the last drop."

This formulation, in terms of the vampire input, takes advantage of the role of blood in the medical/health care input. The link, however, is not part of the relevant cross-space mapping between the vampire input and the health care space: The editorialist is not accusing the hospitals of trying to take human blood away from the patients. Rather, bloodsucking from the vampire space is conventionally mapped to deprivation of resources, money, and energy. In the blend, of course, money and blood are the same, doctors and vampires are the same, and the opportunistic recruitment works splendidly. This is not to say that distinctions between money and blood or vampires and doctors get blurred: All the links to the separate inputs remain dynamically activated. Although bloodletting is incidental topology in the health care input, it ends up being merged with central frame topology (bloodsucking) in the blend.

Call the blend B and the three input spaces M, V, and H (movie making, vampires, and health care, respectively). In the ultimate network configuration, shown in Figure 14.1, the dashed lines stand for cross-space mappings between spaces and the solid lines indicate selective projection to the blend.

This configuration has three cross-space mappings, but we have said nothing yet about generic spaces. Because there are three cross-space mappings, we can have up to three distinct generic spaces. Informally, the schematic structure G_{VH} common to V and H is the oppressor/victim schema. G_{MH}, common to M and H, is a social-status schema: "prominent, important, essential versus low status, unimportant, dispensable." The generic space G_{MV} links real actors to the characters they portray. We discussed this type of generic in Chapter 12 in the section on Drama connectors.

The outer-space vital relations of analogy and drama that link M, V, and H are compressed into uniqueness following the usual principles. The two generic spaces G_{VH} and G_{MH}, which drive the metaphors, are congruent. Though distinct, they can be seen as subcases of an even more schematic generic G of the form "more power/little or no power." Because it fits V and H on the one hand and M and H on the other, G automatically fits V and M, but this is an achieved product of the complex integration network. The Drama Mapping by itself can connect *extras* to *millionaires at a banquet*. But the elaborate Dracula multiple blend brings the Drama Mapping and therefore G_{MV} into congruence with all the other generic spaces: G_{VH}, G_{MH}, and G. Now, *victims, extras*, and *patients* must be aligned, and *vampires, repertory actors,* and *health care professionals* must be aligned.

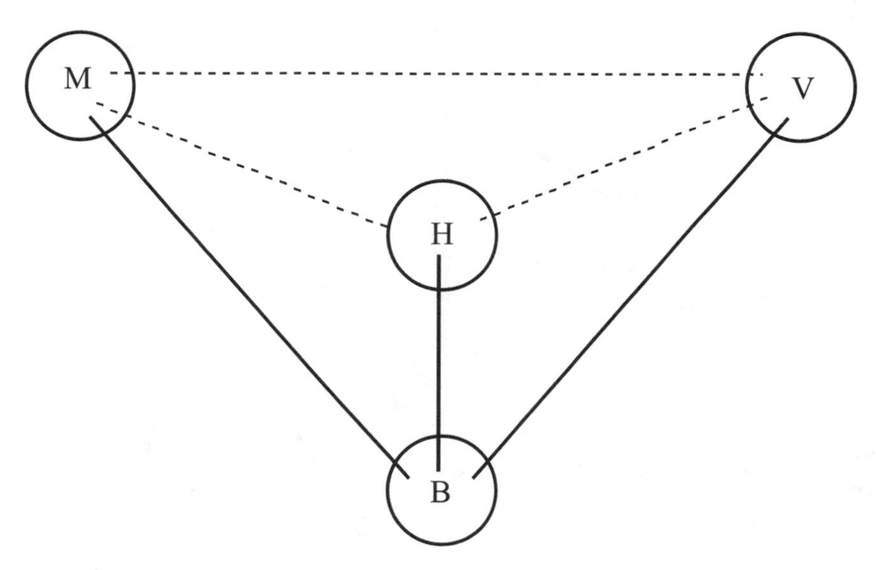

FIGURE 14.1 DRACULA NETWORK

There are additional vital relations between the inputs M, V, and H. Cause-effect links M to V because the film industry has vampire films as one of its products; intentionality links M to V because the filmmakers intend to make these films. The overall structure of the integration network is thus configured as shown in Figure 14.2.

The function of this particular network, rhetorically and conceptually, is to structure H. The writer is saying something about the health care system, not commenting on vampire movies or the movie industry. The notion of a double-scope network, introduced in Chapter 7, is naturally extended here to the notion of a multiple-scope network.

Dracula and His Patients highlights features of generic spaces not previously encountered. Although the network does include a single highly schematic generic space G, which fits all the other spaces, we need to take into account the more specific generic spaces G_{MH} and G_{VH}, which specify the relevant cross-space mappings, and the generic space $G_{MV,}$ which specifies the independent Drama Mapping between M and V. As the network develops, pressure is put on the various generic spaces to align. For example, pressure is put on G_{MV} to make *extras* the counterpart of *victims.* More generally, there is strong pressure to achieve an overall generic space having to do with cruel exploitation of lower agents by higher ones. But that generic space takes some adjustments to achieve because cruelty is not a standard part of G_{MH}: Repertory actors may be indifferent to extras but are not typically viewed as cruel to them. G_{MH} is dynamically

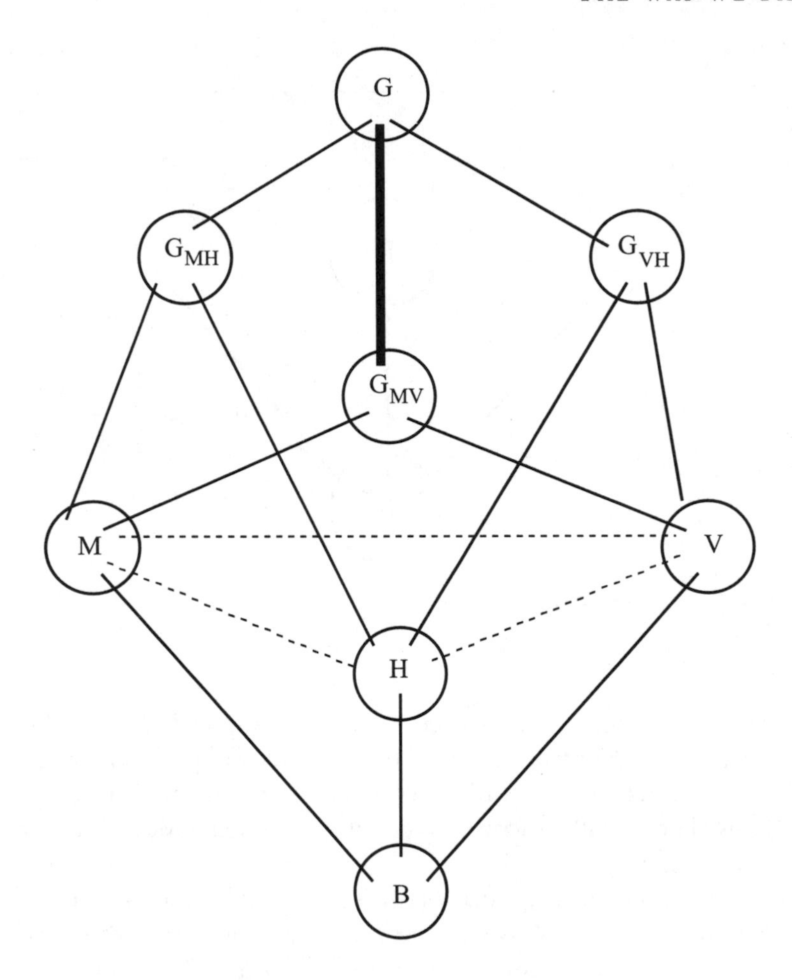

FIGURE 14.2 DRACULA NETWORK WITH GENERIC SPACES

reconstructed in the interest of achieving G. Multiple blends, then, reveal the possibility of elaborate generic space organization, and of pressures between generic spaces, within a single complex conceptual integration network.

PRESIDENT BUSH ON THIRD BASE

Consider the following four-space blend, intended as a joke made by candidate Tom Harkin during the 1992 U.S. presidential campaign:

> George Bush was born on third base and thinks he hit a triple.

One input is baseball and the other is society and one's image of one's relation to society. The generic space contains agents in competition and goals to be reached. The network is double-scope, getting frame structure from both inputs. The baseball situations come from baseball, but the possibility of not knowing how you got where you are is a projection from not understanding why you have your social station. These projections lead to emergent structure in the blend: the possibility of being deluded about how you got on base. This impossible level of stupidity in the baseball input is actually possible in the blend, but still astonishingly stupid.

Now consider this contrived development of the blend:

> The stork dropped George Bush on third base with a silver spoon in his mouth, and he thinks he hit a triple.

Several blends are linked together here to produce a megablend. The conventional multiple blend that delivers the Birth Stork is based on a conventional metaphoric blend of being born as arriving at a location, where the location is typically the world: "He came into the world fifteen years ago." Mark Twain said that he "came in with Halley's comet and would go out with it" (and he did). To get the Birth Stork multiple blend, the conventional blend of Birth as Arrival is itself used as an input, together with other inputs. One is the flight of a stork; another is the very old frame of *air travel*, which covers everything from wing-footed Hermes sent as messenger of Zeus to Aladdin's flight by magic carpet. Yet another is everyday life with a baby. The baby in the blend is on the one hand about to be born, but on the other has features of an older baby: We can see it, it can see the Stork, it smiles, it is already wearing diapers, and it can hold its head up. The inputs, most of them objectively unrelated, are smoothly integrated in the blend into a single scene: The Stork Carrying the Baby (see Figure 14.3). Though impossible in the world as we know it, the scene is easy to understand because it is at human scale and borrows structure from many very familiar scenes.

There is another blend, of life and baseball, which can be manipulated independently, as in "I never got to first base." This is used in "George Bush was born on third base."

Yet another subnetwork blends the hierarchy of stations in life with the hierarchy of dining scenarios. Having a silver spoon in one's mouth belongs to the blended space of this projection.

All three of these blended spaces (Stork Brings the Baby, Life Is Baseball, and Silverware Is Status) are then blended into a hyperblended space, in which being born, the stork, third base, and the silver spoon all reside. Inferences, motivations, and emotions constructed in this hyperblended space can then all be applied to our understanding and feelings about George Bush, his social status, and his candidacy.

This multiple blend expresses something similar to the original joke—being born in a privileged position, having little merit for being in that position, and

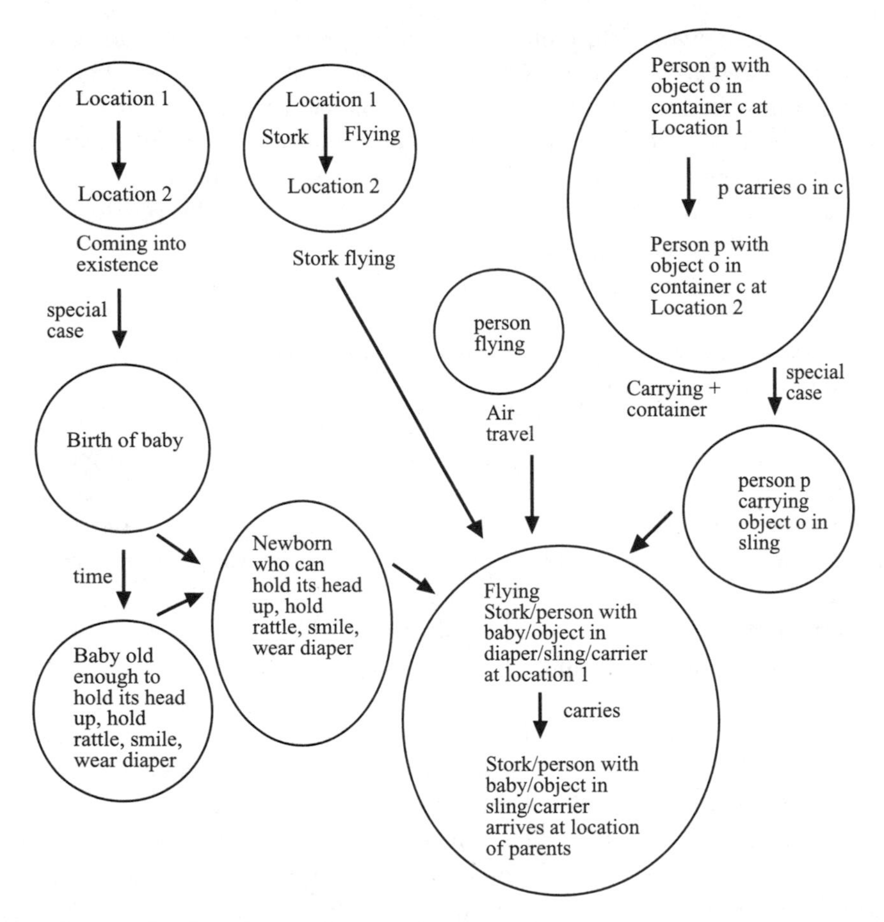

FIGURE 14.3 THE BIRTH STORK NETWORK

being deluded about one's achievements. But it achieves a form of integration that the original formulation lacked. This is because the scene of a stork dropping something on a base is well integrated, as opposed to the scene of someone being born on a base. The blend is more integrated (and not as funny) because the overall scene (however unlikely) is independently conceivable. The blend has a better match both with being born and with being on base because "being born on third base," which was left indeterminate and unexplained in the original blend, has now been framed explicitly so as to accord with both the conventional Stork Brings the Baby blend and the scene of a baseball game: The Stork now flies into the concrete spatial situation of a baseball game, and drops an object on third base, and this counts as birth, even though dropping the baby is not part of any of the three inputs.

Of course, scene integration is not the same as frame integration. The stork scenario disrupts the baseball frame in outlandish ways. Players are not carried around the bases by storks, and they do not have spoons in their mouths. We need to construct a strange and noticeable new possibility for baseball: Nearly everyone has to run the bases, but this person, quite unfairly, began from third base.

Two kinds of cross-space mappings contribute to this complex integration network. The first kind consists of metaphoric mappings—Birth as Arrival, Status as Silverware, Life as Baseball. The second kind consists of partial vital-relation connections between the target spaces of these metaphors—namely, Birth, Life, and Social Status. A Part-Whole connector links Birth to Life; a Cause-Effect connector links social status to social life; a Cause-Effect connector links birth to social status by virtue of the cultural model that assigns a newborn the social status of its parents. Identity connectors connect the person born, the person leading the life, and the person with the social status.

Independent of this multiple blend, Life, Birth, and Status are already integrated for us in a complex cultural "Story of Life." The existence of this integrated Story of Life and the consequent vital-relation connections across Life, Birth, and Status in the multiple integration network give the network a character that was missing from the multiple blend of Dracula and His Patients. There is no conventional integrated story of actors, health care professionals, and vampires that we are expected to know as part of our culture, but there is a conventional story of a person being born with a particular social status and leading a life.

How is this Story of Life handled in the multiple blend? Let us step back and observe that there is a standard rhetorical procedure for telling any complex story that is organized by such a large schematic story. For example, suppose a politician is giving us his autobiography, the story of his ascent from humble beginnings. That story has snapshot moments such as childhood industry, first taste of politics, early work in a worthy cause, and first election. For each of those static snapshots, the story commonly gives us an individual, often metaphoric integration network. So the politician might present the snapshot of childhood industry as "taking the first steps toward responsibility," early political involvement as "developing a taste" for politics, winning the first election as "gaining a seat at the table," and so on. In this rhetorical design, the overall presentation is strewn with individual stand-alone integration networks. There is no further attempt to integrate first steps, tasting, and having a seat at the table.

The Bush-Stork-Baseball-Spoon integration network, by contrast, does a further integration of the apparently disparate inputs to create the coherent scene in which the stork drops the baby onto third base with the spoon in his mouth. To put together that integrated scene, blending takes opportunistic advantage of their all being structured by locations: Third base is a location, arrival is always at a location, and the silver spoon represents a scale of "higher" and "lower" "locations" on the dining register. The integrated scene of "Arriving on Third Base by Stork with a Silver Spoon in the Mouth" is coherent with the Story of Life because its spatial relations match the vital relations between

Birth, Status, and Life in ways that are licensed by conventional metaphor: States are metaphorically locations, so being in a location in life-as-baseball matches being in a state of life and matches having a certain social status; changes of states are metaphorically changes of locations, so changing from one state of life to another matches moving from one base to another in life-as-baseball; being born matches being dropped on the base (the new location); being born matches being sent into the game.

Crucially, some of the essential organizing structure of the conceptual megablend for "Arriving on Third Base by Stork with a Silver Spoon in the Mouth" comes from the vital relations in the Story of Life that connect Birth, Dining, and Life, rather than separately from structure internal to any of those spaces. In the normal unfolding of events in the blend, the stork will drop you at bat, and then, if you have merit, you will run the bases and possibly go all the way to home plate. Therefore, there is crucial *temporal* structure in the blend—being dropped by the stork must precede being on base, just as birth precedes achievement—that comes from the Story of Life rather than from the inputs: In baseball, there is no stork preceding anything, and in the stork input there is no baseball game. There is crucial *intentional* structure in the blend: If the stork drops you directly on a base, this is an intentional act by the stork to *insert you into play in the game that is also social life*. This intention on the part of the stork cannot come from baseball and dining, where there is no stork, or from the input with the stork, where his only intention is to deliver the baby to the mother.

Not surprisingly, an integration network that contains within it several smaller integration networks presents opportunities for conceptual integration of a sort we have not previously emphasized. We have now seen in this integration network compression involving vital relations between inputs in the different integration networks, blending of separate blends (life as baseball, birth as arrival by stork, status as dining) into a megablend, and the use of preexisting overarching integrations (the Story of Life) to guide some of these compressions.

What about generic spaces in multiple integration networks? The many spaces involved (baseball, life, birth, air travel, storks, arrival, status, dining) seem to share no overall generic structure but only local ones. The generic space for life and baseball has an agent trying to succeed in time over various states; the generic space for birth and arrival has an agent who is in one state and then another; the generic space for status and dining has a privilege scale and someone at a spot on that scale.

This massive integration network does, however, offer the chance to recast the Story of Life itself as progress along a path of locations. Each of the individual blends in the network happens to have a structure—sometimes rich, sometimes thin—of moving from one location to another. In life-as-baseball, that structure is obvious and complex, and getting to home plate is good. In birth as arrival by stork, there is the contrast between the arrival location and the locations along the way, and arriving is good. In the scale of dining privilege, moving up from plastic spoons to silver spoons is social success.

Arriving at this overall generic structure for the network is an imaginative achievement that requires work on various parts of the network. All of the individual blended spaces share some structure: In each of them, states are locations, and there is a change from one ranked state/location to another. This structure is therefore a generic space, call it G, for all the blended spaces and for the megablend; but, interestingly, G does not apply to the generic spaces of each of the individual integration networks. For example, this generic space G running over the blends and the megablend has actual locations, not just abstract states. But the generic spaces for life-as-baseball, birth-as-arrival, and status-as-dining do *not* have locations; they have only abstract states. So G does not apply to them. But we can make a more abstract version of G, call it G', that has only change of ranked states. Then G' applies effortlessly to the generic for life-as-baseball and the generic for status-as-dining. Does G', the space with change of ranked states, apply to the generic for birth-as-arrival? It can, but only if we develop for the birth-as-arrival generic space an emphasis on change from a worse state to a better one. This emphasis, though not part of the original birth-as-arrival integration network, can easily be fitted in: The prior state (not yet born, earlier on the path) can be worse than the final state (newborn, at the destination). By this additional work, G' can apply to all the generic spaces in the individual integration networks, and so be an overall generic for the entire multiple integration network. It then applies to all the generics, all the inputs, all the blends, and the megablend. As a corollary, it provides additional structure to the cultural Story of Life that provides vital-relation connections over the input spaces of Birth, Status, and Life.

To recapitulate, in Dracula and His Patients and the Baseball Stork we have seen for the first time how an important generic space can be built up that is not just a local generic space applying to two inputs. We will return to these generic spaces in the Zoom Out section of this chapter.

AS AN UNWANTED CHILD MYSELF . . .

A multiple blend like Dracula with His Patients can look like a three-ring circus, entertaining and impressive, but somehow marginal to real life and reasoning. Yet multiple blends are every bit as crucial in the most serious contexts. Seana Coulson reports the following excerpt from a letter, written by Lee Ezell to the editor of the *Los Angeles Times* in 1992:

> I say thanks that no Planned Parenthood Clinic was available to me in 1963, when, as a virgin teenager, I was raped and became pregnant. The state of California would have been taking advantage of me in my crisis state by offering me this

seemingly easy out. As an unwanted child myself, I decided abortion was too permanent a solution to my temporary problem.

A staggering number of spaces, many of them counterfactual blends, must be constructed if we are to make sense of this seemingly easy argument:

- The counterfactual blend in which Lee's mother did not have the child Lee (because she didn't get pregnant, or didn't have intercourse, or had an abortion . . .).
- The counterfactual blend in which Lee in 1963 was not raped and therefore did not have Julie (the child whom she had as a consequence of being raped, named elsewhere in the letter).
- The space in which it is 1992 and there are Planned Parenthood Clinics providing abortions to teenage rape victims.
- The space in which it is 1963 and Lee is pregnant and there are no Planned Parenthood Clinics and Lee considers having an abortion by some available means but decides against it.
- The counterfactual blend in which it is 1963 and there are Planned Parenthood Clinics and Lee has an abortion at one of them (the "easy out").
- The counterfactual blend in which it is 1963 and no Planned Parenthood is available but Lee still has an abortion.
- The counterfactual blend in which it is 1992 and Lee did have the abortion in 1963 and so Julie does not exist.

There are many vital relation and frame connections across these spaces that will prompt for central blends: Frame-to-Frame, Identity, Counterfactual, Change, Time, Role, and Analogy. Lee's situation in 1963 is clearly analogous to that of her mother when pregnant and to that of teenage rape victims in 1992. Lee identifies deeply with her mother, with teenage rape victims and teenage mothers, and with children in all the spaces, real or counterfactual. She takes into account the unfolding of events over time: the pre-1963 space with her pregnant mother and the 1963 space with the pregnant Lee and the 1992 space where she is a mother with Julie. It is by reasoning over these Time connections that she is able to say that abortion would have been "too permanent a solution" (one whose consequences stretch over many spaces) to "my temporary problem" (a problem that exists only in the 1963 space).

This finely interconnected array of spaces sets the stage for any number of blends that could drive decision and judgment. One such blend has as one input Lee's mother giving birth to Lee, and as the other input, Lee's giving birth to Julie. In this blend, for Lee, who is also Lee's mother, to spare Julie is to spare herself—namely, Lee. The motivations for focusing on that blend, which does drive her reasoning, are obvious.

A maximally contrasting spectacular blend, pointed out by Coulson, has the following spaces: (1) The pre-1963 space in which Lee's mother is pregnant with Lee and has Lee. (2) The counterfactual pre-1963 space Lee's mother would have preferred, in which she is not pregnant with Lee. (3) The counterfactual blend in which Lee in 1963 has an abortion.

Like the expressions "absence," "gap," "missing," and "no milk" discussed in Chapter 11, "unwanted" represents a compression of mutually counterfactual inputs (1) and (2). We blend these to create a new space, (4), in which Lee is an "unwanted" child. Blending (4) and (3) gives yet another blend, the crucial one, in which, as Coulson points out, Lee as mother aborts Lee as unwanted child. On the principle that self-elimination is dispreferred, this blend delivers the inference that the abortion of Julie would have been a bad idea.

Yet further blends are available. If we blend the 1992 space with Planned Parenthood Clinics and the pre-1963 space with Lee's mother and her "unwanted" pregnancy, we can create a counterfactual blend in which Lee's mother aborted Lee, and we can blend *that* blend with the counterfactual blend in which Lee aborts Julie, to create a final bleak blend in which there is only Lee's mother, a missing Lee, and a missing Julie.

THE GRIM REAPER

The representation of "death" as "the Grim Reaper," a sinister, skeleton-like character holding a scythe and wearing a cowl, is an integration involving complex interactions of metaphor and metonymy. The Grim Reaper arises from many spaces: (1) a space with an individual human being dying; (2) a space with an abstract pattern of causal tautology in which an event of a certain kind is caused by an abstract causal element (e.g., Death causes dying, Sleep causes sleeping, Smell causes smell, Sloth causes laziness); (3) a space containing a stereotypical human killer; and (4) a space with reapers in the scenario of harvest (see Figure 14.4). The Grim Reaper resides conceptually in none of the other input spaces. It resides instead in a blend to which we project structure from all these spaces.

At once the least surprising and most surprising space in this array is that of Causal Tautology—least surprising because it is entirely conventional to talk about "death" or "illness" or "love" as general causes that operate across human life, and hardly big news to hear that death causes you to be dead; most surprising because it is shocking to propose that wildly different ways of being dead (through lethal knifing, terminal cancer, extreme old age, automobile accident, starvation) all have a single overarching cause, Death, for which there is no referent in the world and no evidence other than the effects themselves. From the Event, we read off a Cause that is tautologically and exclusively defined in terms of the event category and is referred to by the very terms for that category. These

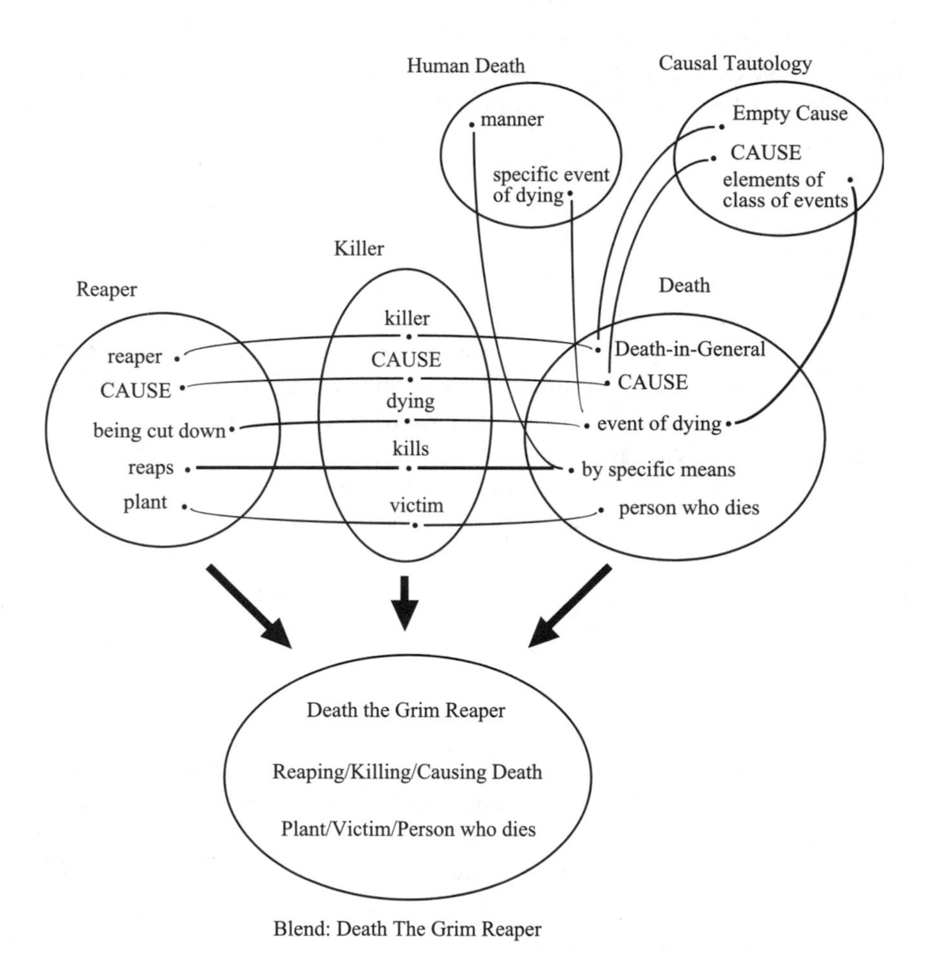

FIGURE 14.4 DEATH, THE GRIM REAPER NETWORK

causes beg the question: If we ask "What caused this death?" and receive the answer "Death," we do not think our question has been answered. We will call such begging-the-question causes "Empty Causes." The construction of empty causes is itself a blending process, in which we blend a space with an event with a space that has a cause and a caused event; in the blend, the original event now has a cause. Interestingly, the features of the cause are projected from the event: Dying is caused by Death, just as the "blueness" of the blue cup is supposedly caused by the "blue" that we create as a feature of the cup.

Not surprisingly, Empty Causes are not like other causes in our world. We think we can see most causes or hunt them down. We can locate the microbes that cause a disease or discover the biological changes that result in old age. But Death, the Empty General Cause, is not available to our perception in the usual ways.

Blending the Causal Tautology with the space for an individual dying gives us a space in which the individual event of human dying is caused by Death the Empty Cause. This is extremely conventional. If we blend again, using this conventional blend as one input and a space with a killer as the other, the result is a new blend that contains an element that is simultaneously Death the Empty Cause and a *killer*. This is personification. The actor and nonactor (*killer* and *Death*) are fused in the blend, as are the two events. There is moreover an action (*killing*) projected from the input space with the killer. Additionally, the person who dies is fused with the victim of the killing. In the blend, Death is a killer who performs an action of killing with the causal result that the victim dies.

We can also activate as an input space the scenario in which a reaper harvests grain. It is easy to connect this scenario to human life because there is already a conventional mapping between stages of plant growth and stages of human life ("He's a young sprout," "He's withering into old age"). The reaper space maps naturally onto the Death-killer blend: The grain maps to the person who dies and the reaper maps to the killer. Now, in the blend, Death-the-killer-reaper causes the death of the person-victim-plant. We could have arrived at the same blend through a different sequence, blending first the reaper scenario with the scene of death to create a blend in which Death is not a killer but instead a good farmer, and then blending that blend with a space that has a killer and a victim.

The Grim Reaper resides in the blend but cannot reside in any of the input spaces. Neither the individual event of human dying nor the input space with the *killer* contains any plants or reapers. And he is too specific to reside in the Causal Tautology space. But he is not in the input with harvests and reapers, either: The stereotype of reapers in that space is incompatible with features of The Grim Reaper.

It is easy to think that the metaphor of death as a reaper is simple and involves little more than a straightforward projection of a source-domain frame of reaping onto a target domain of dying. To see how inadequate this view would be, it is worth going over all the ways in which the emergent structure of reaping in the blend differs from the input of reaping (and *a fortiori* from all the other input spaces).

Human reapers are subject to persuasion and argument, but Death the Empty Cause is beyond persuasion. In the blend, we do not project—from the space of reapers—persuasion, argument, or any other component of human negotiation. Instead, we project the intransigence of Death.

The individual authority of any reaper is unknown: Perhaps he takes his orders from others; perhaps he is a slave. But Death the Empty Cause has authority that, when projected to the blend, yields a reaper who is his own authority. Mortal authorities can command a reaper to stop, but no mortal authority exists who can command The Grim Reaper to stop.

Actual reapers are numerous and essentially interchangeable. But Death the Empty Cause is conceived of as a single cause, which, when projected to the blend, yields a single and definite Death-the-reaper. This explains the definite article: *The* Grim Reaper.

Actual reapers are mortal and are replaced by others. But Death the Empty Cause is neither, and projecting this structure to the blend creates a Death-the-reaper that is eternal: We will be cut down by the same Grim Reaper who cut down our ancestors.

Human reapers are strong, productive, often healthy, and sometimes attractive. But the killer is destructive and unhealthy, and kills *us,* so The Grim Reaper must be unattractive, or "grim." Human reapers work in daylight, harvesting entire fields indiscriminately, indifferent to the identities of individual stalks of grain. But The Grim Reaper comes by dark, he comes for a specific person at a specific time, and he kills. He can stalk you like a killer.

In personification, the appearance and behavior of the person are chosen to express our feelings about the event being personified. If death is grim, then the personification of Death looks grim and behaves grimly.

If we look at the linguistic elements of "Death, The Grim Reaper," we see that they reflect the conceptual blending. The definite article "The" comes from the causal tautology, since it picks out a single general cause. The name "Death" comes from the blending of the causal tautology with the individual event of dying. The adjective "Grim" comes from both the space with the prototypical killer and the space with the individual event of human dying. The noun "Reaper" comes from the space of harvest.

We have described multiple blends ranging from jokes to serious argument, in such genres as newspaper columns, letters of opinion, and personifications. Some are novel, like Dracula and His Patients, while some are entrenched, like the Birth Stork. But all of them use the same operations of conceptual integration, and, perhaps surprisingly, no matter how elaborate the multiple blend, people have no difficulty constructing it. The speed and power of complex conceptual integration shown in "As an Unwanted Child Myself" looks miraculous on analysis. But this is what cognitively modern human beings do all the time. It would be wrong to think that cases of the sort analyzed in detail in this chapter are exotic or arise from exotic mental artifice. One example is the kind of newspaper editorial that we read effortlessly: It is harder to pour our morning coffee than to understand the column. Another is a heartfelt letter to a newspaper that anyone would judge to be clear, simple, and to the point. The Baseball Stork is the kind of joke that political campaigners and late-night comedians can say to a vast audience. And The Grim Reaper is a cultural commonplace for children and adults alike. Although it has taken us fourteen chapters to get to this point, the reality is that most of everyday communication involves multiple blends. Although some multiple blends draw more attention than others, that is not because they are more complex or use more exotic mental instruments. The Baseball Stork is highly noticeable, even ludicrous, but the Unwanted Child, whose multiple blending goes completely unnoticed, is arguably more complex, involving many hidden counterfactuals, acrobatic hyperblends, and cause-effect compression across mutually counterfactual spaces to produce negative elements inside blends. The most striking multiple blends of entertainers and advertisers arise from basic operations of conceptual integration that we use every day.

CHAPTER 14
ZOOM OUT

REPRISE

This chapter has focused on multiple blends.

Question:

- Haven't we seen multiple blends, at least implicitly, earlier in the book?

Our answer:

Certainly. The Computer Desktop has multiple inputs: working at a desk, choosing from lists, giving alphanumeric computer commands at a prompt, looking through windows, changing the size of a window, pointing. The basic integrated manipulation of mouse, arrow, and objects on the screen of the computer desktop is itself a spectacular multiple blend. The desktop interface activity builds on prior successive blends such as that of two very basic inputs—typing a string of symbols and giving someone a command—to create the blended concept of "giving" a computer "commands" by typing.

In fact, just about every blend we have discussed has multiple inputs. Scientific and cultural concepts are the products of successive blending over generations: Our complex numbers example has inputs of number and space which themselves result from elaborate successive blending. A striking case of multiple inputs is the Mythic Race, discussed in Chapter 7, where partial aspects of many previous races were all projected and blended into a single event. In simple mirror networks, like Regatta or the Debate with Kant, the recruitment of an existing frame (racing or debating) to organize and run the blend is technically a projection from a third input space. Blending of multiple identities into one metempsychosis—Cleopatra, Saint Barbara, Queen Elizabeth the First, your great-great-great grandmother (the diva), and Sarah Bernhardt—is of course also multiple blending. The Impotent Smoking Cowboy of Chapter 5 was the product of successive blending, and also involved the multiple blending of implicit counterfactual spaces. Other striking multiple blends we have encountered include Satan Father of Sin (Chapter 8) and the Mindful Pronghorn (Chapter 7).

MOREOVER

The analyses we have seen in this chapter look complicated.

Question:

- But, in fact, aren't these blends even more complicated than the analysis suggests?

Our answer:

Yes. There is even more going on than we have said. Dracula and His Patients has yet another important blend of politicians and gamblers ("What President Clinton did, bravely and brilliantly, I think, was to gamble"), and the excerpt we cited was taken from a newspaper column titled "Best Performance by a Politician" in which the entire political situation was blended with the nominations for the Academy Awards. The newspaper's artist moreover took advantage of the fact that the U.S. Post Office was issuing "celebrity" stamps, each with the face of a great entertainer: The illustration shows two such stamps, one for Elvis Presley, the other for "Bill," in which President Clinton is depicted as an entertainer singing into a microphone.

The Birth Stork blend additionally performs compressions on outer-space vital relations: Features of the later infant are given to the unborn fetus. When the Birth Stork is in flight, the baby has not yet been born, but in the blend the not-yet-born baby looks like an everyday infant.

In general, blends in real life are extremely rich because they are connected to many collective and individual aspects of the situation. Our purpose is not to give an exhaustive presentation of any particular integration network but, rather, to explain the cognitive operations by which they all work.

GLOBAL GENERICS

In this chapter, we have seen for the first time new kinds of generic spaces built within multiple integration networks by abstraction, modification, and elaboration.

Questions:

- Didn't we see megablends in Chapter 8? Why didn't we see this type of generic space in that chapter?

- Can these kinds of generics arise only for multiple integration networks? Can they arise in simplex, mirror, and other kinds of networks?

Our answer:

In Chapter 8, where we discussed Y^n megablends, we skipped these new kinds of generic spaces because we did not yet have the theoretical machinery to analyze them. In a Y^n network, there are of course the usual local generic spaces, built in the usual way, so that each element or relation in the generic space has a counterpart element or relation in each of the two inputs. But the interesting new issue for us here is the possibility of a global generic space for the entire network. We have now seen generic spaces like "The Story of Life" that operate over inputs and outer-space vital relations between them and have a new and very useful property: The structure of the generic space need not apply fully to each individual input. We will call such a space a "Global Generic." Each element or relation in a Global Generic will have a counterpart somewhere in the inputs or outer-space vital relations over which it operates, but not necessarily inside each of the inputs. All of the generic spaces we discussed before this chapter applied fully to each of the two inputs over which they operated. For example, the generic space for Regatta had a boat making an ocean voyage from San Francisco to Boston, and this applied to the space with *Great American II* as well as the space with *Northern Light.*

A Global Generic is an abstract integration over the important elements in the inputs and the important outer-space vital relations between those inputs. It gives an integrated global insight at a schematic level.

A Y^n network has n + 1 inputs, n blended spaces, and n local generic spaces of the usual sort. But we can now point out that it also has a Global Generic that extends over all n + 1 inputs and the outer-space vital relations between them. This Global Generic has (at least) n + 1 elements linearly ordered with a pairwise relation between any two adjacent elements. The Global Generic for "Ann is the boss of the daughter of Max" has a very abstract structure fitting any situation in which there are two simple roles, a compression of those two roles into a complex role, and three elements related to one another by those three roles. That Global Generic fits the situation in which there is the role *boss,* the role *daughter*, the complex role *boss of the daughter of* and the three elements Ann, Max, and some unnamed but specific daughter/employee. It also fits many other situations.

There is another general method of deriving potentially useful generics from integration networks. A blended space is a mental space, and we can always make a more abstract version of a mental space. Consider the blend for "This surgeon is a butcher." The blend has emergent structure, *incompetence*, which is in neither of the inputs. One very abstract generic space fitting this blend has only a person who acts. A less abstract one has an actor and something acted upon. A still less abstract space has an actor and the physical object (living or not) acted upon. A generic space derived in this manner might coincide with the local generic space over the inputs, or be more abstract, or be more specific. Or it might contain abstract structure corresponding to emergent structure in the blend, in which case it will not fit the inputs. For example, a generic for the surgeon-butcher can have a person who acts on a physical object while using methods that are appropriate for a different physical object in a different situation. That generic space does not fit the surgeon input, the butcher input, or the local generic space for the network, but it is highly compatible with the notion of an incompetent person, which is the point of the blend, and we might want to have that generic available for use in other networks. Some of the generic spaces taken from the blend but incompatible with the inputs might be independently available or even conventional. The race frame is obviously a generic for the blended space in Regatta even though it is incompatible with the 1853 space, the 1993 space, and their local generic space. The same is true for generics like *debate* or *dialogue* in the Debate with Kant.

Fifteen

MULTIPLE-SCOPE CREATIVITY

We are double-edged blades, and every time we whet our virtue the return stroke straps our vice.

—*Henry David Thoreau*

IN SIMPLEX, MIRROR, AND single-scope networks, the organizing frame of the blend is borrowed from an input or from a pre-existing pattern. But in double-scope networks we see the new and fascinating phenomenon of innovation, which is unique to cognitively modern human beings. In this chapter, we will investigate the development of new frames in double-scope networks. In the next chapter, we will describe the principles that guide the development of blends, including the development of innovative frames.

ANGER

In 1987, George Lakoff and Zoltán Kövecses did an impressive analysis of metaphoric understandings of anger, in which they revealed the mapping between folk models of heat and folk models of anger: In this mapping, a heated container maps to an angry individual, heat maps to anger, smoke and steam (signs of heat) map to signs of anger, explosion maps to uncontrolled rage. This mapping is reflected in conventional expressions: *He was steaming. She was filled with anger. I had reached the boiling point. I was fuming. He exploded. I blew my top.*

Lakoff and Kövecses also note that this metaphor is based in the folk theory of physiological effects of anger: increased body heat, blood pressure, agitation, redness of face. The Cause-Effect vital relation linking emotions to their physiological effects allows expressions like the following to refer to anger: *He gets hot under the collar. She was red with anger. I almost burst a blood vessel.*

The metaphoric mapping and the vital relation define the following kinds of correspondences:

THE STORY OF EMOTION AND BODY

HEAT INPUT	EMOTION INPUT	BODY INPUT
"physical events"	"emotions"	"physiology"
container	person	person
heat	anger	body heat
steam	sign of anger	perspiration, redness
boiling point	highest degree of emotion	
explode	show extreme anger	acute shaking, loss of physiological control, violent actions

Here we have independently manipulable spaces for the emotion of anger and bodily states. We also have a conventional cultural notion of their relationship, based on correlation: People often do get flushed and shake when they are angry. We will call this notion "The Story of Emotion and Body."

In addition to the metaphoric mapping between Heat and Emotions and the vital-relation connection between Emotions and Body, there is a third partial mapping between Heat and Body. In this mapping, steam as vapor that comes from a container connects to perspiration as liquid that comes from a container, the heat of a physical object connects to body heat, and the shaking of the container connects to the body's trembling.

The three partial mappings set the stage for a conventional multiple blend in which the counterparts in the inputs are fused, yielding, for example, a single element that is heat, anger, and body heat and a different single element that is exploding, reaching extreme anger, and beginning to shake. Once we have this blend, we can run it to develop further emergent structure, and we can recruit other information to the inputs to facilitate its development.

For example, we might say,

> He was so mad I could see smoke coming out of his ears.

This scenario derives from recruiting ears to the Body input and an orifice to the Heat input, and projecting them to the same element in the blend. We now have a new physiological reaction—smoke coming out of the ears—that is inconceivable in the original Body input. In the blend, this reaction is fused with anger. Conventional expressions like "He exploded" can also prompt for new physiological reactions in the blend that are impossible for the Body input itself. In these cases, the notion of physiological correlates of emotion is coming from "The Story of Emotion and Body" inputs, but the specific content of the physiological reaction (smoke, explosion) is coming from the Heat input. This is a multiple-scope network, with a conventional global generic space (The Story of

Emotions and Body) over two of the inputs and their vital relations, and with a systematic compression of those outer-space vital relations to uniqueness in the blend.

The blend remains linked to the inputs. A sentence like "He was so mad I could see smoke coming out of his ears" directly identifies structure in the blend, but inferences—smoke is a sign of great anger—are projected back to corresponding inferences in the Emotion input and the Body input: He was extremely angry and was showing physiological signs of it. (What these signs actually were in the actual human situation is irrelevant.)

Expressions can refer directly to the blend, as in "He exploded. I could see the smoke coming out of his ears." This description, which would be inappropriate for any of the input spaces by itself, coherently picks out the integrated scene of the blend. Additionally, even when the vocabulary is appropriate for one of the input spaces, the blend can often use it in ways that would be ungrammatical for that input: For example, suppose the chef is angry and acts it out by boiling a pressure cooker until it explodes; although "anger" and "explode" apply to this scene, and although we could say the cooker "exploded with force," we cannot say it "exploded with anger." But in the blend, where the anger is pressure and heat and force, we can indeed say "He exploded with anger."

Vocabulary from all three inputs can be combined when referring to the blend, as in "She became red with anger and finally exploded." Again, however, we could not say of a pan heated red by the angry chef that it was "red with anger."

Running the blend can produce elaborate emergent structure, as in "God, was he ever mad. I could see the smoke coming out of his ears—I thought his hat would catch fire!"

There are no burning hats in the heat input or in the anger input. Burning hats are emergent in the blend, which has the frame of somebody on fire. They imply greater heat/anger, greater loss of control, and greater danger.

In multiple-scope networks, we can often pick out patterns beyond the main types. The Anger network has the same pattern as Dracula and His Patients. This might seem surprising, since they treat entirely different conceptual domains, and one is an entrenched cultural model while the other is a novel and highly noticeable blend. But both have three input spaces and a set of vital relations between two of those input spaces. Those vital relations and the two spaces they connect are instances of a Global Generic ("Horror Movie" for Dracula, "The Story of Emotion and Body" for Anger). The vital relations in both cases include Identity connectors. In addition, Dracula has Drama connectors and Anger has Cause-Effect connectors. A superficial difference between the two networks, however, is that in Dracula the topic of the network is the space that is not under the Global Generic ("Health Care" is not under "Horror Movie"), while in Anger the topic of the network is one of the spaces under the

Global Generic ("Emotion" is under "The Story of Emotion and Body"). These parallel networks are illustrated in Figure 15.1.

In some cases—such as a cartoon in which Elmer Fudd turns red, beginning at his feet and climbing thermometer-like to his head, at which point smoke blows out his ears—we consciously set up a world to instantiate the blend, a world in which angry people really do have smoke coming out of their ears. This is like the Regatta network in which a "ghost ship" could (but did not have to) be explicitly brought in. In most cases, however, no such world is set up. The Anger multiple blend is used routinely without any construction of a fantastic world. The cartoons that set up such worlds are merely exploiting the already existing blend. In the same way for the Regatta example, we can say "*Great-American II* is now 4.5 days ahead of *Northern Light*," which prompts for construction of the blend but does not set up a world with ghost ships. The operation of the mappings is the same, but creating a world to go with it makes the blend much more salient to us.

Building a world forces us to make distinctions between structure in the blend that is inferential for the inputs and structure that is not. If Bugs Bunny blows away the smoke in the Elmer Fudd cartoon because he does not like smoke, there may be no inference for the anger input. But one might perhaps say: "I was able to blow the smoke away," with the implication that "I" was able to calm down the angry person. Under that interpretation, if there is no longer any smoke, then there is no fire either, hence no anger. But this is wrong in the heat input and in the everyday world: Blowing away smoke coming out of the heated pot does nothing to the level of heat. In the emotion input, on the contrary, it is assumed that if the signs of anger have disappeared, then the anger has too, and this is the rule that is projected to the blend. The blend violates laws of physics valid in the heat input by taking some of its structure from inputs of emotion and body. This inversion of causal structure is like the one observed in the grave-digging example in Chapter 7, where, in the blend, digging the grave causes death.

THE RETURN OF THE GRIM REAPER

As in the Anger blend, there are compressions in The Grim Reaper network of Cause-Effect vital relations: Death the Empty Cause has many effects far downstream, such as existence of the skeleton and the ceremonies surrounding burial. In the blend, some of these Cause-Effect relations are compressed, but not all the way to uniqueness. For example, the many-step Cause-Effect chain from Death the Empty Cause to the specific death of the specific human being to the corpse to the burial to the decay of the flesh underground to the exhumation to the visible skeleton is compressed to the tightest of vital relations next to Uniqueness—namely, Part-Whole: In the blend, the skeleton is the body of the Grim Reaper.

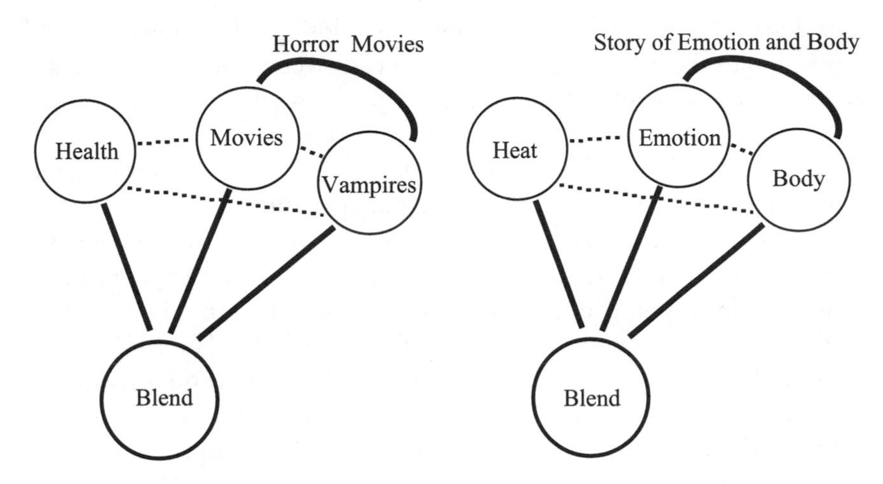

FIGURE 15.1 DRACULA NETWORK VERSUS ANGER NETWORK

Similarly, the many-step chain from Death the Empty Cause to the cowls and robes worn by spiritual attendants at the moment of dying or burial is compressed into another tight vital relation: The cowl and robe are now also an inseparable part of the Grim Reaper.

Vital-relation compression is one of the most powerful governing principles of conceptual integration. The multiple-scope blends of Anger and The Grim Reaper confirm the centrality of vital-relation compression, which we will examine from a general theoretical perspective in the next chapter.

THE GRIM PRINTING PRESS

Elements in inputs are related to each other and, of course, have relations to other knowledge in long-term memory. For example, a newspaper company is related to the published newspaper (its product), the building (its location), and its publicly traded shares. These connections motivate expressions like "The newspaper is on Main Street," "The newspaper was sold for fifty million dollars," and "The newspaper decided to hound the mayor," where the same word, "newspaper," can be used for buildings, companies, and people.

Blends use these connections in creative ways. Consider a cartoon representing a powerful newspaper company about to complete a hostile takeover of a weaker automobile company that it will then eliminate by selling off its assets. The cartoon shows a giant printing press smashing a car. This blend involves metaphor: Input 1 has the stronger and weaker objects, input 2 has the contest between companies, and the cross-space mapping is the basic metaphor that connects stronger objects destroying weaker objects to winning and losing. The strong heavy object is mapped onto the powerful newspaper company; the weaker object is mapped onto the weaker automobile company. In the blend, we find the printing press as the strong heavy object and the car as the weak object.

This is an efficient exploitation of relations: The printing press is a salient instrument of producing newspapers, and cars are the salient products of automobile companies. In the input, the printing press is not an instrument of destruction, but its force dynamics can be associated with a car-smashing machine of the sort used in recycling automobiles. In the blend, the printing press is fused with both the company and the car-smashing machine. The result is a new frame for the blend, in which the stronger object intentionally *defeats* the weaker object. This frame does not fit the input with stronger and weaker object, where there is no intentionality; nor does it fit the input with the businesses, where the competition is not between objects.

What is going on here? The blend must achieve three goals. First, given that the cartoon is a visual medium, the blend must contain some concrete elements that can be represented visually. Second, it must fit the frame of stronger and weaker object. Third, these stronger and weaker objects in the blend must be properly connected to the companies in input 2. These companies, being abstract, cannot in themselves provide the concrete elements needed for the blend. The weaker and stronger objects in input 1 are concrete but not specific and visually representable. But we can exploit relations in the inputs to make the elements in the blend adequate. The printing press and the car are concrete, specific things, associated with the companies, that can be fit into the frame of the stronger object destroying the weaker one. They fit this frame in part because the printing press intrinsically has force-dynamic structure capable of destruction and in part because we are familiar with car-smashing machines. In the blend, two elements are simultaneously (a) two concrete, specific objects; (b) a stronger object destroying a weaker object; and (c) two companies.

Such a blend is clearly creative. The printing press and car have relations in the blend (the press crushes and the car is crushed) that their counterparts in the business input do not have (the press is an instrument of making newspapers and the car is a salient product of the automobile company). The printing press and car in the business input have no counterparts in the input with the stronger and weaker object. Interestingly, the elements that did not project their input-relations (printing press and newspaper) end up being the only objects in the blend.

In the cartoon, the integration of the blend and the match between the relations in the blend and the relations in the inputs are maximized by recruiting special internal connections from the business input. Because the topologies of strong and weak object on the one hand and competing companies on the other will match only at a very abstract level, we find that in addition to the companies, objects closely connected to them are projected to the blend in a way that elaborates the relations in the input with the stronger and weaker objects.

This blend structure is double-scope. The relation of stronger and weaker object comes from the first input, but the intentionality (the printing press intends to crush the car and the car hates it) comes from the business input, where it is attached not to the printing press and the automobile but rather to the respective companies.

What we see in the blend is a tight compression of the inner-space vital relations in the business input. In the business input, the printing press is connected to the newspaper company by part-whole and means and intentionality, and the car is similarly connected to the automobile company by cause-effect (producer-product) and intentionality. In the blend, these inner-space vital relations are compressed all the way into uniqueness: The printing press is the newspaper company and it intends to smash the car, which is the automobile company and which does not want to be smashed.

This analysis shows that conceptual projection is a dynamic process that cannot be adequately represented by a static drawing. Once the conceptual projection is achieved, it may look as if the printing press has always corresponded to the stronger object and the car to the weaker. But in the cross-space mapping, the printing press and the car play no role; they have no counterparts in the input with stronger and weaker object. Rather, the cross-space counterparts are stronger object and newspaper company, weaker object and automobile company. Vital relations in the business input are compressed into uniqueness in the blend: The newspaper company as agent and the printing press as the causal instrument used by the company to create the salient effect are compressed into one element, and the automobile company as agent and the automobile as the salient effect it produces are compressed into one element. The first is also the stronger object and the second the weaker object.

Suppose the cartoon is more elaborate and shows the newspaper magnate operating the printing press to smash the car, which is being driven by the car magnate. The blend structure now contains an adversaries-with-instruments frame in which adversaries fight with opposing instruments, and in which the winner has the superior instrument. Now the printing press and car in the business input have counterparts in the adversaries-with-instruments frame: In the business input, the printing press is a symbol of a capacity for productivity that is an instrument of corporate competition, and the car is a product that is an instrument of corporate competition. The relation between the opposing instruments in the blend now matches the relation between the opposing instruments in the adversaries-with-instruments frame. This frame has the useful property of aligning superiority of instrument with superiority of adversary. Exploiting relations in the business input makes it possible to recruit a frame that increases matches between the relations in the inputs and relations in the blend.

TWO-EDGED SWORD

The phrase "two-edged sword" denotes an argument or strategy that is risky because it may simultaneously help and hurt the user. In the domain of literal swords, two-edged swords are superb weapons, better for stabbing since both edges cut and better for slashing since both edges slash. Their superior performance explains their development and deployment despite the relative difficulty of manufacturing and maintaining them. But a two-edged sword in the blend is quite different: One edge of the sword/argument exclusively helps the user and the other edge exclusively hurts the user. It is not impossible for a literal warrior to be hurt by his own blade, but it is rare, and even then, it doesn't happen that one of the edges always hurts the user while the other always helps. If it did, the two-edged sword would be discarded for the one-edged model.

In this blend, the sword input has two elements that are distinguishable and spatially symmetric, the argument input has two elements that are distinguishable and opposed on intentionality, the generic has two distinguishable symmetric elements, and the blend has two symmetric elements opposed on intentionality. This pattern runs across several similar examples: "The other side of the coin is . . . ," "The flip side is . . . ," "This is a two-way street. . . . " This pattern is so common we hardly notice it. We use it to call opposed political parties "right" and "left." The blends in all these cases have a useful generic space that is incompatible with the local generic space connecting the two inputs. That useful generic space has bilaterally symmetric elements, completely matched spatially but bearing a relationship of intentional opposition.

TRASHCAN BASKETBALL

Seana Coulson asks us to imagine a college dormitory with a wastebasket into which students place or throw their failed efforts—an activity that could easily develop into a game of trashcan basketball, in which the player gets points for throwing the "ball" to land in the "basket." This invention is a conceptual blend that projects selectively from basketball on the one hand and disposing of paper into a wastebasket on the other. Figure 15.2 presents the conceptual integration network for trashcan basketball. The generic space has only agents who throw vaguely spherical objects into a receptacle.

Trashcan basketball does not result from an isolated mapping between a wad of paper and a basketball. Rather, it arises from a whole system of correspondences that speakers can establish between the various domains. Once initial mappings are set up between the basketball and the trash domains, any number of frames from the basketball input (such as dunks, lay-ups, hook shots, one-on-one, and team play) become available for projection into the blend. Moreover, because the physics of interacting with a piece of paper is different from

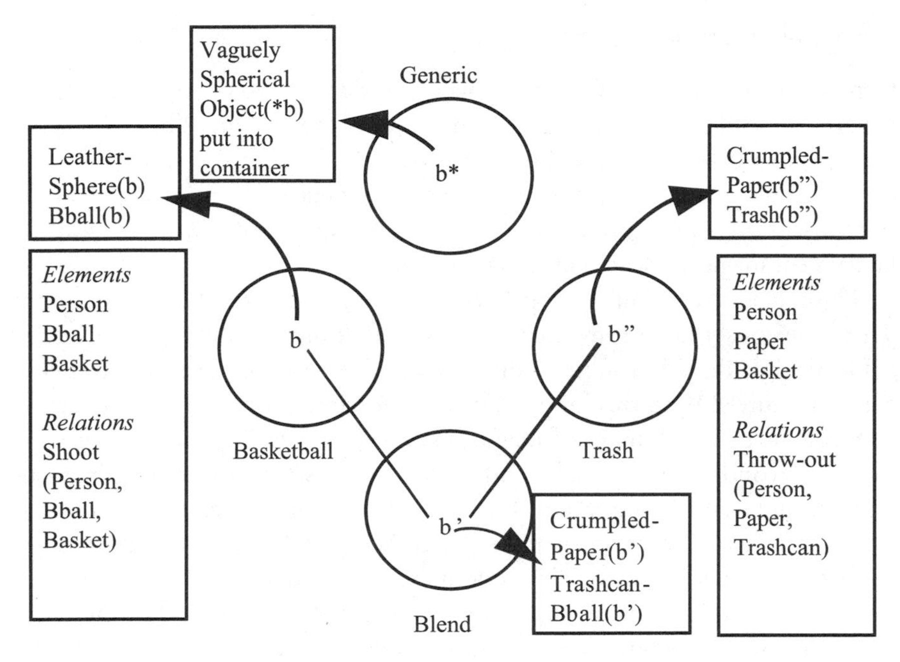

FIGURE 15.2 CONCEPTUAL INTEGRATION NETWORK FOR TRASHCAN BASKETBALL

the physics of interacting with a leather ball, participants in the game will naturally discover differences in shooting or dribbling. These differences lead to emergent structure in the blend under pressure from physical affordances—the walls, the placement of the furniture, the weight of the paper—and social forces of interaction. As the game develops, the players discover this emergent structure. They could not have predicted it in advance, much less have stipulated it in the "rules." The development of the trashcan basketball blend is like the development by mathematicians of the blend for complex numbers. Structures emergent in the blend, such as "multiplication involves addition of angles," for example, are not predictable from the inputs. More generally, crucial scientific leaps involve the discovery of powerful blends that can be run ever more to develop ever more useful emergent structures.

Trashcan basketball does not derive from finding possible local counterpart alignments in the two inputs and then selecting the strongest "match." For example, placing a piece of paper manually in the wastebasket is fine in the input with the wastebasket, and placing a basketball manually in the basketball hoop is a good action in basketball, but placing the wad of paper manually in the wastebasket is too easy to be worth much in trashcan basketball. Despite the extremely strong match between the inputs, that match is not projected to the blend.

DEATH THE MAGICIAN

George Lakoff and Mark Turner originally considered two contrasting personifications of Death, as an evil magician and as a benevolent one. The evil magician makes things disappear forever; in the blend, he is Death the evil magician, who causes people to go out of existence forever. The benevolent magician makes objects disappear and reappear as something else; in the blend, he is Death the benevolent magician, who reincarnates people.

These blends are double-scope, with structure not provided by the inputs. This is obviously so for some elements: In neither input is there a magician who is Death. But there is a subtler point: *Evil* and *benevolent* are themselves emergent structure! Death the Empty Cause is not intentional, and so not evil or benevolent. In the domain of magicians, magicians of any character can perform both kinds of magic tricks, and it might be desirable for some things, such as warts and stains, to disappear. But in the blend, the object that the magician works on is "us." Our permanent disappearance is undesirable, and the return of a person who disappears or dies is desirable. Consequently, the magician who performs the undesirable act counts as evil, and the magician in the other blend, who performs the desirable act, counts as benevolent. In the case of death as eternal, we are left ultimately with an input space that has an evil magician who specializes in one kind of trick; in the case of death as reincarnation, we are left ultimately with a different input space that has a benevolent magician who specializes in a different kind of trick. The construal of the magician in these inputs as evil or benevolent is a backward projection of inferences developed in the blend. In the blend, we compress the desirability or undesirability of the Effect with the character of the Cause.

Sixteen

CONSTITUTIVE AND GOVERNING PRINCIPLES

THROUGHOUT THIS BOOK, WE have explored the ways in which human beings use conceptual integration to create rich and diverse conceptual worlds with such features as sexual fantasies, grammar, complex numbers, personal identity, redemption, and lottery depression. Such a panorama of wildly different human ideas and behaviors raises a question: Does anything go? And if anything goes, is a theory of conceptual integration anything more than a theoretical statement that anything goes?

We point again to the analogy with evolutionary biology. The world of organisms is so rich and diverse that it repeatedly surprises the most sophisticated scientists. Every year, evolutionary phenomena are discovered that escaped even advanced prediction. Now, with genetic engineering, we have, in full conformity with the laws of evolutionary biology, human ears growing from the back of mice, eyes on the legs of grasshoppers. . . . Does anything go?

You might think so. A few decades ago, if we had speculated about fundamental constraints on phenotypes, we might have come up with hypotheses based on overwhelming statistical evidence that, for example, eyes must go on the head. But now we see that eyes can go on legs quite easily. Does this mean that evolutionary biologists have thrown out the constraints of natural and sexual selection?

On the contrary, these monstrous or spectacular deviations show the great power of the theory, and were in fact manufactured by following its guidance. Evolutionary biologists still know that for advanced organisms almost every mutation is lethal. Nonetheless, out of many, many, many new combinations, almost all failures, come enough successes to give us a world of beetles and roses, glowworms and titmice, viruses and naked mole rats.

In crucial respects, the construction of meaning is like the evolution of species. It has coherent principles that operate all the time in an extremely rich mental and cultural world. Many, many, many new integrations are attempted and explored in an individual's backstage cognition, and in interchange by members of a culture, and most of them never go anywhere. But enough survive to provide all the languages, rituals, and innovations we see around us. We need to explore what makes for success versus failure in conceptual integration.

A theory of human cognitive powers must not only account for the richness and variety of human innovation but also show how that innovation is guided. We now turn to these guiding constraints, the constitutive and governing principles of conceptual integration.

In this book we have repeatedly come across the structural and dynamic principles of conceptual integration—partial cross-space mappings, selective projection to the blend, development of emergent structure in the blend. These principles already give a rich and intricate order to the operation. We will call these the *constitutive principles* of conceptual integration. Constitutive principles, in themselves, place extremely strong constraints on any social, cognitive, or physical activity.

For example, in the spirit of John Searle in *Speech Acts*, consider football and language. What counts as football, as opposed to shopping or driving a car, is what conforms to its constitutive rules. In order to be playing football, one must be on the right kind of field, with the right number of players on each side, none of them in forbidden positions, using a certain kind of ball and not using it in any number of ways (such as putting it up your shirt). You must go through a sequence of downs, each with stipulated beginnings and endings. You must turn over the ball in one of the stipulated ways at the end of your four downs, but of course your opponents have a very limited number of ways in which they can take the ball from you before that. You cannot remove your opponent with a spin-kick, as you could in karate. These are extremely forceful guiding constraints. Compared to all the things a human being might be doing at a given moment, the set of things that count as part of playing football is very small. Martian anthropologists trying to investigate the mysterious human activity of football would count it as astounding progress to have discovered the rules of the game.

But knowing the constitutive rules of football will not tell you what you will see when you go to look at any football game, and indeed there are countless events that conform to the rules of football that you will never see in any football game. The cultural development of the game concerns finding the best ways to achieve certain goals within the rules. Falling on the ball every time it is handed to you is perfectly legal, but doing so will not win football games. Sitting down to admire the performances of your opponents is legal, but in general it is a bad strategy. The emergent structure that actually develops for the game of football is far richer than the rules, and further constrains the game. The Martian anthropologist must also try to discover these structures, but unless he has found the constitutive rules and the goals that they define, his hopes are slim indeed. Nonetheless, these two levels of constraint—the *constitutive principles* and the emergent *governing principles*—still do not determine the product.

When we go to see two football teams play, we do not know what will happen, but we know a lot that won't happen and we would be horrified to find that the game was an exact replica of their previous encounter.

Such a twofold pattern of constraint is common in human behavior. Tonal music is no different from football in this respect. The jazz improvisationist, working with several other musicians, is engaged in a behavior that conforms to both principles of tonal harmony and emergent principles that govern the improvisation. Someone who can play a tune may know the principles of tonal harmony but be completely inept at improvisation, because she has not acquired the additional principles that govern that performance.

The same is true of language. The grammatical patterns and vocabulary of a language are constitutive, and those principles limit quite powerfully what can happen in the language, but speakers of the language have also developed a vast additional set of principles governing what to say when and to whom and under what circumstances. Moreover, even a complete knowledge of constitutive and governing principles does not predict what you will hear at your next lunch conversation.

In short, constitutive principles—in the case of conceptual integration, those underlying the network model—already place strong constraints on the relevant processes, but additional governing principles limit their scope much further.

We now turn to a closer examination of these governing principles. They are not all-or-nothing constraints on networks. Rather, they characterize strategies for optimizing emergent structure. Such "other things being equal" principles are called "optimality" principles. Often, satisfying one goes part way toward satisfying another, but governing principles also frequently compete with each other. This relationship is very familiar: In tennis, other things being equal, it is good to hit the ball hard, into the corners, away from the opponent, near the net, and with maximum chance of its landing in the court. Often, a very successful strategy is to follow two principles and hit the ball both into the corner and hard. But hitting the ball hard toward the corner reduces the chance that the ball will land in the court. And if the opponent is in the corner, then following the strategy of hitting the ball into the corner runs afoul of the strategy of hitting it away from the opponent. None of these strategies is constitutive for tennis: You are still playing tennis if you hit the ball softly right to your opponent. In the same way, the governing principles for conceptual integration may lead in different directions and may have lesser or greater force depending on the specific network and the overall goals.

First, we will take up the governing principles for compressing relations. One relation may be compressed into a tighter version of itself. One or more relations may be compressed into another relation. And new compressed relations may be created from scratch in the blend. We will also discuss highlights compressions and provide some overall comments on the governing principles for compression.

Second, we will turn to other governing principles, having to do with topology, pattern completion, integration, heightening of relations, maintenance of connections in networks, perspicuity of the blend, and relevance of structure in the blend for the entire network.

There is one overarching goal driving all of the principles:

- Achieve Human Scale.

The constitutive and governing principles have the effect of creating blended spaces at human scale. The most obvious human-scale situations have direct perception and action in familiar frames that are easily apprehended by human beings: An object falls, someone lifts an object, two people converse, one person goes somewhere. They typically have very few participants, direct intentionality, and immediate bodily effect and are immediately apprehended as coherent.

Once blending achieves a human-scale blend, the blend also counts as human scale, and so can participate in producing other human-scale blends, in a bootstrapping pattern that characterizes much of cultural evolution.

Achievement of a human-scale blend often requires imaginative transformations of elements and structure in an integration network as they are projected to the blend. There are several subgoals worth noting, as follows:

- Compress what is diffuse.
- Obtain global insight.
- Strengthen vital relations.
- Come up with a story.
- Go from Many to One.

COMPRESSION

We have seen time and time again that conceptual integration is a compression tool par excellence. It operates on networks of all types to create compressed blends. These blends have compressed versions of both outer-space relations that connect spaces in the network and inner-space relations that lie within the inputs. Below we review the main kinds of compression. We will look first at compressing a single vital relation.

Compression of a Single Relation

Scaling of a Single Vital Relation. Many vital relations come with a scale. An interval of time, for example, can be long or short. One of the most obvious kinds of compression is simply scaling down. In the ritual of the Baby's Ascent, one input encompasses an entire human life. That temporal interval is shortened, as it is projected to the blend, to equal the amount of time it takes to carry a baby up the stairs. An inner-space vital relation within one of the inputs is thus compressed to a tighter inner-space vital relation in the blend by using the temporal compression already provided by the other input. We see a compression of outer-space vital relations between inputs into tighter inner-space vital relations in the blend in the case of the Bypass, where an interval of decades between the inputs becomes, in the blend, the few minutes before surgery.

In the case of cause and effect, scaling can consist in shortening the causal chain from many steps to few or only one (or in the case of perception, zero, since in the blend the effect is the cause). Scaling of cause and effect can also consist in reducing the number of different types of causal event, as in "digging your own financial grave," where many different financial and social causal actions are compressed in the blend into a single repeated action. The range of effects, of kinds of effects, of causal agents, and of kinds of causal agents may be similarly compressed. Another scaling of cause and effect is to compress a diffuse or fuzzy causation into a sharp one. For Role vital relations, we saw in Chapter 8 how multiple roles (boss and daughter, or secretary and valet) get compressed in the blend into a single composite role.

For Intentionality, complex patterns with either no intentionality, little or diffuse intentionality, or many different intentionalities can be scaled along many scales to give a single, sharp, recognized intentionality (e.g., deceive, attack, question, doubt, believe). Intentionality has a scale. In the Debate with Kant, Regatta, and the Mythic Race, someone (the modern philosopher, the skipper of *Great American II*, el-Guerrouj) is aware in one input space of the participants in the other and has an intentional stance toward them. In the blended space, this intentional stance is tightened into a mutually conscious intentional interaction. El-Guerrouj in the input is merely aware of Roger Bannister in another input. But within the blended space, el-Guerrouj knows he is competing directly with Bannister and intends to make him lose. Consider a politician who blocks a foreign aid bill that would have authorized the expenditure of funds for famine relief. His detractors can say that he is "taking the food out of the mouths of starving children," or just "starving the children," or "depriving the children of food." In the input with the long political process, there is intentionality, but of a diffuse sort. The strength of the intentionality is sharply increased in the blend: The politician is now seen to be directing his action exclusively at the children with the specific purpose in mind of starving them. This is also an example of scaling causality. In the input with the politician, signing the bill is only one of many causes required for feeding the children—the food must also be bought, packaged, shipped successfully, distributed to the right people, and so on. But in the blend, there is only one cause—passing or blocking the bill—for feeding or starving the children. That cause also operates directly, in an immediate human scene with time and space intervals that are at human scale.

The vital relation of change scales very easily, permitting long, complex changes to become quick visible changes of a single object. In the Bypass, the long and complicated process of change through education is compressed into one act of instruction. In the example of the politician blocking the foreign aid bill, the long processes of political change and change in the foreign country become a single act of direct change in taking the food from the child.

Syncopation of a Single Vital Relation. As we noted in Chapter 7, another way to compress a relation into a tighter version of the same relation is through syncopation. We can compress a lifetime not only by scaling it to run very fast but also by dropping out all but a few key moments (being born, meeting Christ, being shot through with arrows, going to heaven). Scaling and syncopation often work together. In the case of the dinosaur evolving into the bird, we have scaling of change and time and number of agents and number of locations, but we also have syncopation: Only a few key moments in that continuous evolution are picked out and conjoined in the blend.

Compression of One or More Vital Relations into Another

A striking general property of blending is that it can compress one vital relation into another. Indeed, there are canonical compressions relating different vital relations. Consider the vital relations Analogy, Disanalogy, Uniqueness, and Change. Analogy is commonly compressed into Uniqueness without Change, and Disanalogy into Uniqueness with Change. We will now explore such hierarchies of compression and the work of blending to replace a vital relation with another, more compressed one.

Analogy, Disanalogy, Change, Identity, Uniqueness. The Pronghorn, for example, compresses analogical relations at various levels into Uniqueness in the blend. The various individual pronghorns over evolutionary time are analogues. The pronghorns of two supposedly distinct periods (ancient and modern) are compressed into a unique ancient pronghorn and a unique modern pronghorn. The two are analogous in many respects and disanalogous in others. The Disanalogy is compressed into a relationship of Identity + Change: We say that the ancient pronghorn "changed" into the modern pronghorn. Finally, the vital relation of Identity is compressed into Uniqueness: In the final blended space, there is only a single pronghorn. We can view Analogy, Disanalogy, Change, Identity, and Uniqueness as relations organized into a hierarchy. Identity and Change are more compressed than Analogy and Disanalogy, and Uniqueness is more compressed than Identity. Quite usefully, when conceptual blending produces a compression of a vital relation, it does not discard the uncompressed vital relation. In the complex network for the Pronghorn, the vital relations of Analogy and Disanalogy still do useful conceptual work, but the blended space gives us a very tight global insight, a platform from which to understand and manipulate the entire complex integration network. All of these vital relations are complex in themselves: The one unique pronghorn in the blend has a complex life, involving change, learning, memory, and experience, with analogies between different moments in that life. The compression yields a conception at human scale: a single animal, changing (or not changing) over the course of a single lifetime. Paradoxically, although this vignette is far more schematic than the complex story of the evolution of pronghorns over millions of years, it is also far richer in humanly meaningful vital relations: Uniqueness, Change of an individual, and Intentionality (learning, memory, intentional action).

Cause-Effect and Uniqueness. We have seen many examples where Cause-Effect is compressed into Uniqueness. The automobile company produces the automobile, but in the blend the company and the automobile are the same thing. In the Impotent Smoking Cowboy network, the cigarette is in the causal chain that leads to the shape of the organ, and in the blend there is a unique shape for the cigarette and the organ. The fundamental Cause-Effect compression in daily life is the compression of the perception and the cause of the perception. We can think about distinguishing the perception and what causes it, but in action, we fuse them in the blend.

Representation, Part-Whole, Uniqueness. The outer-space relation of Representation, connecting a representation to what it represents, can be compressed into Uniqueness in the blend. In a painting of the crucifix, spots of red paint represent the stigmata of Christ, but in the blend, the paint is the blood. A more quotidian example is the use of a photograph of a face to represent a person. Between the photo and the person there are relations of both Representation and Part-Whole, but these are compressed in the blend into Uniqueness. The police officer points at an I.D. photo and says, "Do you know this man?"

Time, Space, Identity, Memory. Our most basic understanding of time is achieved through cultural blends like the sundial, the watch, the calendar. It is customary now to represent time and notions of time by means of static diagrams in space. The Dow Jones Industrial Average, for example, is commonly presented as a graph with value as one axis and time as the other. In general, change over time—reading level achieved, amount of oil spilled from a foundering oil tanker, national debt, popularity in polls—is given graphically. Watches, sundials, and graphs are all material anchors for blends in which Time is compressed into Space.

Another way of using Space to compress Time is to exploit the fact that some locations—the Bastille, Valley Forge, Troy, Bethlehem—are associated with historical events and therefore with times or eras—1789, the American Revolutionary War, the Fall of Hector, the Birth of Christ. Consider the following passage from Norman Mailer's *Armies of the Night*, in which the author is crossing Memorial Bridge to march on the Pentagon:

> He [Mailer] was not used much more than any other American politician, litterateur, or racketeer to the sentiment that his soul was not unclean, but here, walking with Lowell and MacDonald, he felt as if he stepped through some crossing in the reaches of space between this moment, the French Revolution, and the Civil War, as if the ghosts of the Union Dead accompanied them now to the Bastille.

In this passage, Mailer constructs an elaborate blend of "revolutionary" events. The physical place he inhabits, the Memorial Bridge, which connects the Lincoln Memorial to the east with Arlington National Cemetery, once the home of Robert E. Lee, the leader of the Confederate Army, to the west, is an anchor for both his present activities and those of the Union soldiers. The fact that he is part of a mob crossing a bridge in the service of a "great cause" is an analogical anchor to the storming of the Bastille and the crossing of its drawbridge. Compressing these spaces and events entails compressing the times into one moment. In one input, the bridge, the cemetery, and Lee's home are already spatially very close to each other and close to the speaker; this makes it easier to bring the relevant times together as well. That Mailer is walking with Robert Lowell, author of "For the Union Dead," makes it easier to bring the relevant events and times together.

This use of space as a prompt to blend events, intentionality, and times is a basic cultural instrument: We visit the graves of dead relatives, heroes, and martyrs; we visit the towns where Vermeer and Shakespeare were born; we return to our alma mater; we go to chapels or churches to pray even when there is no service, and of course the graves are either in the floor of the church or in the graveyard next to the church. Part of the motivation for these visits is the sense that, if we actually inhabit them, we can more easily integrate our thinking and emotions with the people, cultures, and events associated with them, no matter how ancient. Cultures organize these compressions by designating certain places (the cemetery, the churchyard, the Vietnam Memorial) and certain dates (Memorial Day, All Saints' Day, Easter) as calling for special attention to associated compressions across times and events for the purposes of remembrance. Physical spaces are already attached by memory to sensations and events in our past. A culture does enormous additional work to load these physical spaces with material anchors for memorial purposes (gravestones, relics, plaques). Many other material anchors incidentally become prompts for memory and time compression (such as our personal effects, rooms in a house we once inhabited or inhabit now, cars we have owned).

We have a conception of time as simply a long ordered succession of events separated by "time distances," such that what happened long ago is less and less accessible. That conception of time, which is not unlike the scientific conception of time, has no obvious place for compressions. Yet we have seen again and again that compression of time is conceptually valuable and one of the human imagination's favorite tools. We suggest that there is a fundamental neural basis for these time compressions: The human brain does not for the most part organize events according to the sequence in which they happened or were recorded. Human memory is not a tape that we must rewind to get back to the desired spot. When we go to a place and remember the last time we were in that place, we do not do so by rewinding our memory through the sequence of events between now and then. Simple introspection shows that people cannot predict what thought will come to them a minute from now: Pick up a pen, stub your toe, have a drink, eat a cookie, and "out of the blue" may come a memory from any time in your past, even early childhood. Just as physical space is suffused by culture and memory with blend-prompting powers, our brains, in a very different sense but with equal powers, give us imaginative compressions of things that we know are far apart in time or space.

From an objective viewpoint of time and space, the activities of human memory are bizarre. Why do our memories work in these strange ways? One possible answer to this puzzling question is that memory and conceptual integration evolved to support each other. To do advanced conceptual integration, we need the ability to integrate and compress over inputs that are often very different and highly separated in time and space. We cannot predict which inputs will turn out to be useful, but we do know that useful inputs from many sources need to be activated simultaneously and linked by vital relations. Human memory appears to be superb both at providing simultaneous activation of quite different inputs and at offering good provisional connections between them. Apparently running on autopilot, it often delivers up inputs and connections that have no apparent reason for being activated simultaneously or being connected at all, except that they lead us to quite useful blends.

Other Relations. We have also seen cases where Cause-Effect relations in the inputs are transformed into tighter Part-Whole relations in the blend. In Death the Grim Reaper, the many-step Cause-Effect chain between Death the Empty Cause and the cowl or the skeleton is compressed into a Part-Whole relation. And various outer-space configurations compress into Category in the blend. So, an action can bring a pleasure, and this involves a Cause-Effect connection, but in the blend the action *is* a pleasure: It is an instance of a category. Similarly, something that brings pain becomes a pain itself. Something that brings benefit is a benefit. An action that results in a deception is a deception. And the product of an effort becomes an effort, as in "Here is my effort" said of a charcoal drawing. All of these and indefinitely many more are compressions of Cause-Effect into Category.

Outer-space Intentionality can compress into Category in the blend. If we remember an event, there is a mental space with the person remembering and a mental space with the event remembered. There is an Identity link between the rememberer in one space and the participant in the other. There is an Intentionality link of remembering between them. There is also a Cause-Effect link because the event is causal for the remembering. These Intentionality and Cause-Effect links are compressed in the blend into the category *memory*. We say, "I have a memory of that moment." One can show that *hope*, *desire*, and *belief* also involve compression of Intentionality.

As discussed in Chapter 11, the category *gap* is a compression of Disanalogy and Counterfactuality.

Complex arrays of vital relations can end up being compressed in the blend into the single vital relation of Property. The property *safe* in the blend compresses a complex counterfactual network, yielding expressions like "safe child," "safe jewels," and "safe distance." And the property *likely*, as Eve Sweetser shows, compresses a network of actual and hypothetical spaces, yielding expressions such as "likely candidate," as used to mean a candidate who is likely to give an interview.

Category and Property compression can combine. Consider "guilty pleasure," a phrase used to describe an action (such as eating fatty foods) that brings both pleasure and guilt and that typically brings guilt partly because it brings pleasure. The action is causally related to its results, pleasure and guilt, and the pleasure is causally related to guilt. The action is intentionally related to the pleasure but also to the guilt, since the actor expects the guilt and wittingly indulges nonetheless, perhaps even partly because the action will bring guilt. These outer-space vital relations of Cause-Effect and Intentionality are compressed in the blend: *Guilt* and *pleasure*, which were Effects in the inputs, are now in the blend respectively a Property and a Category. Everyday examples of Property compression are ubiquitous but so conventional that they go unnoticed. A "loud man," for example, is one whose actions sometimes result in our experiencing loud noises. That Cause-Effect relationship is compressed in the blend so that the man acquires an essential Property: loud. "Violent look" and "breakneck speed" are similar examples of compression into a Property.

We saw earlier that "a memory" is already a compression of Intentionality to Category. The phrase "with grateful memories" is on a plaque accompanying a gift to the Center for Advanced Study in the Behavioral Sciences from its class of 1996 Fellows. In one mental space, the person who remembers is grateful. In another is the remembered event that prompts the gratitude. There are Intentionality and Cause-Effect links between these spaces, having to do with remembering. In the blend, the Intentionality and Cause-Effect are compressed into the category *memory*, and the further Cause-Effect link having to do with being grateful is compressed into Property, so the memory itself has the Property of being grateful. This blend has the additional remarkable feature of turning the memory into an intentional gift on the part of the person being thanked. We can thus express our gratitude to someone for memories even if that person did not know us and did not notice us at the time.

Achieving Inner-Space Scalability. Individual mental spaces have inner-space vital relations that scale easily: Time, Space, Change, Similarity, Property, and Part-Whole. But in full integration networks, there are vital relations that are only outer-space and nonscalable: Representation, Analogy, Disanalogy, Identity. To achieve a human-friendly integrated blend, we often compress such outer-space relations into inner-space relations in the blend that are scalable. This is a very general principle of compression in integration networks. To achieve human-scale blends, we need to achieve scalable vital relations. Converting nonscalable outer-space relations to scalable vital relations is a general mechanism for achieving human-scale blends. Some of this compression is so conventional and entrenched that it is hard to notice. For example, we have an outer-space vital relation of Representation between a person and the label for the person, such as a name. In the blend, the name becomes a part and a property of the person. So we see that our everyday frame for thinking of a person is already a remarkable compression of outer-space relations to inner-space relations. The outer-space role-value connections between *colonel* and Oliver North or between *father* and Paul become, in the blend, property relations. We can even compress a role-value connection to a part-whole relation in the blend, as when we put on a nametag or wear a military uniform with stripes, bars, epaulets, or medals that signal our rank. Cognition is embodied, and the spectacular intellectual feats that human beings perform depend upon being able to anchor the integration networks in blends at human scale, using the vital relations that are employed in perception and action.

Creating a New Relation Through Compression

As we discussed briefly above in the passage on scaling, conceptual integration often creates a relation in the blend where there was none in either the inputs or the connections between them. Roger Bannister, for example, in the Mythic Race, although he is aware of competitors in his own input space, has no outer-space relation of Intentionality to the input with el-Guerrouj: He is not in competition with el-Guerrouj or trying to defeat him. But in the blend, Bannister is aware that he is in a competition with el-Guerrouj and aware that he loses to el-Guerrouj. There is no intentional connection in the inputs from Bannister to el-Guerrouj, but there is a tight and poignant one in the blend. In the Mindful Pronghorn, the Cause-Effect connection of adaptation is compressed into a relation of change for the individual pronghorn in the blend, who goes from being young and slow to being older and faster; and that relation of change is incorporated into a richer human scenario of learning, an intentional scenario. In the scientific account of evolution, there is a stage of adaptation where the pronghorns change over time so that later pronghorns are on average faster, and then there is a stage of inheritance where that speed is transmitted genetically down through generations. Both stages have Cause-Effect relations, the first through adaptation and the second through continued genetic transmission. In the blend, those two kinds of outer-space cause-effect vital relation are compressed into two kinds of intentionality: learning and memory. It is quite an imaginative feat that the two-stage scenario of evolution has been fit nicely into a two-stage model of life for an individual: First you learn something and can do it, and after that you can still do it because you remember how. Similarly, in the drawing of the dinosaur evolving into a bird, the dinosaur is driven to evolve by his desire to catch the dragonfly. When an effect of compression is created for the blend by constructing a new vital relation for it, we will call this "Compression by Creation."

The Highlights Compression

We have seen global generic spaces that present a "Story of Life," a "Story of Birth," or other "stories" that consist of a number of key events and participants connected by strong vital relations extending over mental spaces. Blends, as we mentioned, can provide integrated versions of these stories. Death The Grim Reaper compresses into a single scenario the various stages of the "Story of Death." That generic story contains such stages as being about to die, expiring, burial, decay of the corpse, and, much later, the visible abiding result—the skeleton. The highlights in the overarching story are compressed into simultaneous highlights in the blend. The Grim Reaper blend has the arrival of the Grim Reaper for being about to die, the scythe for the expiring, the cowl for the burial, and the skeleton for the final result.

What we see here is a general pressure for blended spaces and global generics to reflect the structure and highlights of the overarching stories. This is possible only because vital relations of one type can be compressed into vital relations of another—for example, Cause-Effect to Part-Whole in the Grim Reaper. We are now in a position to see some remarkable aspects of Time compression in a blend like the Grim Reaper. What seems like an essential structure of the overarching Story, the order of events through time, is absent from the blended space. The highlights compression transforms the temporal and causal chain into a part-whole structure where everything is simultaneous.

Borrowed Compression

Cross-cutting all these kinds of compression is the distinction between borrowing a compression that already exists in one input and compressing outer-space relations between inputs. Sometimes one of the inputs already has a tightly integrated scenario that is projected to give a tight compression in the blend. For example, in "digging your own financial grave," the *digging* input already has a compression of agent, action, time, space, and cause-effect, which is projected to the blend. Much of the frame structure of the blend comes from the *financial* input, but that structure is placed inside the compressed scenario provided by the *digging* input. The ritual of the Baby's Ascent also uses the tight integration provided by one of the inputs—carrying the baby up the stairs—to give compression to the blend.

Optimality and the Bubble Chamber of the Brain

Like other principles that we will see, those for compression are optimality principles. Because they compete among themselves and with other principles and goals, they are only partly satisfied in any network.

As we have noted, there is a useful analogy between biological evolution and the construction of new blends. Every organism faces competing values and constraints: It may be good to be powerful but also good to be fast, and good to have low nutritional requirements. But these values pull in opposing directions. Evolution does not work by somehow conceiving the optimal design of an organism for a niche and then constructing that organism. Rather, interactions happen and products result, and those products are selected for or against. A particularly fit organism for a niche might not come up simply because the mutations needed for it did not happen. Similarly, the brain can be thought of as a bubble chamber of mental spaces: New mental spaces are formed all the time out of old ones. We surmise that the brain is constantly constructing very many blends, and that only some of them are selected out for further development and application. Even fewer become available to consciousness. A "culture," which includes a large collection of brains, is an even larger bubble chamber for evolving candidate blends, testing them, discarding or cultivating them, and promoting and disseminating some of them. Yet despite the teeming activity of the brain's bubble chamber and the vastly larger bubble chamber of a culture, only some blends will come up, only some inputs will be activated, only some conceptual mutations will happen. Many could have happened in principle, but did not. We see in the history of mathematics and science, for example, that absolutely useful blends can take centuries to bubble up, even though, once available, they are easily intelligible and memorable. Moreover, in such a system, optimization is not absolute: When an organism or blend is entirely good enough for the purposes at hand, relative to other organisms or blends, it is not additionally required to be the best one *possible.* Evolution does not look through all possible worlds; it selects only from what has come up. In the same way, conceptual integration employs whatever happens to have been activated in individual brains and collections of brains, rather than reviewing everything that could have been activated. In blending, as in evolution, good enough is good enough: The system does not have infinite time or infinite facts or activations, and it is strongly biased by its present state. Once evolution has a pronghorn that outruns its predators, it doesn't have any motivation to evolve even greater speed for the pronghorn. Evolution does not come up simultaneously with the fastest possible lion and the fastest possible gazelle, but it does optimize lions and gazelles, not in any absolute sense, but relative to the available ecology and the random mutations that occur. In the same way, success in blending does not consist in meeting every possible governing principle in the strongest possible way—a logical impossibility given that the principles often conflict. In a system of many optimality principles, success is a matter not of satisfying each of them but of satisfying *all* of them *enough.*

Converting to Human Scale

At the beginning of this chapter, we said that the governing principles are driven by one overarching goal: Achieve Human Scale. Human beings are evolved and culturally supported to deal with reality at human scale—that is, through direct action and perception inside familiar frames, typically involving few participants and direct intentionality. The familiar falls into natural and comfortable ranges. Certain ranges of temporal distance, spatial proximity, intentional relation, and direct cause-effect relation are human-friendly. Other things being equal, it is good for a blend to belong to these ranges. Transforming vital relations by compression or creating new ones, properly done according to the governing principles, will make the blend more human-friendly. For example, we can deal easily with strong and simple intentionality in interaction, so adding intentional relations to a scene of interaction can help make the blend more intelligible, vivid, memorable, and useful to us as human beings. We have seen many cases of converting to human scale. The complex evolutionary story of pronghorns is converted to the human-scale story of a single animal who learns and remembers. The diffuse and intricate notion of a person's life is converted to the human-scale event of going up the stairs in the ritual of the Baby's Ascent. And the far-off and abstract connections between educating children you don't know and technical competence on which your life depends is converted to the human-scale event of children getting ready to do your bypass.

The principles of conceptual integration—constitutive and governing—have been discovered through analysis of empirical data in many domains. These principles, with all their intricacies and technical mechanisms, conspire to achieve the goal

- Achieve Human Scale

with its noteworthy subgoals:

- Compress what is diffuse.
- Obtain global insight.
- Strengthen vital relations.
- Come up with a story.
- Go from Many to One.

The fact that blends achieve impressive compressions is one we have amply discussed. We have also seen how the blended space of a complex and diffuse network could be structured by a human-scale scenario. For example, the Mindful Pronghorn achieves a blended scene at human scale in which a single pronghorn runs from predators. This scene compresses complex and diffuse structures that operate in evolutionary time. The ritual of the Baby's Ascent gives a very simple scene with very few agents and events—an adult carries a baby up the stairs—that compresses over very many events and agents involved in at least two human lives. The Impotent Smoking Cowboy presents a highly compressed blend at human scale with a single agent, event, and time—a man smoking a bent cigarette. Such blends produce impressions of global insight. It seems that the construction of the network with a blend at human scale and the appropriate connections to a complex array of mental spaces is what generates the impression of global insight. Another consequence of the principles of conceptual integration is the strengthening of vital relations, either by making new ones, or by intensifying existing ones, or by converting vital relations of one type into vital relations of another. We have also seen many cases in which the blend itself provides a simple story for the entire network—as when, in Regatta, there is a simple story of a race in the blend. Stories can also be provided by global generics of the sort we saw in discussions of the "Story of Life" and the "Story of Death." Finally, nearly every network we have seen goes from many elements in the inputs to one or few in the blend. For example, the very many agents and events involved in going bankrupt through investing become one agent performing a single action of digging repeatedly in digging one's own grave. And the millions of pronghorns involved in the evolutionary history of the American pronghorn become in the blend a single pronghorn with learning and memory.

These goals are not independent. Compression is a way to achieve human scale, and by the same token achieving human scale will produce compression. Strengthening vital relations is also part of achieving human scale, and a scenario at that scale typically involves a simple story. The human scale is the level at which it is natural for us to have the impression that we have direct, reliable, and comprehensive understanding. This is why achieving a blend at human scale will induce the feeling of global insight. The compression and scale of the blend make it cognitively more tractable to deal with, more manipulable, and since it is tied to the complex network, its manipulation gives mastery of a diffuse network, which creates a feeling of global conceptual mastery and insight. Going from many elements and relations in the network to few in the blend also helps achieve human scale, compression, and stories, because a simple human-scale scenario, with a minimal number of agents in a local spatial region and a small temporal interval, is what we are set up to engage with perceptually and behaviorally.

Governing Principles for Compression

We can now state several governing principles for compression. They all maximize compression in the network; that is, they lead to compression of vital relations as they are projected to the blend.

- *Borrowing for compression.* When one input has an existing tight coherence at human scale but the other one does not, the tight human-scale coherence can be projected to the blend with the effect that the other input is compressed as it is projected to the blend. Examples: "digging your own financial grave," "He digested the book," the ritual of the Baby's Ascent.
- *Single-relation compression by scaling.* Some inner-space or outer-space vital relations can be scaled to a more compressed version of the same vital relation in the blend. Time, Space, Change, Part-Whole, Intentionality, and Cause-Effect can all be scaled, whether from outer-space to inner-space or within a single space. Also, within a single space, Property and Similarity can be scaled. The scale of Similarity is correlated with the number of shared properties and the property scales themselves. Something can be more or less *blue*, and two things can be similar if both are blue, dissimilar if one is blue and the other red, and if both blue, more or less similar according to their blueness.
- *Single-relation compression by syncopation.* Diffuse structure in an input or across inputs can be compressed as it is projected to the blend by dropping out all but a few key elements.
- *Compression of one vital relation into another.* A relation of one type can be compressed into a relation of a different type.
- *Scalability.* The overarching goal of compression is to achieve human scale in a single mental space. In a single space, elements satisfy Uniqueness by definition, and other vital relations—Time, Space, Change, Cause-Effect, Part-Whole, Property, Similarity, and Intentionality—are scalable. Non-scalable relations—Analogy, Disanalogy, Identity, Representation—can accordingly be compressed to scalable ones.
- *Creation by compression.* Adding new vital relations to a space can help it achieve human scale. Compression can create a vital relation for the blend that is not in the inputs.

- *Highlights compression.* Distributed elements in an overarching Story can be compressed into a simultaneous arrangement in the blend by such instruments as compression to Category, compression to Property, and syncopation over detail.

Figures 16.1 and 16.2 illustrate two of the compression hierarchies for which we currently have evidence.

TOPOLOGY

In Chapter 6, we discussed organizing relations within and between input spaces. We called the first "inner-space topology" and the second "outer-space topology," and saw that an essential part of the topology of a mental space is defined by its vital relations. In a network with input spaces and outer-space connections, various possibilities exist for projecting inner-space and outer-space topologies to the blend. The default possibility is that a relation is projected without change. For example, the distance traveled by *Great American II* from San Francisco is not changed as it is projected to the blend.

All other possibilities involve some difference between the blend and the inner-space or outer-space relations for the inputs.

The first such possibility is that a relation has no counterpart in the blend. It might seem intuitively as if nonprojection of some topology to the blend is necessarily detrimental, since it is a loss of information. On the contrary, the presence of only some topology in the blend emphasizes that particular topology and so can provide a *better* understanding of its importance. A good example of this, discussed in Chapter 10, is provided by the two alternative displays for the airplane cockpit.

A second possibility is that the relation is projected to the same relation in the blend but is scaled. In the ritual of the Baby's Ascent, time in the input with the entire life is scaled as it is projected to time in the blend.

Third, there may be syncopation in projecting the relation to the blend. Syncopation preserves ordering but leaves out all but certain highlights.

Fourth, a relation may be compressed into another relation (as in Analogy to Uniqueness). Although there are manifest differences between the relations along a cline of this type, such clines are basic in human conceptual systems. For example, if we see a person once and then again a year later, we can effortlessly draw analogy and disanalogy between what we see on the two occasions. We automatically and effortlessly invoke a process of change, even though we did not see it. We think the change left some features invariant (the analogical parts) but changed others (the disanalogous parts). That relationship of change is unconsciously compressed into a personal identity: We think we are perceiving the "identical" person, even if in the last year all his hair has gone grey. The cause-effect relation

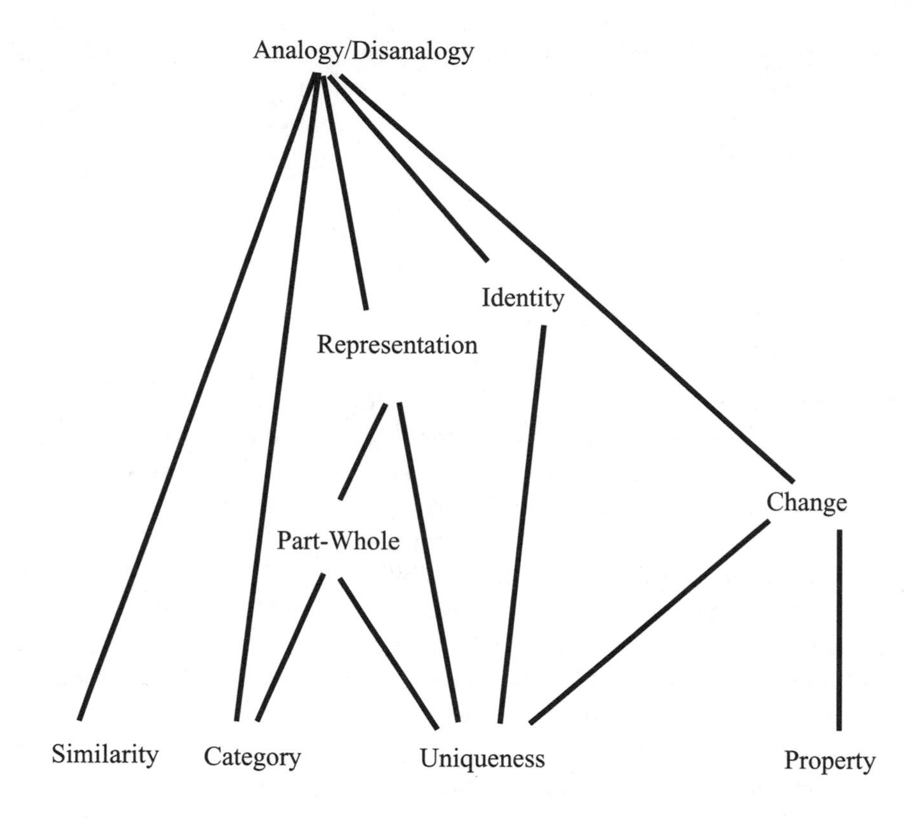

FIGURE 16.1 COMPRESSION HIERARCHY FOR ANALOGY/DISANALOGY

between Death and the skeleton is changed into a part-whole relation in The Grim Reaper. This is a standard strategy for iconographic representations of Empty Causes—Famine and War, for example.

Though it may seem obvious, this mental process should not be taken for granted; it can be distorted in neurological pathologies. In Capgras' delusion, a rare neuropsychological pathology, a patient views people who are in fact his close relatives and friends to be perfect impostors, exactly analogous to his real relatives and friends in every detail, but not identical to them. Additionally, as we have discussed before, compressing cause-effect into uniqueness (in the perception of the blue cup, in hearing the roar of the tiger, in perceiving someone's expression as "violent") is a very natural and useful compression for human beings to make. In most cases it goes without notice, but it can be exploited in creative blends, as we saw in Chapter 15 in the example where the corporation that produces a certain kind of automobile becomes that automobile in the blend.

Fifth, a relation in one input can be the inverse of that relation in the other. The blend takes the relation from one space but its compression from the

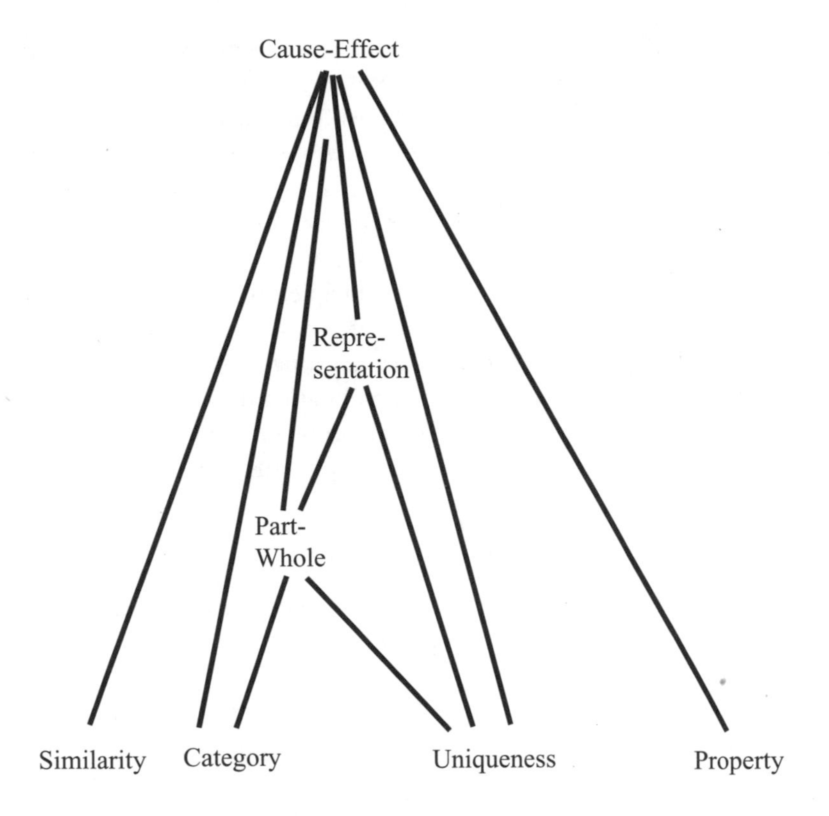

FIGURE 16.2 COMPRESSION HIERARCHY FOR CAUSE/EFFECT

other. For example, in "digging your own financial grave," the direction of causality in the *investment* input is the inverse of the direction of causality in the *grave* input. The *investment* input has the preferred topology, but the *grave* input has the preferred compression. We take for the blend the tight integration of the *grave* input but the causal direction of the *investment* input.

The five ways we have seen of aligning the topologies of the blend and of the inputs represent strategies for preserving organizing topology while optimizing compression. They are all based on the guiding principle of Topology.

> *Topology Principle:* Other things being equal, set up the blend and the inputs so that useful topology in the inputs and their outer-space relations is reflected by inner-space relations in the blend.

The guiding principles of Topology and of Compression often conflict. The constitutive principles of conceptual integration do not by themselves forbid compressing everything that is in the inputs and in their outer-space relations into a single undifferentiated element in the blend. Compression pulls in that direction, but such a severe compression produces a blend with virtually no counterparts for the topology of the inputs and their outer-space relations. The Topology Principle resists Compression that eliminates important topology. Similarly, the constitutive principles of conceptual integration do not by themselves forbid projecting every bit of the topology of the input spaces and their outer-space relations to the blend with no change and no fusion. But such a blend will typically not be at human scale. The Compression principles therefore resist the Topology Principle's preference for the preservation of topology. An appropriate balance must be struck between them for any integration network. The widely exploited patterns of compression are a way of accommodating these competing demands. The less compressed topology in the inputs can be inferred from the more compressed topology in the blend if you know the patterns.

PATTERN COMPLETION

In many cases of blending—such as the Debate with Kant and Regatta—an existing tightly integrated frame is recruited to the blend to give it tight integration. In the version of the Grim Printing Press where the newspaper CEO is turning the press and the automobile CEO is in the driver's seat, we recruit to the blend a frame in which opponents use instruments against each other and win or lose by virtue of having the better or worse instrument. These recruitments structure the elements in a blend into an integrated pattern. The Mindful Pronghorn, for example, has a frame for the blend with learning and memory, and these relations serve in the network as the compressed versions of the important outer-space vital relations of Cause-Effect, Time, and Change (adaptation and inheritance) between the inputs with the various pronghorns. There is a general guiding principle for Pattern Completion.

> *Pattern Completion Principle:* Other things being equal, complete elements in the blend by using existing integrated patterns as additional inputs. Other things being equal, use a completing frame that has relations that can be the compressed versions of the important outer-space vital relations between the inputs.

INTEGRATION

Integration—one of the mind's three *I*'s—is the main subject of this book. The impulse to achieve integrated blends is an overarching principle of human cognition.

> *Integration Principle:* Achieve an integrated blend.

Since the essence of a conceptual integration network is to project from many different and sometimes clashing inputs into a single blended space, integration in that space is a considerable achievement, not something implicit in the inputs. Integration in the blended space allows its manipulation as a unit, makes it more memorable, and enables the thinker to run the blend without constant reference to the other spaces in the network. Integration helps bring the blend to human scale, and thereby also increase the possibility for further useful recruitments to the blend from the range of our knowledge that is already at that human scale.

As we just saw, the inputs often have opposed topology. Projecting these topologies to the blend could create a disintegrated space. It is a corollary of the Integration Principle that in such cases, selections and adjustments must be made to avoid a disintegrated blend. For example, in the Debate with Kant, the language, German, from the first input, is not projected because integration in the debate frame requires a single language.

PROMOTING VITAL RELATIONS

We saw in the Mindful Pronghorn that new vital relations of intentionality (learning, memory) come up in the blend. It is a common occurrence for blends to develop internal vital relations. Naturally, this maximization of vital relations cooperates with the recruitment of new frames to the blend and with the conversion of blends to human scale. There is also a general principle of maximizing reflection between the outer-space vital relations and the vital relations in the blend, which can lead to building new outer-space vital relations. The Toblerone ad we discussed in Chapter 7 does both. The viewer of the ad sees chocolate, the pyramids, and the expression "Ancient Tobleronism?" This is the basis for building a much richer blended space by analogy with conventional cases of monuments built in honor of somebody or something. That blended space is accordingly unpacked to a familiar mapping scheme in which a person who does great deeds comes to be admired by followers and later, perhaps after his death, is given a statue in his likeness. To complete the conceptual integration network for Toblerone, the outer-space vital relations need to be considerably proliferated: The chocolate was available in antiquity and was so good that a cult of admirers erected monumental likenesses in its honor. The end result of this remarkable cognitive work is an integration network with many supplementary vital relations in the blend and in the outer-space connections. A similar example is Hiding the Penny, discussed in Chapter 7, in which someone's psychology during a romantic relationship later in life is explained by reference to his childhood habit of hiding things so well he could not find them. The analogical connection between "hiding the penny" and "hiding your love" becomes a causal relationship of either the "It Is Written" or the "Early Habits Persist" variety.

> *Maximization of Vital Relations Principle:* Other things being equal, maximize vital relations in the network. In particular, maximize the vital relations in the blended space and reflect them in outer-space vital relations.

As we will see, the relative weight of a guiding principle can depend on purpose. Where the purpose of the blend is to reveal something about the relation between the inputs, maximizing vital relations takes on special importance since that revelation can depend on new outer-space vital relations, such as the exalted status of Toblerone in antiquity. Where the purpose of the blend is, by contrast, to develop a hypothetical space markedly different from the inputs, the principle has less weight. For example, someone who is wondering whether to become a stockbroker might form a blend of herself and a friend who is a stockbroker, so that, in the blend, she is a stockbroker and has emergent feelings and judgments about her new occupation. The purpose of that blend is not to explore the comparison between herself and her friend as they are now, and therefore it does not prompt for novel outer-space vital relations. Under such a purpose, the guiding principle of maximizing outer-space vital relations has less weight. Still, it exerts a latent pressure on the network and can lead to conceptual consequences. Although exploration of comparison was not its purpose, the blend may still lead, by virtue of the principle, to comparison with her friend and new feelings toward her friend, such as admiration, envy, jealousy, or contempt. We often reject invitations to fuse two elements, even when the purpose is clearly not comparison, because there is always latent possibility for tighter connection between the inputs. For example, blending an oppressed minority with an endangered species may have good intentions but it will be rejected as insulting.

The process of maximizing vital relations in the blend and reflecting them in outer-space vital relations between the spaces is a way of intensifying the structure in the blend and of intensifying the overall connection between input spaces. Another way of doing this is to intensify vital relations already available. We commonly find networks involving human action where the form of causation and intentionality in the blend is sharper, simpler, and stronger in the blend than it is in the inputs. An example is the politician who blocks the foreign aid bill and is accused of "taking the food out of the mouths of starving children." This network also compresses the many agents involved in the political process into one.

> *Intensification of Vital Relations Principle:* Other things being equal, intensify vital relations.

WEB

> *Web Principle:* Other things being equal, manipulating the blend as a unit must maintain the web of appropriate connections to the input spaces easily and without additional surveillance or computation.

This principle reflects the fact that even when we are focusing on only one space in the network, the entire network is implicated. It is because good connections in the web are maintained unconsciously that some work in the blend can have automatic effects in the inputs or across the inputs. For example, the Buddhist Monk's meeting himself in the blend has immediate consequences for the inputs, because the time of day is always identical in the blend and the two inputs, and a monk's location at a time of day is identical to the location of his counterpart at that time of day in the respective input. This strict identity between blend and inputs of the progress of clock time and of the monk's location at a particular clock time gives very good satisfaction of the Web Principle.

When Topology is satisfied, it helps satisfy Web, because users typically assume that the web connections are those provided by projections of topology. But these two principles are distinct and need not be aligned; they can compete. Topology pushes in the direction of maximizing topology between an input and the blend. But Web is about maintaining the *appropriate* connections between spaces, so it can push in the direction of limiting the topological connections. In using a blend as the basis for thought and action, we must remain clear about how inputs and blend do *not* match topologically. For example, the Computer Desktop circa 1995 had no use for pointing with the finger at a screen, giving voice commands, or scribbling information on the outside of a folder (like "Discard after April 15"). These topologies are important in the input spaces of giving interpersonal commands, working in actual offices, and indicating a selection from a list (as when we indicate to the waiter which wine we want from a list of names we cannot pronounce). But they are not available in the blend. Speaking into the mouse as if it were a microphone (as a human visitor from the future does in one of the Star Trek films) is funny to someone who knows that the topology of vocal communication is not projected to the Computer Desktop blend. Such a "natural mistake" comes from using topology in the blend that is available only in the inputs. Interestingly, the desktop interface is currently evolving to include more of these topologies.

The visitor from the future projected too much topology from an input to the blend. In the other direction, from blend to input, we must not project back from the blend emergent topology that is inappropriate for the inputs. This is what happens when obsessive viewers of soap operas attack actors in the street who play evil villains in the soap operas. There is emergent topology in the blend: The evil person has the actor's appearance. That unification of character and person is not to be projected back to the real-life input with the actor.

In both directions, the mistakes come from maximizing topology in violation of the Web Principle.

A corollary of the Web Principle is that we should not disconnect valuable web connections to the inputs. The Computer Desktop has web connections to the space of computer operations, in which all shifts of focus require only a simple click. For example, if a user is running five different applications on the desktop and wants to see only one of them, he can click "Hide others" (conversely, "Show others"); to see a given document partially occluded by another, he need only click anywhere on the desired document. But in the space of offices, to hide everything on his desk except the one thing he wishes to focus on would require complex physical operations. Projecting all these operations to the blend would sever its useful web connections to the input of computer operation. Function guides competition here. The web connection to "change of focus" in the computer operations input is important because the desktop interface is designed to run a computer. If its function were to simulate an office working environment, then the complexity of the physical operations would be maintained at the expense of computing efficiency.

Web combines with Integration to force novel integrations in the blend. For instance, in the metaphor "digging one's own grave," the blend's causal, temporal, and intentional structures (*actor is unaware of his actions, sufficient repetitions of that action bring about the failure*) come to the blend from the input space of mistakes and failure. This web connection is crucial to the reasoning, but it would be destroyed if we instead projected the commonplace structure of the *grave* input (death followed by conscious grave-digging by somebody else). In the Nixon-in-France example, we project to the blend Nixon, but not his U.S. citizenship, which would prevent him from being president of France, thus cutting off a crucial web link from the blend to the second input.

UNPACKING

> *The Unpacking Principle:* Other things being equal, the blend all by itself should prompt for the reconstruction of the entire network.

One of the powers of the blend is that it carries in itself the germ of the entire network. If one already has the entire network active, then running the blend gives inferences and consequences for the rest of the network. But if the entire network has not yet been built or has been forgotten, or if relevant portions of it are not active in the moment of thinking, then the blend does good work in prompting for those activations. Part of the blend's power to provide global insight lies in its utility as a mnemonic device—in cases where we have knowledge of the network and merely need to retrieve and activate it—or as a triggering device, carrying small compressions that guide us to unpack them into full-blown parts of the network. While the discourse environment will sometimes set up the inputs and connections before the blend, in other cases—as when we see a billboard—we may be presented only with a material anchor for the blended space. Unpacking is often facilitated by disintegrations and incongruities in the blended space. The Impotent Smoking Cowboy billboard, for example, gives us just the cowboy with the drooping cigarette, and the phrase "Warning: Smoking Causes Impotence." The incongruity of the drooping cigarette itself prompts us to construe this representation as something more than a simple picture of a cowboy smoking a cigarette. The cigarette "unpacks" to both the normal cigarette in the space with the virile cowboy and the drooping organ in the space with sexual performance. In fact, it might be better to say that at first we recognize a space with incongruities and that those incongruities prompt us to take the space as a blend and look for its inputs. In the same way, the Debate with Kant blend satisfies Unpacking because it has someone from the eighteenth century and someone from the twenty-first in the same space. In these cases, the scene presented to us has disintegrations that prompt us to take it as having disparate inputs. Similarly, "He was so mad I could see smoke coming out of his ears" presents a scene that is integrated at the highest level but has explicit prompts for *anger, heat,* and *bodily physiology.* Prompts for making the right connections often take the form of tight compressions, as in The Grim Reaper, whose body, clothing, and tools prompt for skeletons, funerals, reaping, and killing.

As Douglas Hofstadter points out, the Unpacking Principle is not purely one of structure within the network but, more broadly, one of communication, since the unpacking possibilities offered by the blended space will depend on what is already active in the context of communication.

RELEVANCE

> *The Relevance Principle:* Other things being equal, an element in the blend should have relevance, including relevance for establishing links to other spaces and for running the blend. Conversely, an outer-space relation between the inputs that is important for the purpose of the network should have a corresponding compression in the blend.

Participants in communication are under general pressures to make their communications relevant. When a blend is used in communication, it is subject to these general pressures, but part of its relevance derives from its location and function in the network. An element in the blend can fulfill the general expectation of relevance by indicating its connections to other spaces or indicating the lines along which the blend is to develop. Speaker and listener are both aware of this fact, and it guides their construction and interpretation of the network. The expectation of relevance encourages the listener to seek connections that maximize the relevance of the element *for the network*, and it encourages the speaker to include in the blend elements that prompt for the right network connections, but also to exclude elements that might prompt for unwanted connections. We will call this principle "Network Relevance." Network Relevance can be satisfied for an element in the blend if it can be successfully taken as a prompt for Unpacking. Consider The Grim Reaper. Once Death the Empty Cause is personified, it must have a shape, body, and manner, and may easily have clothes. But these aspects are not enforced by the frames of either the event of death or the killer. As we saw in the discussion of highlights compression, these elements are well-chosen Part-Whole compressions that guide unpacking and prompt for vital relations in the network such as Cause-Effect. Interpreting them as prompts for unpacking gives them Network Relevance.

It is a general feature of the presentation of self in everyday life that one's clothes, accessories, manner, expression, gesture, and appearance have relevance. They can thus prompt for blends in which they have Network Relevance. The sullen high-school student dressed in a dog collar and fatigues is implying not that he is a soldier (or a dog) but, rather, that parts of his character are received by projection from trained acts of aggression and violence.

The Relevance Principle pressures networks to have relationships in the blend that are compressions of important outer-space relations between the inputs. We have seen this at work in many networks. For example, in Regatta, the space with *Northern Light* is the reference space: The sailors of *Great American II* know about it and have ambitions with respect to it, and that knowledge is causal for their activities. These outer-space relations of Intentionality and Cause-Effect need, by the Relevance Principle, to have compressions in the blend, which they do because of the intentional and causal structure of the frame of *race*.

RECURSION

One crucial corollary of the overarching goal of blending to Achieve Human Scale is that a blended space from one network can often be used as an input to another blending network. Once blending delivers a new blend at human scale, that new blend is a potential instrument for achieving yet more compression to human scale. We saw that the complex multiple blend of The Grim Reaper produces a human-scale situation with a person, The Grim Reaper. There is no reason in principle why The Grim Reaper would have to fit an existing category like *person*. It could be the case that double-scope networks typically produce elements belonging to entirely novel categories. A complex network with many compressions delivers The Grim Reaper blend. The Grim Reaper, as a person, can be part of any input that includes persons. In particular, personification blends can now have The Grim Reaper as the input with the person. We can personify a hawk as The Grim Hawk, complete with skeleton, scythe, and cowl but also with wings and beak, who is a divine force in the world of mice. This kind of recursion happens routinely in the development of science and mathematics. If we start with a wave such as we see at the seashore, and then consider sound, and recognize that sound, though a different phenomenon, still has longitudinal motion in a medium, we can make a new blended category *wave* that now includes various kinds of "longitudinal" waves. That new category *wave* can be an input to a further blending network, whose other input has electromagnetic phenomena. The blended space in the new network now has a category *wave* that includes "electromagnetic" waves.

The history of the concept of number has seen many successive blends, where at each stage a blended concept of number serves as the input to a new integration network, whose blended space has yet a newer concept of number. If we already have counting numbers 1, 2, 3, . . . , they can serve as an input to a network whose other input has proportions of these numbers. The cross-space mapping will be a one-to-many partial correspondence, where each counting number n maps onto any ratio of counting numbers r and s where r has n parts, each of magnitude s. By selective projection, all proportions are projected into the blend from the space of proportions, and all counting numbers from the space of counting numbers. Equivalent proportions are projected onto the same element in the blend. Operations like multiplication are emergent in the blend. The blend has a new category whose elements are projections of widely disparate elements in the inputs, and yet, *in the blend,* the category for all these elements is *number* again. They are all just numbers. For example, there is an element in the blend that is the projection of the ratio of 9 to 3, and the projection of the ratio 12 to 4, and the ratio of the projection 333 to 111, which is also the projection of the counting number 3. In other words, that element in the blend is the projection of an infinity of input elements that are in quite different categories. Another element is the projection of the ratio of 9 to 5 and 18 to 10 and 27 to 15 and so on, but is the projection of no counting number at all. Although the proportions themselves are not linearly ordered, the number created in the blend by projections such as the projection from the proportion of 550 to 900 turns out to be smaller than the number created by projection from the proportion of 2 to 3, even though the numbers of the first proportion are very large and the numbers of the second are very small. On the one hand, the blend, which creates the new concept *number* including rational numbers, is a great mathematical achievement. On the other hand, the blend delivers *numbers*, an old category. The category that is the output of the blend has the same name as the category that was the input to the blend, and is felt to be the same, even though the inner structure of that category has been dramatically altered. Again, because *number* is a human-scale notion, the blend has human scale. But when we look inside that notion of number, we see that it is now more complicated, and attached to various diffuse input spaces. The fact that the same category, *number*, organizes both the input and the output is what introduces recursion in the successive blending process, just as recursion is introduced in the Grim Reaper network by the fact that *person* organizes both an input and the output. What is particularly dramatic about the history of the concept of number is that through many successive blends, the category *number* is always an input and an output. *Number*, meaning rational numbers, comes to be an input to an integration network whose output is *number*, meaning real numbers. And as we saw in Chapter 13, *number* meaning *real numbers* blends with two-dimensional space to produce *number*, meaning complex numbers. At each stage the organizing category is felt to be the same in both the input and the output; yet at each stage the internal structure of the category is different in the blend than it was in the input. Can such recursion go on indefinitely? On the one hand, the category of the output is the same as the category of the input and is at human scale in that sense, so the blended space at any step is a candidate for yet further blending. But on the other hand, the internal structure of the category at each step is associated with more and more decompressions in the rest of the network, and that makes the network less tractable.

HOW THEY COOPERATE AND COMPETE

We have seen throughout this chapter how constitutive and governing principles cooperate: Compression helps human scale, human scale helps getting a story, getting a story helps global insight, going from Many to One helps the blend achieve human scale. We have also seen how the governing principles may compete. Compression competes with topology, since topology is a pressure to preserve various distinctions and elements while compression works in the countervailing direction. Similarly, integration competes with unpacking since absolute integration leaves a blend that carries no sign of its distinctive inputs. For example, a picture of the Annunciation, in which the Angel of the Annunciation tells the Virgin Mary that she is to be the Mother of God, might look like two young women having a conversation in a bedroom, in which case we would not be prompted to unpack it to two quite different inputs—one of which has a girl in her bedroom, and the other the eternal divine relationship of God, Christ, the Holy Spirit, and the Mother of God.

HOW GOVERNING PRINCIPLES WORK IN TYPES OF NETWORK

With any conceptual integration network, we must resolve the lines of cooperation and competition presented by the various principles. This is a difficult matter, but one that human beings are uniquely evolved to handle. The various canonical types of network negotiate among the governing principles in different ways.

Mirror Networks

Mirror networks have special characteristics that suit some governing principles well and preclude others.

We saw in some detail in Chapter 7 how mirror networks provide compressions. Like any network, a mirror network can use Scaling Compression, Syncopation Compression, Compression from one type of vital relation to another, and Creation of vital relations. On the other hand, mirror networks have no role for Borrowing Compression since both inputs and the blend share the same compressed frame.

It is a special virtue of mirror networks that they easily satisfy Topology, Integration, and Web simultaneously. The sharing of the organizing frame automatically transfers a rich topology from space to space. Integration is provided in the blend by the shared frame and its elaboration. This elaborated frame is often already a common, rich, and integrated frame, like *race* or *debate* or *encounter*. The sharing of the frame throughout the network automatically preserves the Web connections between spaces.

When the counterpart elements in the inputs to a mirror network are fused in the blend, the organizing frame of the inputs will exhaust the elements of the blend and there will be no Pattern Completion: Blend and inputs will have the same organizing frame. We saw such a case in Chapter 8: You Are My Long-Lost Daughter. There are also cases where some counterpart elements from the inputs are not fused but the organizing frame for the inputs can nonetheless accommodate their composition in the blend, as in the Successor Skipper. Often, however, the multiplicity of nonfused elements in a blend in a mirror network creates relationships that can be recognized as partial patterns to be completed by bringing in a fuller frame, as when the nonfusing of the two philosophers in the Debate with Kant produces a blend of two philosophers musing on the same problem, which can be completed with the frame for *debate*.

The integration provided by the shared frame and its elaboration would seem to work against Unpacking, by providing a blend that is a straightforward instance of the shared frame. But a space can be well integrated at one level of detail and poorly integrated at a finer level. For instance, a race between two boats is a fine integration, but at the more detailed level, the race between an 1853 clipper and a 1993 catamaran is a poor integration.

Thus it is possible for a blend to satisfy Integration by being well integrated at one level and yet to have a weaker integration at a more detailed level. As we saw in Chapter 7, the spaces in a mirror network share an organizing frame but not necessarily structure at lower levels. In a mirror network, Unpacking is often provided by just such poor integration at lower levels, which provides a signal to connect elements in the blend to counterparts in different input spaces in which their counterparts are well integrated. The blend in Regatta, for instance, is integrated at the level of the organizing frame, but the contrast of clipper ship and small catamaran produces poor integration at a lower level, and this serves as an aid to Unpacking. When we encounter the blend, the poor integration prompts us to project the clipper ship and the catamaran back to different inputs. In fact, the phrase "ghost of *Northern Light*" simultaneously evokes backward projection to an input and emergent structure in the blend. The word "ghost" signals vital relations between the input spaces, and in the blend it is a compression of those vital relations. The sailors in the 1993 space know about the history of *Northern Light*. That is a link of Intentionality through memory. "Ghost" prompts for the construction of these outer-space vital relations and also for the creation of an element in the blend. This of course satisfies Unpacking.

The category *Ghost* is a powerful compression of Intentionality (in the form of memory), Time, Representation, Identity, Counterfactuality, and Cause-Effect. It compresses Cause-Effect because the historical event is causal for the memory. It compresses Identity between the content of the memory and the historical space. It compresses the Counterfactual links between the blend and each of the inputs. The ghost in the blend is naturally decompressed into a Time link between the two inputs. The Intentionality link of remembering that connects the inputs is often supplemented with a Representation link, making the memory a representation of the past event. The particular ghost in the blend compresses that representation link into a directly perceived instance of the category *ghost,* and that instance is identical to the remembered thing in the historical input. The features of *ghost* that we have noted for Regatta are general and by no means restricted to mirror networks. This remarkable category is found in cultures all over the world, and perhaps in all cultures.

A mirror network provides a straightforward satisfaction of Relevance to the extent that an element in the blend is, by virtue of the shared frame, automatically connected to its counterparts in the inputs. But crucially, an extended frame recruited or constructed for the blend is also subject to Relevance. The relations in that frame that are not in the shared frame are assumed to be compressions of outer-space connections between the inputs, as when memory in the Mindful Pronghorn compresses unchanged inheritance between the ancient and modern pronghorn inputs.

Single-Scope Networks

In a single-scope network, such as the portrayal of business competitors metaphorically as boxers, Integration in the blend is automatically satisfied because the blend inherits a compressed organizing frame from one of the inputs—in this case, *boxing*. For the same reason, Topology is satisfied between the blend and the input that provides the organizing frame. But Topology is also satisfied between the blend and the other input because the conventional metaphor of competition as physical combat has aligned the relevant topologies of the two inputs. Thus, when an element in the blend inherits topology from an element in either input that is involved in the cross-space metaphoric mapping, the topology it inherits is automatically compatible, by virtue of the existing conventional metaphor, with the topology of that input element's counterpart in the other input.

Single-scope networks also give us straightforward Borrowing for Compression and Scaling Compression. The input that provides the frame is typically integrated at "human scale" (e.g., two men fighting). When projected into the blended space, that frame becomes a massive compression of the focus input—in this case, the business rivalry. We find scaling of time down to the duration of a fight, the number of agents compressed to two, the types of actions reduced to blows and dodges, and the causal chain scaled down to hitting or missing.

Web is similarly satisfied by this shared topology provided by the borrowed frame and the metaphoric mapping. Unpacking is provided just as it was for a mirror network: Although the blend is integrated at the frame level, it is disintegrated at the more specific level. Suppose, for example, that the competitors are represented in a cartoon as boxing in business suits. The lack of integration between business suits and boxing prompts us to unpack the blend to two different spaces. In the same way, if we know that "Murdoch" and "Iacocca" refer to businessmen and not boxers, then their use in the sentence "Murdoch knocked out Iacocca" directs us to the lower levels of the input of businessmen, and this helps satisfy Unpacking.

In the case of single-scope or mirror networks with a custom-built source domain, as in Hiding the Penny, or You Are My Long-Lost Daughter, or the various Redemption examples, there are many natural opportunities to strengthen the outer-space vital relations. For example, in Hiding the Penny the analogy between the inputs becomes a Cause-Effect relationship, under the Early Habits Persist reading. These single-scope networks also offer natural opportunities for syncopation, so that the distinctions between the input spaces drop out of consideration in the blend, where the fusion can produce an intensified personal identity that overshadows any of the other events in life. In You Are My Long-Lost Daughter, daughters in the inputs related by outer-space analogy are compressed into identity. In Hiding the Penny, events in the inputs related by outer-space analogy are compressed into essential identity.

Double-Scope Networks

In a double-scope network, Compression, Topology, Integration, and Web are not automatically satisfied: It is necessary to use a frame that has been developed specifically for the blend and that has central emergent structure. (This may be why double-scope networks—such as the Computer Desktop, Complex Numbers, and Digging Your Own Grave—are typically thought of as more creative, at least until they become entrenched.) In these networks, then, we expect to see increasing competition between governing principles and increasingly many opportunities for failure to satisfy them.

The Computer Desktop illustrates many of these competitions and opportunities. We stress that failing to satisfy a governing principle does not necessarily mean that the resulting blend fails; on the contrary, constructing a useful double-scope blend often depends upon finding a suitable way to relax governing principles. First let us consider an aspect of the Computer Desktop blend in which Topology clashes with Integration, and Integration of the blend wins. The purpose of this blend is to provide an integrated conceptual space that can serve as the basis for integrated action. The basic integrative principle of the Computer Desktop is that everything is on the two-dimensional computer screen. But in the input space of real office work, the trashcan is not on one's desk. Topology would place the trashcan off the desktop in the computer interface blend; but because this would destroy the internal integration of the blend, the computer screen has the trashcan on the desktop. Integration of the blend in this case can be achieved only by relaxing the Topology Principle. Also, as Ricardo Maldonado has pointed out, the computer trashcan essentially never fills up. This is another case where Integration wins out over Topology.

There are at least two reasons why we are content to relax Topology in this way. First, the topology being dropped is incidental to the cross-space mapping: The three-dimensionality of the office and the position of trashcans under desks has no counterpart in the cross-space mapping to the input of computer operation; neither does the capacity of the trashcan, since any folder (of which the trashcan is one) has as its capacity the entire unused capacity of the system. Second, as we have mentioned, the purpose of this blend is to develop a conceptual basis for extended action, not to draw conclusions about the input space of offices. In a case like the Buddhist Monk, where the purpose is to draw conclusions about topology of input spaces—specifically, coincidence of locations and times—relaxing Topology is likely to allow inferences in the blend that would project wrongly or not at all back to the input, and so defeat the purpose of the blend.

A satisfactory blend is achieved by finding an effective relative weighting of the governing principles for the purpose at hand. A good relative weighting for the Buddhist Monk would not be good for the Computer Desktop, and vice versa. Real failure occurs when the weighting is not achieved and the massive violation in the blend defeats its purposes. The most noticeable such failure for the Computer Desktop is the use of the trashcan both as the container of what is to be deleted and as the instrument of ejecting floppy disks. This failure involves failures of Integration, Topology, and Web. It provides good compression, but in this case the compression blends two operations that should have been kept separate so as to preserve essential topology, and to preserve integration in the blend.

There are violations of the Integration Principle. Using the trashcan for both deletion and ejection violates Integration in three ways. First, in the frame elaborated for the blend, the dual roles of the trashcan are contradictory, since one ejects the floppy disk to keep it rather than discard it. Second, in the frame elaborated for the blend, all other operations of dragging one icon to another have the result that the first is *contained* in the second with the unique exception of dragging the floppy to the trashcan. Third, for all other manipulations of icons on the desktop, the result is a *computation*, but in this case it is a physical *interaction* at the level of hardware.

This dual use of the trashcan also violates Topology. In the input of office spaces, putting an object in a folder or in the trashcan results in containment, and this topology is projected to the blend. The trashcan on the desktop is like any icon that represents a metaphoric container: If we drag a file to a folder icon or to the trashcan icon, the file is deposited there, and this is the topology of the input of office spaces. But putting the floppy disk icon into the trashcan icon so as to eject it violates the projection of topology from the input of offices. It also violates topology by not preserving the relation in input 2 (the space of real offices) that items transferred to the trashcan are unwanted and destined to become nonretrievable.

The operation also violates Web. The very process of ejecting floppy disks from the computer desktop creates nonoptimal web connections, since the floppy is sometimes "inside" the world of computer operations and sometimes "inside" the world of the real office.

The design of word-processing programs for the desktop interface demonstrates other forms of competition among governing principles. The command sequence Select–Copy–Paste on word-processing applications violates both Topology and Web. It violates Topology because in the input where text is actually copied by scribes or xerox machines, copying (after selection) is a one-step operation. There is no pasting and no clipboard. Properties specific to the Integration in the blend make it convenient to decompose this operation into two steps, but they do not map topologically onto corresponding operations in the input of "real copying."

The labels "Copy" and "Paste" chosen for these two operations also violate Web: The Copy operation in the blend (which actually produces no visible change in the text) does not correspond to the Copy operation in the input (which does produce visible change); the Paste operation, which does produce change, is closer to "copying" in the input, but the label "Paste" suggests a counterpart (pasting) that is not even part of the copying process.

Not surprisingly, these flaws in the blend lead to mistakes by novice users. They click Copy instead of Paste, or try sequences like Select–Select Insertion Point–Copy. This fails miserably because the first selection (not marked for copying) is lost when they click the second selection, and, anyway, Copy at that point is the wrong command. Mistakes like this are interesting because they are an effort by the user to maintain optimal Topology and Web connections. If double selection were possible on the blended interface (as it is, in terms of attention, in the Input), Copy and Paste could easily be reintegrated into a single process operating on both selections, and the attempted sequence would be viable. In fact, the application being used to type the present text has a keyboard command (with no counterpart in the menus) that comes closer to this conception.

The "Cut and Paste" method of moving text is a less severe violation, because the projected operations from the "office" input are plausible and properly web-connected. But it does add conceptual complexity to what is more easily conceived of as simple unitary "moving." Recent versions of the application we are using to type this manuscript have added the possibility of selecting and dragging text directly to the appropriate location. The portion of text does not actually "move" (only the arrow does) until the mouse is unclicked.

Despite these failures, the Computer Desktop blend draws rich and effective structure from familiar frames, and users are able to use it in a rudimentary fashion very quickly and to learn the elaborated frame, warts and all. The nonoptimality creates difficulty for novices, who are reluctant to put the floppy disk in the trashcan since by topology it should then be lost, but advanced users forget this difficulty and learn a less optimal but more elaborate blend.

The fact that in double-scope networks the organizing frame of the blend is not available by extension from the organizing frame of either input increases chances of nonoptimality and of competition between the governing principles, but it also offers opportunity for creativity. Pressure to satisfy governing principles in highly complex double-scope networks has historically given rise to some of the most fundamental and ingenious scientific discoveries.

The development of complex numbers is a case in point. The complex-number blend turns out to be a double-scope network. Some key elements in each input have no counterparts in the basic cross-space mapping: The operation of multiplication for numbers has no counterpart in the geometry input, and the angles in the geometry input have no counterparts in the number input. The blend, however, inherits both the multiplication operation from the frame of the "number" input and the angle from the frame of the "geometric" input. This is already enough to make it a double-scope network, since multiplication in the blend has topology from the second input while angle in the blend has topology from the first input. But beyond this, multiplication includes addition of angles as one of its constitutive components. This fact was discovered only by running the blend; it turned out to be a highly unexpected essential property of the new concept of number. Indeed, the pressures to satisfy the governing principles in this double-scope network led to an important mathematical discovery.

Let us consider some of these competitions in the construction of complex numbers. As we mentioned in Chapter 13, there are actually three inputs to the complex-number blend. Two of them are already superbly rich and well integrated: The two-dimensional geometric plane and the real numbers. The third input is a badly unintegrated space: the real numbers and, in addition, some strange, "impossible" elements that are useful in doing numerical calculations but do not seem to qualify as actual numbers. Consider a typical mental challenge faced by a mathematician as great as Euler and as late as 1768, in what Morris Kline calls "the best algebra text of the eighteenth century":

> Because all conceivable numbers are either greater than zero or less than zero or equal to zero, then it is clear that the square roots of negative numbers cannot be included among the possible numbers [real numbers]. Consequently we must say that these are impossible numbers. And this circumstance leads us to the concept of such numbers, which by their nature are impossible, and ordinarily are called imaginary or fancied numbers, because they exist only in the imagination.

Euler is driven to this reasoning because he requires well-ordering to be projected from the input of real numbers into any concept of "possible" number. The impossibility of well-ordering these elements automatically makes them something less than full numbers. But in the modern blend of complex numbers, well-ordering is indeed not projected to the blend. 1 and the square root of negative 1 are not equal, yet it is not the case that one of them is "less than" the other. Euler is insisting on too thorough a satisfaction of Topology with respect to real numbers. It turns out that the successful blend finally achieved had to relax that particular governing principle.

It is tempting to think that only wildly acrobatic mathematical constructs like imaginary numbers could require such a profound relaxation of Topology, committing criterial properties of number to the flames. On the contrary, the same sorts of struggles and the same kinds of relaxation were necessary to grant the status of "number" to irrational numbers, zero, and negative numbers. Blaise Pascal himself judged that irrational numbers had no existence independent of continuous geometrical magnitude. Consider the illuminating reasoning of the great theologian and mathematician Antoine Arnauld, a close friend of Pascal's in the mid-seventeenth century, about the turmoil involved in thinking of negative "numbers" as actual numbers:

> Arnauld questioned that $-1:1 = 1:-1$ because, he said, -1 is less than $+1$; hence, How could a smaller be to a greater as a greater is to a smaller? The problem was discussed by many men. In 1712 Leibniz agreed that there was a valid objection

> but argued that one can calculate with such proportions because their form is correct, just as one calculates with imaginary quantities.

Arnauld, Leibniz, and others were struggling to find an integrated notion of number that would include "negative numbers" and yet retain all the properties deemed crucial for what everyone already accepted as numbers. For those traditional numbers, it was certainly true that if a <b, a/b could not be equal to b/a. In fact, a/b had to be smaller than b/a. Well, –1 is less than +1 in anybody's conception, but –1/1 = 1/–1. To accept "negative numbers" as full numbers that could be involved in division required relaxing what looked like crucial topology from the input space of numbers. In the blend, this relaxation yields a fantastic emergent structure, a new concept of division. Dividing two numbers is now a matter of determining the ratio of their *magnitudes* and assigning a *sign* to that ratio based on other operations not previously invented.

The struggle to achieve the conceptual blend was not eliminated by achieving perfect formal routines. As early as the sixteenth century, Raphael Bombelli had perfected formal routines for both negative and imaginary numbers. In fact, it was the success of these formal routines that created the mathematicians' dilemma. John Wallis in the seventeenth century had developed formal routines for placing imaginary numbers into correspondence with the two-dimensional plane, and associated operations on such numbers with geometric constructions. Amazingly, mathematicians used the formal routines even while refusing conceptual status to the nontraditional numbers.

Given the blends that we have been taught in our schools and universities, and that we take for granted, it is easy to be blind to the great creativity involved in the invention of the concept of number and the centuries-long struggle to incorporate the notion of negative and imaginary numbers into the category *number*. That struggle was a titanic competition between governing principles. These advances in number theory were not a mere matter of extending what one already knew about number by adding a few new numbers. On the contrary, basic notions like *division, greater than,* and *multiplication* had to be thoroughly reconceived for all numbers in order to create a successful new blend.

Many potential blends come up in the struggle to achieve an effective one. To give just one example, Wallis, by careful projection from the input space of real numbers, achieved a blend in which negative numbers are both less than zero and greater than infinity. Like Arnauld's reasoning, his logic is quite crisp and admirable:

> In his *Arithmetica Infinitorum* (Arithmetic of Infinitesimals, 1655), he argued that since the ratio *a*/0, when *a* is positive, is infinite, then, when the denominator is changed to a negative number, as in *a*/*b* with *b* negative, the ratio should be greater than *a*/*0* because the denominator is smaller. Thus the ratio must be greater than infinity.

Wallis's reasoning is simple. For usual numbers (i.e., what we call positive numbers), as *b* moves toward zero, *a/b* gets larger and larger, so that *a*/0 is infinite. Continuing this procedure, when *b* is even smaller than zero (i.e., negative), shouldn't *a/b* continue to become larger (i.e., larger than infinity)? This is a fine blend, but not the mathematically most successful one. In the modern blend we use now, we give up the criterial properties of division that Arnauld and Wallis wanted to maintain.

We take the modern blend as almost self-evident, with the negatives nicely trailing the positives on an infinite line, but this apparent simplicity is deceptive from a cognitive, cultural, and historical point of view.

Jeff Lansing has pointed out other marvelous examples of important scientific blends leading to discovery (by Fourier, Maxwell, and Faraday), which suggests that this is a general process. We emphasize that this type of creativity is possible by virtue of the competition of governing principles and the power of blending to accommodate them. Douglas Hofstadter has also analyzed the remarkable analogies in a number of scientific discoveries in physics in the twentieth century; we would add to his analysis that the analogies in question are themselves components of creative conceptual blending networks.

Finally, Unpacking is relatively easy to satisfy in the double-scope networks since key elements in the blend cannot all be projected back to the same organizing frame of one of the inputs. For example, in Digging Your Own Grave, the gravedigger is responsible for the death, and this structure cannot be provided by the single organizing frame of digging graves. Thus the blend must be unpacked to the organizing frames of different inputs.

SUMMARY

One of our major tasks for the future is to investigate and clarify the constitutive and governing principles of conceptual integration. The principles we have already discovered turn out to be simple to state, but they interact to produce a rich world of products. We recapitulate these principles below, both as a summary of the progress made so far and as a prompt for further research. This is a rich area for further exploration.

CONSTITUTIVE PRINCIPLES

- Matching and counterpart connections
- Generic space
- Blending
- Selective projection
- Emergent meaning
  - Composition
  - Completion
  - Elaboration

GOVERNING PRINCIPLES FOR COMPRESSION

Borrowing for compression
Single-relation compression by scaling
Single-relation compression by syncopation
Compression of one vital relation into another
Scalability
Creation by compression
Highlights compression

OTHER GOVERNING PRINCIPLES

The Topology Principle
The Pattern Completion Principle
The Integration Principle
The Maximization of Vital Relations Principle
The Intensification of Vital Relations Principle
The Web Principle
The Unpacking Principle
The Relevance Principle

OVERARCHING GOALS

Achieve Human Scale.

Noteworthy subgoals:

Compress what is diffuse.
Obtain global insight.
Strengthen vital relations.
Come up with a story.
Go from Many to One.

CHAPTER 16
ZOOM OUT

HUMAN SCALE

We have seen governing principles for maximizing compression.

Question:

- But don't we think of tiny or very brief things by "exploding" them?

Our answer:

Yes. For example, we can understand atoms better by thinking of pinheads revolving around an orange. In that sense, we are scaling up rather than scaling down. But the net effect is still to bring us to human scale, and also to compress very many different events into an intelligible scenario. Similarly, we can understand events in high-energy physics that happen in a few nanoseconds better if we use a blend where they happen in a few seconds. Again, the effect is to compress many different events into a scenario at human scale. Compression is not a matter of absolute size, and is not uniformly achieved by scaling down. In some cases it requires scaling up.

TIME WARPS THROUGH COMPRESSION

So time can be shortened or lengthened to make a human scale blend.

Question:

- What else can happen to time in the blend?

Our answer:

A highlights compression gives at a shot the elements of a story that unfolds in time. The painting of the Annunciation in the Mérode Altarpiece shows the breath of God blowing toward the Virgin Mary's abdomen, and traveling on that breath is a homunculus already carrying his own miniature cross. In Rogier van der Weyden's Louvre *Annunciation*, a medallion hanging from Mary's bed depicts the Resurrection. These two paintings prompt for conceptual blends that include the story of Christ from the moment when His birth is announced to Mary to the time of His Crucifixion and Resurrection. In these scenes, the carrying of the cross and the resurrection from the tomb are presented as facts before the birth of Christ. The blend in these cases is a human-scale scene. All at once, the blend is a single momentary highlight in a larger global story, but it includes the other highlights of the larger global story in a single compression. A crucial highlight such as the Resurrection, which in the larger story happens much later than the Annunciation, is now in the blend a cause of the medallion present at the Annunciation.

The medallion looks so natural in the painting that it is easy to overlook the remarkable system of compressions to human scale that it is carrying. There is a causal link between the space of the annunciation and the space of the resurrection. There is an identity link between Christ in the womb of Mary in the Annunciation and Christ at the Resurrection. There is a representation link between the Resurrection and the medallion. And of course there are time and space links between these spaces. All of these imperceptible outer-space links are compressed into a visible object in the blend. The medallion in the blend is like the blue hockey stick we discussed in Chapter 10. It, too, is a visible object in the display that compresses outer-space relations of time, space, cause-effect, identity, and representation.

It might seem as if such a time warp is possible only in blends that represent stories that transcend human time, stories of eternal truths or divinity. But in fact it is a standard operation of symbolic highlights compressions. In Vermeer's *Arnolfini Wedding*, the couple, holding hands, is presented in a bedroom, in front of the bed, with many extra elements, such as a small dog at their feet. The image of the bride is sometimes construed as indicating pregnancy. If we interpret this painting as a snapshot in time, the pregnancy can pose a problem. But in a different interpretation, it is a highlights compression. The story of marriage includes the house, the bedroom, the holding of hands, the wedding clothes, the fidelity (represented by the loyal dog), and, of course, the pregnancy and the children. The highlights compression giving pregnancy at the moment of the wedding is another temporal inversion through compression of cause and effect. Again, the wedding is a momentary highlight in a larger story and includes other highlights in a single compression.

The Grim Reaper gives a similar time warp through highlights compression. It is one moment—the moment of impending death—in a larger story of dying, and it includes other highlights in a single compression. Inevitably, there is a time warp in the blend. For example, the costume for the religious attendants at the funeral is already on the Grim Reaper before he appears to announce the impending death.

SCALING AND CREATING INTENTIONALITY

We have seen that achieving human scale can involve constructing or intensifying Intentionality in the blend.

Questions:

- Is there a system to constructing Intentionality in the blend?
- Are there limits to this magic?

Our answers:

There is a system and an order to constructing Intentionality in the blend, but that system is more powerful than we have so far discussed. Consider first the case where intentional actions are already in the input spaces. Earlier, we discussed the political cartoon in which the politician who vetoes the foreign aid is represented as taking food out of the mouths of starving children. In the input with the veto, the politician is already entirely intentional. In the input with the frame for taking something away from someone, the agents are already entirely intentional. Borrowing the frame for taking something away from someone provides compression in the blend and has the effect of heightening the intentionality of the politician.

Now consider the case where the topic input has only an event without intentionality. The biological event of death does not involve intentionality. But "Death took him" prompts for a blend in which the events are now actions of an intentional agent. This is a very general pattern of meaning construction, discussed in *More Than Cool Reason* and elsewhere as EVENTS ARE ACTIONS. Potentially, it allows any event to be construed in the blend as intentional. Intensifying existing intentionality and attributing intentionality to nonintentional events are not two different kinds of operation. They are both cases where a blend integrates the intentionality from one input with events projected from the topic input. The distinction is only that in one case the event in the topic input has relatively weak intentional structure of its own, and in the other it has no intentional structure. EVENTS ARE ACTIONS, then, is a systematic way of constructing blends in accordance with the governing principles that intensify and create vital relations.

What if something is not even an event? Can blending give it intentionality? The answer is yes, through a precise conceptual procedure called "fictive motion." We will return to this topic in detail in Chapter 17. Fictive motion blends a dynamic scenario of motion with a static situation so that the static situation can be conceived and described as having motion. Some well-known examples are "The fence runs all the way down to the river," "The mountain range goes all the way to Canada," and "The road traces a winding path through the mountains." The dynamic input contributes a moving trajector on a path, which is mapped onto a relevant dimension of the static object in the other input. Thus, in the blend we have a trajector moving along the relevant dimension of the static object.

Such a blend, in which the static scene now has motion, can be an input to a new blend constructed according to EVENTS ARE ACTIONS, in which the static scene now has not only motion but intentional structure for that motion. Consider the attested sentence "Trees climb the hills toward the Golan and descend to test their resolve near the desert." The line of trees up the hill is a static scene that acquires motion by a fictive-motion blend. Blending that nonintentional fictive motion blend with an input that has an intentional agent gives a new blend in which the trees "climb." Fictive motion on the one hand and EVENTS ARE ACTIONS on the other are systems of blending that can be combined into a complex system in which nonevents can be given intentional structure. This system is an exceptionally good way of satisfying the Intensification of Vital Relations Principle and the Maximization of Vital Relations Principle, as well as the goals of achieving human scale and getting a story.

CONSTRAINTS ASSOCIATED WITH THE CONSTITUTIVE AND GOVERNING PRINCIPLES

We have seen that the constitutive and governing principles can be satisfied in very many ways. Indeed, everywhere we have looked, we have found impressive blends that provide superb satisfactions of these principles.

Questions:

- You seem to be finding blends everywhere. Does this mean everything is a blend?
- If so, isn't your theory vacuous because it applies to everything? If blending is everywhere and in everything, how could one say anything focused about it?

Our answers:

Brains can put together elements in very many ways other than blending. When we see a table next to a chair, we are organizing them as spatially adjacent, but we are not blending the table and the chair. We may assume that the chair is for sitting at the table; we may think they were manufactured to go together and were sold together. In this case, we are putting them together in a number of different frames, but we are not matching the inputs, doing selective projection, and blending the chair and the table. When we remember that we left our house, went to the furniture store, and bought the chair and table, we are organizing events as temporally adjacent, but we are not blending our leaving the house with our entering the store, or our entering the store with our buying the furniture. We may further organize the events into a coherent single scenario (leaving the house to go to a furniture store in order to buy the chair and table), but again this brings mental spaces together without blending the sequential events. We may categorize table and chair together as things that are sold in furniture stores or as kitchen furniture, but again we are not blending chair and table. We may look at a restaurant and see first a chair and then a table, so that they are linked in a perceptual sequence, but again without blending. And so on. We have no reason to think that other species cannot perform these organizing operations involving categorization, spatial and temporal sequencing, and episodic memory.

All of these organizational operations are available to put together a skier skiing and a waiter waiting on tables. For example, the airborne skier out of control might land on the waiter at the lodge. Someone might work as a waiter at a lodge and ski during her days off. We might see first a skier skiing down the hill and then the waiter bringing us our martini. We might do the accounting for a lodge that has to take out insurance on both the ski instructor skiing and the waiter waiting on tables. None of these conceptual organizations establishes a cross-space counterpart mapping between the skier skiing and the waiter waiting for the purposes of projecting them to a blend.

Yet as we have seen, there is another way to put them together—into the novice who improves his skiing by making a provisional blend of his skiing and the waiter's carrying a tray.

Of all the ways in which the brain can put two things together, conceptual blending is a relatively small subset. The constitutive principles of conceptual blending already place a very strong set of limiting constraints on mental organizations.

But as small as it is compared to the brain's full range of possibilities, the set of mental organizations that fits the constitutive principles is huge compared to the subset that also meets the governing principles. It is easy to come up with any number of blends fitting the constitutive principles that no one would ever construct because they run counter to the governing principles. For example, it is easy to construct a blend that satisfies the constitutive principles but not the governing principles for compression. Recall the case where we said of some politician's vetoing a foreign aid bill that "he's taking the food out of the mouths of starving children." Here we see a cross-space mapping between two inputs and a compressed blend. But suppose we want to talk about someone's actually snatching food from a child. Why not use the same inputs and the same cross-space mapping to say that the snatcher is "vetoing a bill that would provide foreign aid to a number of countries"? This blend fits the constitutive principles perfectly, but it is entirely alien to us. It takes something that is already compressed and at human scale with a clear story and recasts it as something diffuse with many agents and many patients and no single clear goal. It violates Borrowing for Compression, Scaling of Time and Space for Compression, Syncopating Compression, the Integration Principle, the Maximization of Vital Relations Principle, and the Intensification of Vital Relations Principle. It also violates overarching goals: Achieve Human Scale, Compress What Is Diffuse, Strengthen Vital Relations, and Go from Many to One.

Consider the mirror network You Are My Long-Lost Daughter. Nothing in the constitutive principles prevents us from keeping the cross-space mapping between frame counterparts but projecting those frame elements in the inputs onto different frame elements in the blend. So the father could say to the daughter, "I am your long-lost daughter." This is nearly unintelligible, but not because it violates constitutive principles. It completely violates Topology in a case where the topology does most of the good work. As a result, it ends up violating Web.

The Buddhist Monk is a mirror network in which the shared frames makes it very easy to keep the essential topology of space-time collocation. We have seen a blend that gives an easy global insight into the solution. But now suppose we chose different projections of the moments in the inputs onto the moments in the blend. Suppose that there is a one-to-one mapping between moments in the inputs and moments in the blend, but that there is no additional order to the mapping. So, for example, dawn in the inputs might be mapped to 11 A.M. in the blend, and 11 A.M. in the inputs might be mapped to 9:30 A.M. in the blend. Then it is still true that in the blend, the monk meets himself as the two monks hop back and forth along the path, and this does indeed, mathematically, prove that there is a spot on the path that the monk inhabits at the same time of day in the two different inputs, but no one can get that inference easily out of a blend with a hopping monk. The hopping monk blend fits the constitutive principles just as well as the original Buddhist Monk blend, but it violates Topology (in a particularly perverse manner) and as a result destroys Web almost completely and gives no global insight. It also precludes the pattern completion that comes in the original blend from recruiting the common frame of *two people approaching each other along a path and meeting each other.*

We leave it as an exercise for the reader to recall a good integration network but then, keeping the same inputs, to violate the governing principles to produce horrid expressions like "If the *Titanic* had been President Clinton, it would have been unanimously vilified in the press, instantly impeached by a unanimous Congress, tried summarily in the Senate, and promptly kicked out of office." This is a blend. It fits the constitutive principles. Yet it violates the governing principles so badly that it is hard to believe that it could be the structural analogue to the good Clinton-*Titanic* blend, although it is. We have only switched the topic spaces, but that is enough to force massive decompression.

Seventeen

FORM AND MEANING

> Linguistics is arguably the most hotly contested property in the academic realm. It is soaked with the blood of poets, theologians, philosophers, philologists, psychologists, biologists, and neurologists, along with whatever blood can be got out of grammarians.
>
> —*Russ Rymer*

THROUGHOUT THIS BOOK, WE have emphasized the importance of compression and of maximizing and intensifying vital relations. Double-scope capacity provides human beings with the ability to do remarkable compressions, and every language provides a systematic array for types of compression. Various highly useful compression patterns become conventional and are associated with specific grammatical forms. Having studied one example of this in Chapter 8, we will now consider the general cognitive phenomenon of associating a compression pattern with a linguistic form.

DOUBLE-SCOPE COMPRESSION IN A TWO-WORD NUTSHELL

We can bring two things together mentally in various ways. Blending them is one subset of those ways, and the blends that satisfy the governing principles are a much smaller subset. An even smaller subset consists of those core compression patterns that are entrenched in a culture. The next subset down consists of those entrenched compression patterns that have associated grammatical forms.

It is easy to think that a simple form corresponds to a simple meaning. But as we saw, blending can perform massive compressions and express them in simple forms. So, by virtue of the power of compression and decompression, a simple form can prompt for the construction of an extremely complicated meaning.

Some of the apparently simplest forms in the language consist of putting two words together: Noun-Noun compounds like "boat house," Adjective-Noun combinations like "angry man," Noun-Adjective combinations like "child-safe or "sugar-free."

Through decompression, we can construct a variety of complicated integration networks for any given phrase using "safe," such as "dolphin-safe," "shark-safe," or "child-safe." In all such cases, there is successive blending. We first blend the mental space for the current situation—involving dolphins, sharks, or children—with an abstract frame of *danger*. This yields a specific counterfactual mental space in which *dolphin, shark,* or *child* is assigned to a role in the *danger* frame. This mental space of specific harm is disanalogous to the mental space for the current situation. These two disanalogous spaces are inputs to a new blend, in which the disanalogy is compressed into the property *safe.*

"Dolphin-safe," as it currently appears on cans of tuna, means that measures were taken to avoid harming dolphins during the harvesting of the tuna. "Shark-safe," as applied to, say, swimming, refers to conditions under which swimmers are not vulnerable to attack by sharks. "Child-safe," as applied to rooms, means the rooms are free of typical dangers for children. In every case, from simple forms the understander must construct elaborate integration networks.

How one does this mental work may differ from case to case. In "dolphin-safe tuna," the dolphins are in the role of potential victims. But in "dolphin-safe diving," when said of mine-seeking human divers who are protected by dolphins who are not themselves at risk, the dolphins are in the role of agents of the safety. Contrastingly, "dolphin-safe diving," when said of diving that imitates the way dolphins swim and is therefore safe, uses the manner of swimming associated with dolphins. If we assume that dolphins eat goldfish, then "dolphin-safe goldfish" would cast dolphins in the role of predators. Genetic engineers who are concerned not to produce anything resembling a dolphin might refer to a technique that is known never to lead to a dolphin embryo as "dolphin-safe." The dolphin here does not fill the role of victim, victimizer, causal agent, or role model. In a world in which the most humiliating thing for a shark is to resemble a dolphin, behaviors that are unquestionably sharkly might be called "dolphin-safe." And of course, a compositional theory of meaning immune to our dolphin examples would be hailed as "dolphin-safe."

If the adjective "safe" comes before instead of after the noun "dolphin," as we saw in Chapter 1, then we find another multiplicity of potential meanings. "Safe dolphin" can mean a dolphin who is protected, a dolphin who will not inflict the kind of harm other dolphins might cause, the role of swimmer-at-the-front-of-the-school whose responsibility it is to keep the rest of the school safe from running into obstructions, or the decoy dolphin robot that behaves in such a way as to communicate to other dolphins a situation of complete safety and so lulls them into being caught.

A more specific blending pattern for the compounding of a noun with "safe" arises when the noun refers to an endangered species. Now, we can talk about "turtle-safe nets" or "salamander-safe landscaping" or "hawk-safe agriculture." In order to make appropriate sense of these phrases, one must know the compression pattern chosen by the culture for thinking about endangered species, and also know that it is associated with the Noun-safe form where the Noun picks out an endangered species. Without this knowledge, the meaning cannot be predicted compositionally from the noun and the adjective. What the culture ends up with in this case is a very powerful compression—a maximally simple two-word form that points to an integrated blend and that satisfies Unpacking by evoking mapping schemes and counterfactual spaces that go with "safe." The particular culture that has entrenched this compression has double-scope creativity, a language (English), ecological concerns, food-packing corporations, and grocery stores. Double-scope creativity is universal for our species. The English language offers a syntactic Noun-Adjective form and entrenches certain blend types that go with it, like those prompted for by Noun-safe. The ecological concerns are relatively recent, certainly as applied to dolphins. They make salient and important the scenario of dolphins being harmed by fishing methods, and make it desirable that the scenario be counterfactual. Consequently, the scenario is a good candidate for Noun-safe expression, and a good candidate for marketing departments seeking to induce shoppers to buy their products. This convergence of syntax and commerce creates the compression that appears on cans of tuna fish: "dolphin-safe."

Examples like "dolphin-safe" are useful because they highlight, in a transparent and uncontroversial way, the nature of the blending process. Furthermore, they abound. Think of "cruelty-free" on bottles of shampoo, or the variety of noncompositional integrations running across "waterproof," "tamper-proof," "foolproof," and "child-proof," or "talent pool," "gene pool," "swimming pool," "football pool," and "betting pool."

Familiar compositions like "dirt-brown," "pencil-thin," "red pencil," and "green house" work in the identical fashion, but because they are very deeply entrenched, it is possible to misinterpret them as somehow different in their operations from the examples above. Following Charles Travis, we observe that noncompositional conceptual integration is just as necessary in these "core" cases. As we have said, "red pencil" can mean a pencil whose wood has been painted red on the outside, a pencil that leaves a red mark (the lead is red, or the chemical in the pencil reacts with the paper to produce red, or . . .), a pencil used to record the activities of the team dressed in red, and a pencil smeared with lipstick—not to mention pencils used only for recording deficits. For a set of houses that differ only in location and in the color of the kitchen linoleum, "green house" can mean the house with green linoleum, where "green linoleum" means the one with spots that are green, where "green spots" means spots created with a green pencil, where "green pencil" means. . . .

The scenarios needed for these integrated meanings are no simpler than those needed for "dolphin-safe" and "fool-proof." The cognitive capacities needed to construct these integrated meanings are the same as those needed to interpret the supposedly exotic examples, and these cognitive capacities apply as well to the supposedly central examples like "green house," used to mean a house whose exterior walls, exclusive of windows, shutters, trim, porches, flashing, foundations, and fascia boards, are mostly green on the weather surface. That some interpretations stand out more than others—especially when the phrases are taken in isolation—stems from the existence of strong defaults. This difference has to do with the conceptual and linguistic defaults most likely to be activated in any given situation, not with the mechanisms of integration.

If we look across the central cases like "red pencil" and "government bond," we find that some of the relevant defaults are provided by cultural frames with rich structure, others by generic roles that run across many frames, others by the local situation at the moment of utterance. This last case includes elicitation by linguists and philosophers: The subject is asked to judge an expression in a supposedly context-free way, but in fact must construct a minimal context in which to interpret it. Such minimal contexts typically use the strongest defaults.

How do we go from the linguistic units to the conceptual elements or from the conceptual elements to the linguistic units? In the case of nominal compounds, the formal unit names two elements in two different spaces, and directs the understander to find the rest. We will call these conceptual elements the *named elements*. Consider "land yacht" as a reference to a large, luxurious automobile. Clearly, "land" and "yacht" come from different domains: Yachts are associated with water as opposed to land. "Land yacht" gives us land from one space and yacht from another, and asks us to perform a mapping between these spaces. In this mapping, yacht corresponds to luxury car, land corresponds to water, driver corresponds to skipper, and the road for the car corresponds to the course for the boat.

Figure 17.1 shows how the conceptual blend depends on building an analogical mapping, and how, in the corresponding integrated syntactic form "land yacht," "land" and "yacht" name elements that are not counterparts in the mapping. "Land yacht" now names the new element in the blend, even though it names nothing in either of the inputs, and even though *land* is not a counterpart of *yacht*. Formal expression, in this case a two-word combination, prompts for the construction of the blend and provides a way of naming part of the emergent structure. The general conceptual and linguistic patterns are the same in the earlier "prairie schooner."

Consider "Language is fossil poetry." "Fossil poetry" works just like "land yacht": *Fossil* comes from the domain of paleontology and *poetry* from the domain of expression. In the mapping, poetry corresponds to the living organism, while language corresponds to the fossil of that organism. The conceptual elements named in the integrated syntactic form "fossil poetry" are not counterparts in the conceptual mapping.

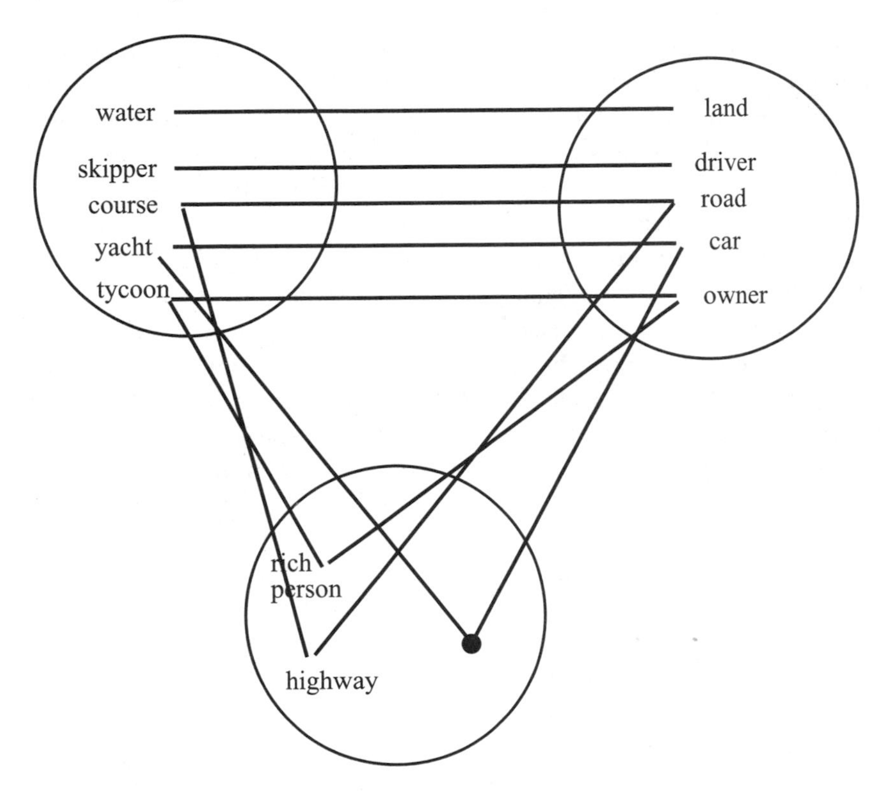

FIGURE 17.1 LAND YACHT

Now let us look at the kinds of compressions that can be provided by such two-word Noun-Noun nutshells. Consider "jail bait," a phrase used to refer to an underage girl whom an adult man finds sexually attractive. "Jail" comes from the domain of human criminality, while "bait" comes from the domain of fishing or trapping. In the mapping between them, attraction to the girl corresponds to attraction to the bait, initiating sex corresponds to swallowing the bait, and ending up in jail (for sex with a minor) corresponds to being caught. The conceptual elements named in the integrated syntactic form "jail bait" are not counterparts in the conceptual mapping. Here, obviously, we are prompted to borrow the compressions and intensities of the *fishing* frame for the purpose of compressing the *sex with a minor* frame and intensifying many of its vital relations. For example, the causal chain in the *sex with a minor* frame, which runs from perception to incarceration, can be long and diffuse, whereas the *fishing* frame has direct human-scale causation: A single bodily action results in immediately being caught. There is extraordinary emergent structure. In the blend, the man is not to blame. In the space of fishing, the fish does not know that the bait is bait. In the space with the man and the minor, the man certainly does know about laws and jail and he recognizes that sex with the girl is legally forbidden. But in the blend, he is blameless for the action, indeed even the prime victim, even though he understands the law, the prohibition, the possible punishment, and the reasons for it. The "jail-bait" blend may acquire further emergent structure through the Intensification of Vital Relations Principle. In the fishing space, the intentionality is in the fisherman's attempt to trick the fish and catch it. In the other input (*sex with a minor*) the intentionality is in the man's attraction. There is no counterpart for the fisherman in the space with the minor, but it is nevertheless possible to project something like the fisherman's intentionality into the blend. One corresponding interpretation holds the girl herself responsible for what befalls the man. Another might bring in the Devil. Yet another might bring in the injustice of society and its laws.

"Jail bait" is an example of a two-word nutshell that prompts for compression through borrowing of the inner-space relationships in the *fishing* frame. In that compression, Time is scaled down and a diffuse interpersonal interaction with many actions is compressed into a single action—swallowing the bait. This compression can create relations in the blend, such as the attribution of intentionality to the young woman. There is also highlights compression: The sequence in the human story of perception, greeting, seduction, doing the deed, having it become known, being arrested and tried and sentenced and jailed is all compressed in the blend into seeing and doing, where, because taking the bait is automatically taking the hook, there is no separation between committing the act and being punished. This is an intense Cause-Effect compression. In the blend, the Effect is literally in the Cause because the hook is literally inside the bait. "Jail bait" is said as *advice:* The compression is meant to focus the man on the Effect by making it part of the Cause, thus giving him powerful global insight.

By contrast, we have seen many cases where it is *outer-space* vital relations between the inputs that are compressed in the blend, as in "caffeine headache," "money problem," and "nicotine fit," analyzed in Chapter 11, in which the Disanalogy between the inputs is compressed into a Property in the blend. For example, money problems are a certain kind of problem, the ones caused by absence of money. Again, we see the simplest possible linguistic form prompting for remarkably complicated integration networks. Communicating through simple grammatical forms is possible because cognitively modern human beings can bring to bear on those forms all of double-scope integration and its governing principles and overarching goals. The language itself does not have to carry such operations as compression or pattern completion because human brains supply those operations at no linguistic cost.

One common aspect of these compounds is that someone attempting to "unpack" the linguistic form does not begin from the assumption that the named elements are necessarily conceptual counterparts. When presented with such a linguistic form, we cannot predict, *a priori*, the relationship between the named elements. Notice that the generic roles of these elements are different in "land yacht," "fossil poetry," and "jail bait." "Land" is a locative, "fossil" is a product of a process, and "jail" is a result. "Yacht" is a means, "poetry" an activity and its product, and "bait" an instrument.

Now consider "boat house." The same operations are involved. As in "land yacht," we have a connection between the two different spaces of land and water: Houses are associated with land, boats with water. In the mapping between them, the residents of the house correspond to the boats, the house itself corresponds to a protective shelter for storing the boats, and leaving the house corresponds to being launched. "Boat" and "house" name elements that are not counterparts in this mapping.

Of course, there is no restriction that prevents the named elements from being counterparts. Consider "house boat," which again evokes two different spaces of land and water. In the space of land, the resident lives in the house; in the space of water, the sailor is aboard the boat. In the formal integration "boat house," "boat" and "house" *are not* conceptual counterparts; but in "house boat," the boat and the house *are* conceptual counterparts, and they map onto a single element in the blend. Similarly, "jail house" evokes a domain of domestic residence and a domain of criminal punishment. In the mapping between them, the jail and the house are conceptual counterparts, and they map onto a single element in the blend. This corresponds to the possibility we saw in Chapter 8 for "Y-of" expressions like "city of London" and "burden of guilt" where Y and Z are counterparts. As Christine Brooke-Rose shows in great detail, Noun-Phrase-of-Noun-Phrase can name metaphoric counterparts, such as "fire of love." Charles Fillmore gives the example "One needn't throw out *the baby of personal morality* with *the bathwater of traditional religion.*" These counterparts need not be metaphoric: "the nation of England," "the island of Kopipi," "the feature of decompositionality," "the condition of despair."

In all of these cases, including those in which the syntactic form names elements that are blended conceptually—"house boat" and "jail house"—the blend is both less and more than the composition of the input spaces. In "land yacht," we ignore that yachts have cooking and sleeping facilities and require no manufactured course. Yet the blend contains more than the inputs: For example, the inputs may supply the knowledge that we are dealing with a vehicle, but not that it is a car as opposed to something else, or that many specific features that we link with luxury cars belong to the land yacht: electric windows, leather upholstery, opera windows, and suspension built for comfort rather than handling.

In "fossil poetry," on the one hand, we ignore that fossils are typically associated with extinct species, and generally that poetry is not physical or biological. On the other hand, the conception of language as a derivative of poetry, which is the central inference of the blend, is absent from the input space. In "jail bait," we ignore that someone intends to lure the fish while perhaps no one intends to lure the man. We ignore that the man is neither a fish nor (in the case where he merely admires) a criminal. In the blend, we make use of the particular social frame according to which the world is full of pitfalls and traps for the man, teasing him with what it forbids. In the space with the fish, fish are not capable of such a perspective, while in the space of criminal action, the world does not necessarily tempt people to commit crimes.

The situation is no different when the named elements happen to be conceptual counterparts. In "house boat," we ignore that houses have yards and are stationary or that boats are designed principally for travel. We also know many things from background knowledge about house boats that are not derivable from the inputs. We know that a house boat cannot be simply a regular boat put on land that happens to have people living in it, or a regular boat at mooring that someone has been living in; but there is nothing in blending or the use of language to prompt for integration networks that would forbid these meanings.

We see in these examples the falsity of the general view that conceptual structure is "encoded" by the speaker into a linguistic structure, and that the linguistic structure is "decoded" by the hearer back into a conceptual structure. An expression provides only sparse and efficient prompts for constructing a conceptual structure.

The problem, then, is to find the relations between formally integrated linguistic structure on the one hand and conceptually integrated structures built by the speaker or retrieved by the hearer on the other. In general, we will find that the conceptual integration is detailed and intricate, while the formal integration gives only the briefest indication of a point from which the hearer must begin constructing this conceptual integration.

BLACK KETTLES AND BROWN COWS

Another deceptively simple construction is Adjective-Noun, again a two-word nutshell. We have already seen with "safe" that the adjective prompts for a specific complex mapping scheme and conceptual blend. While nominal compounds give named elements, one from each input, and leave the blending to us, the adjective in an Adjective-Noun compound brings its own mapping scheme for the resulting integration network. The mapping schemes are especially distinctive for adjectives such as "safe," "likely," "possible," "eligible," and "fake." As we discussed, the mapping scheme for "safe" includes a general scenario in which something is harmed, a specific real scenario, the application of the harm scenario to the real situation to produce a counterfactual scenario in which some particular thing in the real situation is harmed, and creation of a blend in which there is now "absence of harm."

Eve Sweetser considers the case in which "likely candidate" means not someone likely to become a political candidate or succeed as a political candidate but, for example, a political candidate likely to grant an interview. As she writes, "So long as we can think up a scenario relative to the candidate in question, and evaluate that scenario for likelihood, *likely candidate* can mean the candidate who figures in the scenario we have labeled as likely." On her analysis, conceiving of such a scenario and evaluating it consists of finding a blend of the frame for likelihood, conceived of as probability of occurrence in a sequence, and the frame for candidate. Like "safe," "likely" prompts for a blend. "That is a likely event" decompresses into two contrasting outcome spaces, one expected and with the event, the other not expected and where the event has no counterpart. In the blend, the event has the property *likely*. "Likely person" this time decompresses into a series of events that share the same frame and are disanalogous only with respect to the value of some role, and marks one of those events as *likely* in the sense we just discussed. For example, if the frame is *giving an interview*, "likely person" can be interpreted as indicating that the contrasting event space that has that person in the role of interviewee is an expected one. The property *likely* in this case is a double compression of contrasting expectations and disanalogy on the value of a role. If the noun modified by "likely" evokes a particular frame, that frame may be selected as the frame that organizes all the contrasting events. "Likely candidate" can therefore mean "someone likely to become a candidate" or "a candidate likely to be chosen or elected." But, as Sweetser shows, other frames can be used. "Likely candidate" can be used to mean a candidate likely to give an interview; "likely suspect" is similar, meaning alternatively "a person likely to be deemed a suspect" or "a suspect likely to be guilty" or "a suspect likely to be convicted." But the scenario of likelihood in the blend need not be the one evoked by the noun. Sweetser's reading, "a candidate likely to grant an interview," does not use the likelihood of either becoming a candidate or being chosen. If every year the governor pardons someone in the jail house, we might bet on who is the "likely suspect." Similarly, "possible textbook" may refer to a textbook that might possibly be chosen as required reading in a college course.

Just as the different meanings of "safe" may go unnoticed, so may the different meanings of "possible" and "likely." But from a logical standpoint, a "possible textbook" in the sense of an existing textbook that might be adopted is not the same as a "possible textbook" in the sense of a textbook that might be written or of an existing popular book that could serve as a textbook.

Seana Coulson has shown that the adjective "fake" carries an especially elaborate blending and mapping scheme. It calls for two input spaces with a Disanalogy connector, such that an element in one space is "real" but in the other space its counterpart is not. For example, the "fake" of "fake gun" prompts a mapping between an actual scenario with an agent and an instrument, and a counterfactual scenario in which the counterpart of that instrument is a real gun and some other participants react accordingly. In the blend, the reactions and beliefs of other participants are projected from the counterfactual scenario, while the nature of the object and the beliefs of the gunman are projected from the actual scenario. The entire situation is now construed through the blend, and the language for picking out the element in the blend can be used to pick out its counterpart in the actual input space. In the blend, the counterfactual vital relation of this interplay between the two perspectives is compressed into a Property of the element identified by "gun" (through projection from the "real" input). In the case of fake flowers meant to decorate a dining table, there is an input space in which the objects are flowers and a space in which they are not, because they are silk or rubber or for some other reason. There is a Disanalogy connector between the flowers in one space and the silk or rubber object in the other space. This relation is compressed into a property of the object in the blend. It is now a "fake flower." Note there need be no indication that any of the participants believes the object is a flower. It can be treated as a flower, perhaps for aesthetic purposes, by projection from one space, and as not a flower by projection from the other space. One might enjoy it but not have to water it. "Fake money" requires two spaces, one in which the object is money and the other in which it is just paper, for example. There are many possible projections to the blend. For example, the fake money might not be legal currency but still have monetary value because it is issued by a casino for in-house use. Alternatively, it could be used to fool someone by making him a worthless payment. In that case, the space where it is money is also the belief space of the person fooled, and the space where it is paper is the belief space of the crook.

Adjectives like "safe," "likely," "possible," and "fake" compress complex outer-space vital relations between inputs into properties of elements in the blend. But so do less remarkable adjectives like "little" and "big." First, it is well known that such adjectives are "relative." Something is little or big only relative to a standard. Furthermore, the notion of little or big relative to a standard presupposes comparison with the standard and, in general, comparison of two objects. Following Euclid, it is natural to think of this comparison as consisting of bringing the two objects into an encounter with each other—superimposing them, in Euclid's case, or putting them next to each other and looking at the result. Whether executed mentally or physically, this encounter is a blend, with composition and emergent structure. In the blend, one is part of the other in the case of superposition, or the top of one is lower than the top of the other in the case of placing them next to each other, or some similar relationship emerges from the encounter. In the case of superposition, for example, the outer-space disanalogy between the two objects corresponds to the inner-space relationship of part-whole. In the same way, we can compare three objects of different sizes, 1, 2, and 3, so that in the blend, 1 is part of 2 and 2 is part of 3.

These part-whole relationships in the blend correspond to the outer-space relations in which 1 is smaller than 2 and 2 is smaller than 3. With this blend in hand as one input, and the general schema of a category and its central example as the other input, we make a new blend in which we have three members of the category and the central example is 2. This blend is a general schema of comparison of three elements of some category, and in this blend, 1 is little, 2 is standard, and 3 is big. The outer-space disanalogy relations in the network have as their counterpart the inner-space properties of little, standard, and big.

For these properties to actually apply to anything, two further blends are required. In the first, the general schema of little, standard, and big is one input and the other input is a specific category (e.g., elephants, butterflies, planets) and its central example. In this blend, there is a standard prototypical elephant, for example, and also a metric for elephants on which this one is the standard size. To judge the size of a particular elephant, we need one more blend, which has as one input the standard elephant and its spot on the elephant metric, and, as the other input, the particular elephant. In the blend, the particular elephant will be little, standard, or big.

This imaginative compression of relations into properties follows lines we have seen many times. The first integration network, which gives us 1 as part of 2 and 2 as part of 3 in the blend, is a mirror network: The input spaces have the same frame, in which there is a particular type of object with a specific size. The next integration network, which blends the 1–2–3 blend with the general frame for a category and its central example, is a double-scope network: In the blend, for example, 2 is both the central example and the container for 1 and part of 3. The third integration network is a double-scope network where the two organizing frames are completely compatible: The general *little-standard-big* frame, which already includes the structure of category and prototype, is blended with a particular category and its prototype. The last integration network is a simplex network, in which the particular category, its prototype, and the *little-standard-big* metric for that category are blended with a particular element of that category.

Because *little* and *big* strike us as simple properties, it is natural to think that the conceptual work underlying their meaning must be simple. But just as the simple perception of a color is caused by extremely complex perceptual work, so is the simple recognition of something as little or big.

And brown cows? It's the same story, a story partly told already by Charles Travis for black kettles and green apples. Suppose we are looking at a kettle and Charles Travis says "This kettle is black" and we all agree that it is indeed black. But then Charles wipes the soot off the kettle and we see that it is green, so it isn't black. But is it really green? Charles scrapes off the green paint, and we see that the underlying metal is black, so maybe the kettle is black after all. But now as the sunlight hits it, we see that it is actually dark brown. Charles gives us purple glasses, and now we see that the kettle is black. The moral of this first part of

Travis's story is that there is no absolute sense in which the property *black* applies to the kettle or not. Particular circumstances and contextual presuppositions make it appropriate or inappropriate to call the kettle black.

The second part of the story is that we can call the kettle black if it has a black decorative design; if it has a black spot while all the other kettles have green spots; if it is one of two identical kettles with no black color on their surfaces and it is on the black oven instead of the white one; if it is the one that came in the black box; if it was manufactured by a company owned entirely by blacks; and so on. Each time "black" is used in these cases, we feel that it is indicating a Property of a kettle. And at the same time, we see from the examples above that there is no such stable property in any absolute sense that could be defined for the kettle once and for all independently of anything else. How can this be? The answer is that "black" is like the other adjectives we saw. It prompts for an integration network and compresses some outer-space relations into the inner-space Property *black* in the blend.

Note that we focus exclusively on cases where it seems that the color black has been assigned literally to an element, and not on metaphoric or metonymic uses such as "black magic" or "black arts." In what looks like strict color assignment, "black" followed by a noun prompts for a mapping scheme in which one space has colors (in particular, the color black) and the other space has the element picked out by the noun. We are to find a cross-space mapping between the color black and something salient in the other space. Typically, the noun might pick out an object one of whose visible parts has a color that is close to black. In the integration network, that color is mapped onto the color black in the space of basic colors. By selective projection, only the color black from the color input is projected to the blend, while the object and its environment are projected from the other input. Thus, in the blend, the object itself has the Property *black* to the exclusion of any other colors. And as the black spot example shows, this is possible regardless of the actual objective amount of other colors on the object.

But parts of an object are only one kind of salient aspect of an element in a space. As the oven example shows, what the object is sitting on can be salient. Similarly, what it contains can be salient (a red cup containing black paint can be a "black cup" next to a red cup containing white paint). Also, the producer of the object can be salient, so the kettle produced by blacks can be the "black kettle." But the element in the space evoked by the noun does not have to be a prototypical object. We can refer to a "black sky" where "black sky" can mean that a few ominous clouds are dark enough to be mapped onto black in the basic color space.

Evidently, color adjectives prompt for specific complex integration networks in which outer-space connections correspond to color Properties in the blend. This blending scheme explains why the red of "red ball," "red lipstick," "red hair," and "red fox" can be very different shades. The color of the fox, for instance, is mapped onto *red* because the mammalian fur colors are mapped onto the basic colors black, red, white, brown, yellow, and grey, and the fact that these fur colors are basic for mammalian fur is mapped onto the fact that black, red, white, brown, yellow, and grey are basic for color in general. This outer-space mapping between the inputs is compressed in the blend so that the fox does indeed have the Property *red.* In the blend, the color of the fox is now a central instance of the color category *red.* A fox spray-painted candy-apple red could still be called a "red fox," and in the blend would have the Property *red,* but now this red would count as a nonstandard instance of the category *red.* In the blend, the color of the everyday red fox comes from the input of furry animals, but its status as basic and red comes from the input of basic color terms.

So, the interpretation of "brown cow" turns out to involve an integration network, and "little brown cow" turns out to involve all the major types along the continuum of integration networks, from simplex to mirror to double-scope. This is not as remarkable as it might seem. Semanticists who have looked closely at size and color terms have produced ever more complicated analyses of how they work. The most sophisticated analysis of size terms is perhaps Ronald Langacker's, which recognizes the same degrees of complexities that we do. The folk notion that simplicity of form, conventionality of vocabulary, and frequency of use indicate simplicity of underlying conceptual organization has been thoroughly recognized as a fallacy.

That blending is a component of many diverse constructions should be no more surprising than that chemical bonds are ubiquitous in chemicals. In pointing out the role of blending, we are not claiming that the phenomena in question reduce to just blending.

FORMAL BLENDING

Novel conceptual blends do not generally need novel forms of expression. A language already has all of the grammatical forms it needs to express almost any conceptual blend. For example, all the two-word expressions we have considered in this chapter use existing syntax: compound forms like Noun-Noun, Adjective-Noun, and Noun-Adjective, as well as particular existing nouns or adjectives filling those forms. The concept *land yacht* may be a new blend, but the phrase uses existing grammar and vocabulary to prompt for the *land yacht* integration network.

Forms are mental elements, and they can be blended just like any mental elements. Sometimes, this blending will align with the blending of conceptual structure to which the forms attach. We can see this pressure to achieve formal organization to express conceptual blending across a range of constructions, from morphemes to sentences. Consider single-word integrations like "Chunnel," referring to the tunnel that runs under the English Channel. Clearly, there is a conceptual construction that integrates structure from both the abstract frame of a tunnel and the specific frame of the body of water between England and France. This integrated unit can serve as the site for integrating a great range of knowledge, from the relevant geology to problems of engineering, from the history of relations between England and France to issues of quarantine, disease, and ecology. Its corresponding grammatical form, "the tunnel under the English Channel," is already tightly integrated. English makes available an even more compact compound-noun construction, "the Channel tunnel." By fortuitous accident, a further integration of form is possible, given the phonemes in "Channel" and "tunnel." This integration is a formal blend, triggered by a partial phonological and orthographic mapping between the two words. Pressure to integrate produces (in the case of English) "Chunnel"; the corresponding accidents are lacking in French, leaving as the most integrated form "tunnel sous La Manche." This shows another important aspect of integration: It is opportunistic. The fact that this opportunism depends in any specific case upon apparently peripheral accidents can lead to the mistaken view that the operation itself is peripheral. But in actuality, it is precisely through opportunistic exploitation of accidents that the most central events and structures can arise. Evolution teaches us that this is not paradoxical.

Suzanne Kemmer reports the following example of integration:

> "The Whiners' most common complaint is that they've been relegated to what Mr. Coupland calls 'McJobs.'"

"Mc" evokes a space of fast food and employment in that industry. "Jobs" evokes the more general frame of seeking employment. They have roles in common (workers, employers, wages, benefits, possibilities for advancement, etc.), providing a generic foundation for the blend. Two input spaces are set up, one for aspects of McWorld and another for seeking employment. A straightforward mapping links common roles. But that mapping does not in itself provide the central inference of "McJobs." Specific aspects of jobs at McDonalds—no prestige, no chance of advancement, no challenge, no future, boredom, a certain kind of social stigma—are blended with the more general notion of low-level service jobs. Absent this blend, we would be free to associate low-level service jobs with other stereotypes—altruistic and even saintly devotion to others, climbing of the social ladder, small-town serenity and routine, freedom from avarice and grueling ambition. So this simple blend brings in analogical mapping, the construction of a generic space, and, in the blend, such new categories as the McJob, the type of person who has the McJob, and the pay scale for a McJob. One purpose is to bring inferences from the blend to the conception of crucial realities—such as the plight of young people in the modern economy—and to influence legislation and government policy. The blend's power and efficiency seem to derive from its homogeneous internal structure and its corresponding formal compression into a single word. The striking thing about this unit is that it creates new conceptual structure while all the conceptual engineering that went into building it can be retrieved by a member of the relevant linguistic and cultural community from the single word "McJobs." We see that the blend satisfies governing principles at the conceptual and formal levels: It offers maximal formal and conceptual Compression and Integration and satisfies both Web and Unpacking.

Formal blending can occur independent of whether there is any background conceptual blending. An ad on the back of a bus for the Del Mar racetrack, whose post time is 2:00 P.M., reads: "Hunch hour. 2 P.M." To get the pun, we must access both "hunch" and "lunch hour" simultaneously. This means doing pattern completion from "hunch hour" to "lunch hour." It means doing a partial mapping of "hunch" to "lunch hour": "Hunch" and "lunch" are noun phrases that differ only in their initial phoneme. It means projecting both the "unch" of "hunch" and the "unch" of "lunch" onto the "unch" of "hunch hour," and projecting the nominal compound structure (N1 N2) of "lunch hour" onto the corresponding nominal compound structure of "hunch hour." And it means projecting the common noun status of the counterparts ("hunch" and "lunch") onto the noun status of "hunch" in the blend.

Suppose one speaker in a group wonders aloud whether the new mall down the road is still open at 9 P.M., and another speaker responds, "Well, they should be, since everyone knows about *Amahl and the Night Visitors*." Here, "Amahl" in the title of the opera is blended with the noun-phrase "a mall," but there is no conceptual blend. Or consider the caption in *Latitude 38* accompanying pictures of the 1994 Vallejo boat race, which had two legs, with the upwind leg particularly sunny: "Vallejo 94—Two legs, sunny side up." This caption requires formal blending, partial projection, mapping between forms, pattern completion, and so on, but no conceptual blending: The race is not blended with a particular breakfast dish.

Sometimes, the formal blend parallels the conceptual blend very closely. *The Atlanta Constitution* of February 17, 1994, carried a front-page caption reading "Out On a Limbaugh," followed by a summary of the story to be found inside: "Critics put the squeeze on Florida's citrus industry for its $1 million deal with broadcaster Rush Limbaugh." To get the punning effect of "Out On a Limbaugh," we must access "out on a limb" and "Limbaugh" simultaneously. Behind this formal blend is a conceptual blend with two input spaces: one with an agent who climbs out on a tree limb, another with the deal between the Florida citrus industry and Rush Limbaugh. Just as *limb* and *Limbaugh* are blended conceptually, so "limb" and "Limbaugh" are blended formally. In this particular unusual case, conceptual counterparts that are conceptually blended have formal expressions that are formally blended. In fact, the formally blended element refers to the conceptually blended element. The formal blend, as is standard for a blend, contains formal structure that is not calculable from the formal inputs. Let us look at these formal inputs. "Out On a limb" has an indefinite article with a common noun. "Limbaugh" is a proper surname. Although a proper surname in English can become a common noun indicating one of a group of people with that surname ("She's a Kennedy," "She's the poorest Kennedy") or one of a group of people analogically equivalent to a particular person ("He's an Einstein"), this is not what is going on here. Nor is "a Limbaugh" a case of an indefinite article used with a proper noun referring to an unknown person who happens to have that label, as in "There is a Fidelia Cumquat on the telephone for you." In "Out On a Limbaugh," "Limbaugh" has not become a common noun, referring to namesakes or analogues of Limbaugh; nor is it picking out an unknown person. On the contrary, it is picking out an extremely well-known particular person. Nonetheless "Limbaugh" follows an indefinite article; this is a property of its counterpart "limb" in the other input to the blend. As a result, the formal blend has new syntactico-semantic structure—namely, indefinite article + proper name associated with a known person.

In the Limbaugh example, we see formal blending and conceptual blending in close parallel, but this mirroring is very rare. We have already seen cases of formal blending that have no corresponding blend at the conceptual level. Conversely, most conceptual blending happens with no corresponding formal blend. For example, "Look before you leap" is a standard prompt for metaphorical blends that has no formal blending. There are even cases where conceptual blending and formal blending are both at work but the formal blend runs contrary to the conceptual blend. For example, contestants on the BBC game show "My Word" were challenged to come up with an intelligible expression identical to the title of any popular show song except for a single letter in the last word—for example, "When you wore a pink carnation/and I wore a big red nose." One contestant's response was "Why can't a woman/be more like a mat?" (The original, from *My Fair Lady*, is "Why can't a woman/be more like a man?") At the formal level, "man" and "mat" are blended, but at the conceptual level, *man* is to be blended not with *mat* but with its opposing element: *person who walks on the mat.*

The impulse to do formal blending for its own sake, and the corollary disposition to find conceptual blends behind the formal blends, is evident in many jokes, such as the following anonymous obituary, posted widely on the Internet:

> Veteran Pillsbury spokesperson, the Pillsbury Doughboy, died yesterday of a severe yeast infection and complications from repeated pokes to the belly. He was 71. Doughboy was buried in a slightly greased coffin. Dozens of celebrities turned out, including Mrs. Butterworth, the California Raisins, Hungry Jack, Betty Crocker, the Hostess Twinkies, Captain Crunch, and many others. The graveside was piled high with flours as longtime friend, Aunt Jemima, delivered the eulogy,

describing Doughboy as a man who "never knew how much he was kneaded." Doughboy rose quickly in show business, but his later life was filled with many turnovers. He was not considered a very smart cookie, wasting much of his dough on half-baked schemes. Despite being a little flaky at times, he even still, as a crusty old man, was considered a roll model for millions. Doughboy is survived by his second wife, Play Dough. They have two children, and one in the oven. The funeral was held at 3:50 for about 20 minutes.

COMPLEX GRAMMATICAL STRUCTURES

We have already seen in detail how many grammatical constructions work:

- single words like "safe."
- resultative clausal constructions like "He boiled the pan dry" and "She bled him dry."
- Y-of networks like "Ann is the boss of the daughter of Max."
- nominal compounds like "boat house," "house boat," and "jail bait."
- Adjective-Noun compounds like "guilty pleasures," "likely candidate," and "red ball."
- morphological combinations in a single word like "Chunnel."

In all these cases we saw that the construction in the language had a stable syntactic pattern that prompts for a specific blending scheme. The blending scheme carries with it a particular kind of compression. In Chapter 8, we studied "of" constructions such as "Prayer is the echo of the darkness of the soul" and showed that they prompt systematic blending schemes that deliver compressed megablends over multiple input spaces. And in Chapter 9, we saw that "I boiled the pan dry" compresses a long, diffuse chain of events into a blend in which a single agent performs a single act that causes a result for a single object.

A language is a powerful culturally developed means of creating and transmitting blending schemes. As we also saw in Chapter 9, the capacity for language depends intricately on the capacity for blending and compression. The patterns we find in a language are the surface manifestation of blending schemes that have emerged within a culture and that have wide applicability. Over the last 50,000 years, cultures have developed many systems for saving people from the work of inventing all the useful blending schemes from scratch. The most obvious and perhaps most powerful way cultures provide children with useful blending schemes is through language.

In linguistics, the forms in language have been studied under the term "syntax." It turns out that, however one looks at them, these language forms are exceptionally complex, much more so than we ever realize. So, notoriously, the study of syntax is also complex. Yet that study is essentially incomplete if we do not simultaneously study the blending schemes for which these language forms prompt. Even when the language forms are simple—as in "jail bait"—the corresponding integration networks can be very complicated.

We will now analyze the central role of blending in some more elaborate constructions. This will give an idea of the general inseparability of language, blending, and compression. Language has elaborate formal patterns because it prompts for powerful blending schemes. But the formal patterns and blending schemes are so deeply entrenched as to be almost invisible to consciousness.

CAUSED MOTION

One of the most common and familiar human scenes involves moving an object: We throw it, kick it, hurl it, or nudge it, and it moves in a direction and lands somewhere. This is the "caused-motion" scene. It contains an agent who does something, and that act causes an object to move. There are verbs whose entire job is to indicate some specific version of the caused-motion scene, like "throw," "hurl," and "toss." Many, perhaps all, languages have such verbs and a pattern into which they fit. In English, that pattern is Subject-Verb-Object-Place, as in "*Jack threw the ball over the fence*." But in English, unlike most languages, this pattern can also be used with verbs that in themselves do not express caused motion, verbs like "walk," "sneeze," and "point." So we get "I walked him into the room," "He sneezed the napkin off the table," and "I pointed him toward the door." This pattern can even accommodate verbs that involve no physical motion at all—verbs like "tease," "talk," and "read." So we get "They teased him out of his senses," "I will talk you through the procedure," and "I read him to sleep." In these examples, the subject performs some action and this causes the object to "move" either literally or metaphorically, in a "direction" either literal or metaphoric. The caused-motion construction has been studied in great detail by Adele Goldberg.

In all of these cases, the form is prompting for an integration network that has the caused-motion scene as one input and some other diffuse scene (involving walking, or throwing, or sneezing, or talking, or reading, etc.) as the other input. In the case of "He sneezed the napkin off the table," the diffuse input consists of a sequence of events in which there is a person, a napkin, and a table; the person sneezes; the sneeze moves the ambient air; the air stream impacts the napkin; and the napkin, because it is light, moves under pressure from the air stream, typically reaching the edge of the table where gravity causes it to move in a quasi-parabolic path (except for air resistance) to land on the floor, which it cannot penetrate, because, after all, it is only a napkin. In the diffuse input, we have an action (sneezing), an agent, and a motion by an object (the napkin) in a direction. The action is causally related to the motion. In the compressed caused-motion input, we have an agent, an action-motion, an object, and a direction. Conceptually, there is a natural mapping from the caused-motion scene to the diffuse input: The agent maps onto the agent, the object onto the object, the direction onto the direction, and the action-motion onto any of a number of distributed candidates—the action, the causal relation, or the motion.

In the compressed input, there is a syntactic form associated with the conceptual compression. In the diffuse input, particular words like "sneeze" and "napkin" and "off" and "table" are associated with individual events and elements. In the full integration network, the conceptual compression and the syntactic form come from the compressed input, while some individual words come from the diffuse input through selective projection. In the case of "He sneezed the napkin off the table," we integrate the conceptual structure of the caused-motion scene with many elements and events from the diffuse input: the agent, the action, the causation, the object, the motion, and the direction. The single action-motion in the caused-motion input maps to at least three different elements—the action, the causal link, and the motion in the diffuse input. We also integrate the caused-motion syntax (Subject-Verb-Object-Place) with a few of the words available for the diffuse input—in this case, a word for the agent ("he"), a word for the action ("sneeze"), and some words for the object and the direction.

But notice that we do not bring in a word for the causal link from the diffuse input, or a word for the motion of the object. Because the mapping from the action-motion to the diffuse input is not one-to-one, we have many possibilities for projecting words from the diffuse input. We might project a word for the causal link rather than the action or the motion, as in "Sarge let the tanks into the compound." This example typically evokes a military situation in which Sarge's permission is needed for the tanks to enter the compound. The sentence does not specify the particular causal action performed by Sarge (waving his hand, signing a paper, giving a verbal OK by telephone) or the motion of the tanks (being carried in on trucks, airlifted by helicopters, moving on their own power). We might project a word for the motion of the object, as in "He rolled the barrels into the warehouse," where it is the barrel that rolls, not the agent. The sentence does not specify the particular causal action (pushing the barrels, kicking them down a ramp, pressing a button to release the queue of barrels, . . .), nor does it specify a causal link. In fact, the single integrated action-motion in the caused-motion input includes manner of both action and motion, as we see in specifically caused-motion verbs like "throw," "push," and "hurl," which indicate something about both the manner of the action and the manner of the motion of the object. So the single action-motion connects to the manner of the action and the manner of the motion in the diffuse input as well, and we can project words for those manners into the blend, as in "He floated the boat to me" and "He wiggled the nail out of the hole." The scenes evoked by these examples include action, motion, and manner, but the words that are projected, "floating" and "wiggling," do not themselves require motion along a path or external action. They focus on the manner of being.

Such constructions offer ready-made and powerful blending schemes. A tightly compressed frame and a corresponding syntactic form from one input can be recruited into a blended space linked to a diffuse input. Constructing a network based on that scheme for a particular case depends crucially on being able to construct a generic space that applies to the two inputs. In the case of the caused-motion examples we have seen, this generic space has agent-action, object-motion, and direction. This description also fits the very striking example "They prayed the two boys home." Here the blend is performing an extreme compression: The scene with the prayer and the boys contains many causal steps over an expanse of time, and intermediate agents, with relatively weak or vague causality, but in the blend there is a single action that is directly causal for the boys' coming home.

As we have seen many times in this book, metaphoric mappings provide one of the standard ways of locating a cross-space mapping between inputs and constructing an integration network. If the diffuse input has causation of change of states, then there is already an existing template for blending states and locations and changes of states with changes of locations, and that template can be recruited wholesale to provide much of the cross-space mapping and much of the projection to the blend. For example, "I pulled him out of his depression" prompts for a network where one input is the caused-motion scene with its syntax and the other input has a complicated interpersonal causation involving a change of psychological state. We automatically construct much of this particular caused-motion integration network by recruiting the states-locations blending network. "Pull" is a prototypical caused-motion verb, but the analysis works identically for "I talked him out of his depression" or "He drank himself into oblivion."

We have seen cases where the cross-space mapping is a bundle of Role-Value mappings in a simplex network (as in "He threw the ball over the fence," where *throw* is a value for the role *causal action-motion*), and cases where the cross-space mapping is analogical or metaphorical ("He drank himself into oblivion"). The caused-motion network also operates in interesting ways when there are counterfactual connections. Adele Goldberg offers the example "Pat blocked Chris out of the room"; others include "We barred him from the building," "John forbade him from participating," and, with the metaphorical connection, "We kept him out of trouble." Such examples show that the compressed input is more general than just caused motion. The more general frame is that of an agent exerting a force and of an object undergoing a force in the direction of a goal. In the simple case, the agent exerts force on the object in the direction of the goal, as in "We moved the wolf to the door." In this case, "move" indicates the application of a force by the agent, "to" indicates that the force is in the direction of the goal, and "door" indicates the goal. But alternatively, the force exerted by the agent can oppose the movement of the object, as in "We kept the wolf from the door." Here, the verb "keep" indicates that a certain force is applied in opposition to an existing force, "from" indicates that this applied force is in the direction away from the goal, and "door" indicates the goal.

We see, then, that the general compressed input in a "caused-motion" network is force-dynamic and applies equally well to the caused-motion and blocked-motion examples. The blocked-motion examples have an implicit counterfactual space: The verb "keep" implies that the agent's force opposes the object's force, so if the agent's force disappears, the object moves to the goal.

EMERGENT SYNTAX

In the caused-motion construction, the syntactic component comes entirely from the compressed input space of integrated caused motion, while words come from the space of the events associated with the causal sequence. But there are other constructions in which the syntactic form used for the blend does not come entirely from one space. Part of it comes from one space, part of it from the other, and part of it develops specifically for the blend. In this case, the blend has emergent syntax relative to the inputs. Consider causatives in French, which are formed using the verb *faire* ("do"):

NP	V	V	NP
Pierre	fait	manger	Paul.
[Pierre	makes	eat	Paul]

meaning "Pierre feeds Paul."
["Paul" is the agent of "manger"]

Pierre	fait	envoyer	le paquet.
NP	V	V	NP
[Pierre	makes	send	the package]

meaning "Pierre has the package sent."
["le paquet" is the object of "envoyer"]

Pierre	fait	manger	la soupe	à Paul.
NP	V	V	NP	à NP
[Pierre	makes	eat	the soup	to Paul]

meaning "Pierre feeds Paul the soup."
[Paul is the agent of "manger" and "la soupe" is the object]

Through these double-verb forms, French provides its speakers a way to evoke an integration network that delivers a compressed, human-scale scene in which at least two agents (Pierre and Paul), a causal action, a causal link, and a caused action (eat) are integrated into one event. Now, the French language already has single-verb forms that are good for parts of this integration. For example, "Jean fait le pain" is fine for evoking a scene in which Jean performs a causal action involving another element (le pain). And "Paul mange la soupe" is fine for evoking a scene in which Paul performs some action on another element (la soupe). French also has several basic single-verb clausal constructions into which one can put these single-verb forms. But how shall we express the scene in which Pierre does something that causes Paul to eat the soup? None of the basic single-verb clausal constructions quite serves. French offers three complex blends for doing the job. Each has as one input one of three compressed basic single-verb clausal constructions, and as the other input the diffuse chain of causal events with intermediate agents that we want to compress. The blend takes much of its clausal syntax from the compressed first input, but, and this is crucial, it has additional, emergent syntax. In that syntax there are now two verbs. Moreover, there are novel positions for clitic pronouns (such as *le, lui,* and *se*, as in "Paul *se* fait tuer par Jean," meaning "Paul has himself killed by Jean," versus "Paul fait *se* tuer Jean," meaning "Paul makes Jean kill himself") and various complements (as in "Paul fait envoyer le paquet à Marie à Jean," meaning "Paul has the packet sent by Jean to Marie"). In these double-verb causatives we see double-scope integrations at the conceptual level: The conceptual frame for the basic construction does not match in a one-to-one fashion the complex and diffuse causal chain in the other input. We also see double-scope integrations at the formal level, delivering new, emergent syntactic forms for expressing the blend. In the caused-motion construction, the syntactic form came entirely from the compressed input. Here, in the double-verb causatives, the syntactic form comes only partly from the compressed input. Some words and their grammatical categories come from the other, diffuse input. And the full syntactic form is emergent in the blend.

The double-verb causative constructions may be a response to an integration problem faced by humans everywhere: One person does something that is causal for someone else's action. A blending template achieves appropriate compressions in these instances, and language forms can prompt for these compressions. Many languages have arrived independently at the solution of double-verb causative constructions, using blends like the ones in French but having different emergent syntax. The development of these language forms is an example of the way in which human cultures, using the cognitively modern capacity for double-scope integration, evolve blending templates, including language forms, that are transmitted to subsequent generations. Our principal argument in Chapter 9 was that language could not be developed as an efficient set of forms without the capacity for double-scope integration. Here we make that point more explicit by highlighting the crucial role of blending in a few particular constructions. Languages, with their grammars, are great cultural achievements produced and perpetually transformed through the exploitation of the capacity for double-scope integration.

A different response, one that uses the same conceptual blends but that exploits morphology to create corresponding formal blends, can be seen in Semitic languages. Nili Mandelblit has given the most thorough and elegant analysis of the role of blending in a particular grammatical system in her study of the verbal system of Hebrew.

Hebrew verbs all consist of (a) a skeleton of consonants (the "root"), slotted into (b) some vowel pattern, or prefix + vowel pattern. Such a vowel or prefix + vowel pattern is called a *binyan*. The plural is *binyanim*. The consonants carry the "core meaning" of the verb. There are seven major *binyanim* in Hebrew (the capital C's stand for the root consonants to be inserted): *CaCaC, CuCaC, CiCeC, niCCaC, hiCCiC, huCCaC,* and *hitCaCeC*. Consider, for example, the root *r.?.h.* (where ? stands for a glottal consonant), with the meaning "see." Here are five forms of this verb with different *binyanim*:

- CaCaC + *r.?.h* [*ra?a*] "to see"
- niCCaC + *r.?.h* [*nir?a*] "to be seen," or "to seem"
- hiCCiC + *r.?.h* [*her?a*] "to show"
- huCCaC + *r.?.h* [*hur?a*] "to be shown"
- hitCaCeC + *r.?.h* [*hitra?a*] "to see each other"

Mandelblit shows that each *binyan* prompts for a particular blending schema. Consider, for example, the *binyan* hiCCiC (termed *hif'il*), which suggests causation, and the root verb r-u-c, meaning "run." They are blended as follows:

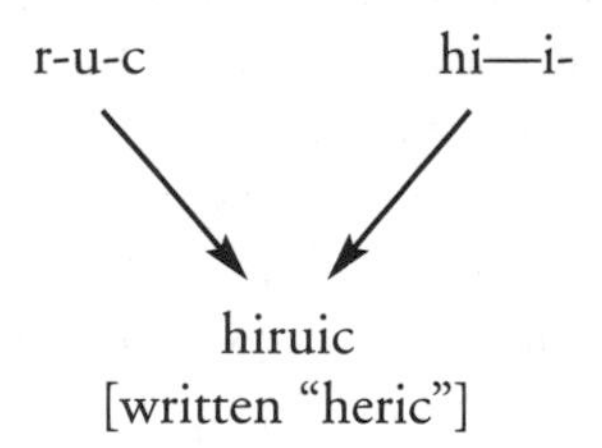

This formally blended verb occurs in the following sentence:

hamefaked	*heric*	*et*	*haxayalim.*
the-commander	run-hif'il$_{past}$	direct-object-marker	the-soldiers.

"The commander made the soldiers run."

This sentence integrates a whole causal sequence of events: There is the causing event of the commander acting on the soldiers and the resulting (or effected) event of the soldiers running. We see from this sentence that Hebrew has a way of prompting for a compression of both a causing event and a caused event in a single-verb clausal construction. The verb is "heric," and the single-verb clausal construction is the Hebrew Transitive, whose language form is

NounPhrase-Verb-*et*-NounPhrase.

Figure 17.2 depicts the integration network evoked by this sentence.

This is a *hif'il* sentence in which Hebrew is employing exactly the same strategy for conceptual and grammatical compression that we saw in French: Use the compression of a basic single-verb construction in order to compress an elaborate, more diffuse causal sequence of events. But while the strategy expressed itself superficially in French through new emergent syntax, it is expressed in Hebrew through novel morphology. Using the formal blending of morphological patterns, Hebrew is able to produce a single verb that fits perfectly into the transitive construction, so that the grammatical compression of the compressed input is borrowed without modification for the blend.

HUMAN-SCALE CONSTRUCTIONS FOR SPACE, FORCE, AND MOTION

We have seen many examples in which diffuse and complex ranges of conceptual structure give rise to the construction of human scale blends: "digging your own grave," "boiling the pan dry," "snatching food from a child's mouth," "guilty pleasures," "grateful memories," "driving someone out of his mind," and so on. This achievement of human scale is not simplification *per se*, because in every case we have the full network attached to the blend. The blend provides global insight into a human-scale pattern, but it remains connected to all the inputs.

Ostensibly simple conceptual structures also give rise to human-scale blends. In these cases, the blended space looks at first blush to be more complex than the inputs. But the overarching goal, as we saw in Chapter 16, is not to augment or reduce complexity for its own sake but, rather, to converge on human-scale patterns. For example, it might seem that there is nothing simpler at the conceptual level than recognizing that an object has a shape and location and is contiguous to other objects. Using mathematical or computational representations, we could express that knowledge with diagrams or with Cartesian coordinates and functions or with indicators of relative position.

Certainly, human beings use indicators of relative position and diagrams, and we can develop a sophisticated ability to use coordinates and functions; but we also have a standard way of using conceptual blends to give insight into location, shape, and contiguity. In this case, the blended spaces seem to increase the complexity of the description gratuitously by involving superfluous dimensions and structures. Len Talmy's example, "The mountain range goes all the way from Mexico to Canada," has as its purpose to give us global insight into the

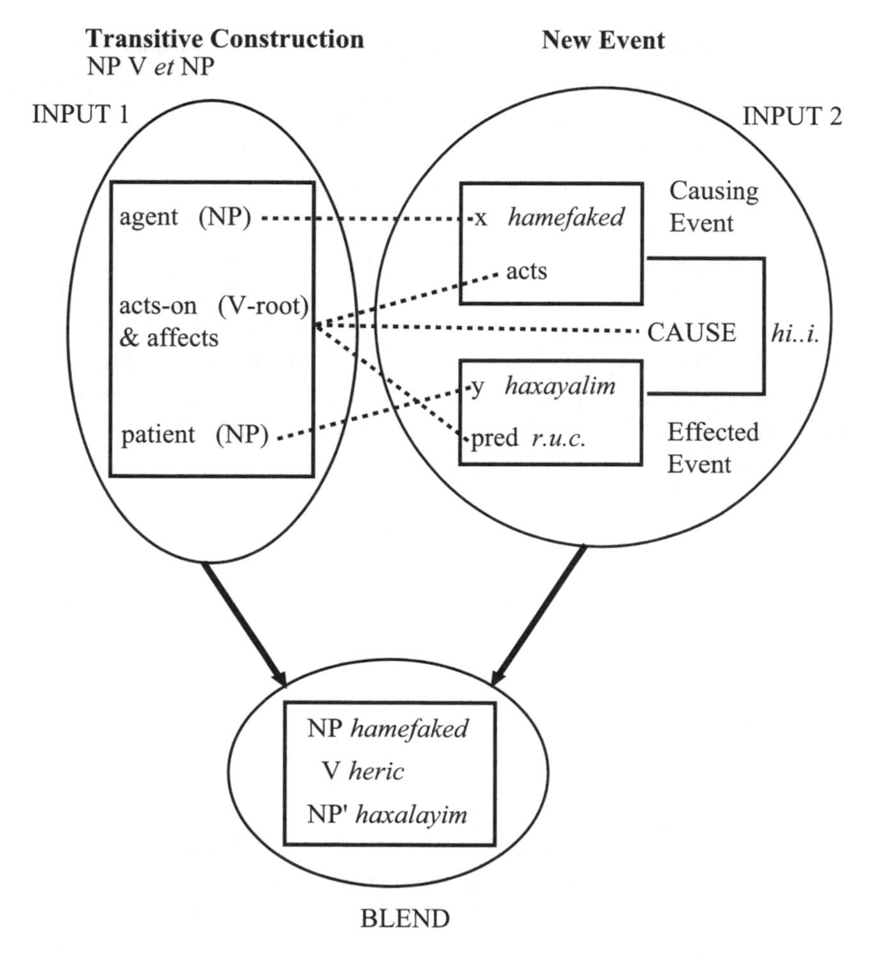

FIGURE 17.2 NETWORK FOR A HEBREW CAUSATIVE CONSTRUCTION

location of the mountain range and its spatial relationship to Mexico, Canada, and the United States. It presents a static scene using motion: *goes all the way from . . . to . . . .* Far from being unusual, this is a standard strategy in many and perhaps all languages. As Talmy writes, "Most observers can agree that languages systematically and extensively refer to stationary circumstances with forms and constructions whose basic reference is to motion." "The mountain range goes all the way from Mexico to Canada" is a basic idiomatic expression that does not feel either figurative or complex. Yet the addition of a dimension of motion to a static scene confronts us with a paradox: We now have something that is more complex and also patently false, since the mountain range is not really moving along a path from Mexico to Canada.

Why would human cognition and language work in this bizarre way? The answer is that we have the overarching goal of achieving human scale, and the operation of conceptual integration accomplishes that by projecting motion to the blended space. We also have grammatical constructions that prompt for just such integration networks, and we can use those constructions to describe the static scene, provided that we use the motion input. The integration network has, in one input, the static situation of the mountain range. In the other, we have the more general frame of a trajector moving along a path from a beginning point to an end point. The cross-space mapping connects the trajectory to a relevant dimension of the static object. In the case of the mountain range, this dimension is the horizontal. In the mountain range blend, the horizontal dimension of the static mountain range is now a trajectory, and there is additionally a trajector, projected from the motion input, that moves in time along the trajectory, from Mexico to Canada. The grammatical construction that sets up this blend normally assigns the trajector to the subject position and the movement to the verb ("Marie goes to the store"). But in the blend the construction uses the label for the trajectory in the subject position: "*The mountain range* goes from Mexico to Canada." Of course, the mountain range moves in neither the inputs nor the blend. But because the trajector in the motion input has no counterpart in the static input, there is no label available for it in the static scene. Using the label of the trajectory (*mountain range*) in subject position with a movement verb has the virtue of presenting as topic what we are actually talking about (the mountain range) and evoking the moving trajector metonymically.

The blended space now has a human-scale scene of a trajector moving in human-scale time along a human-scale path. Space and time have been scaled down, and a simple, ideal path has been created along which there is motion. Understanding the dynamic scene in the blend allows one to do the correct projection back to the static input with the mountain range, and to construct the appropriate relevant configurations of the range and the countries. In most respects, the blended space is more complicated than the static scene. It has all the aspects of dynamic motion, including time-space coordinates and contiguous positions in time. It can include the global form of the trajectory and the manner of motion, as in "The mountain range goes straight/curves/meanders/winds/skips from Mexico to Canada." In spite of or, rather, because of this added complexity, the blended space gives global insight at human scale into the static configurations.

Human action, with motion and intentionality in physical space and time, is a basic human-scale structure. The fictive-motion blends take states as inputs and add motion and possibly intentionality to produce the blended space. Adding motion transforms a state into an event. Adding intentionality transforms an event into an action. The blended space, with action, event, and state, is more complex from a formal point of view than the input with just a state, but the added complexity crucially enhances the human-scale quality of the scene. It is more congenial for human beings to process a full, dynamic, intentional human-scale action than it is to process one apparently simple component of it.

There are many ways to project motion to the blend in order to enhance its human-scale quality. Talmy provides an insightful taxonomy of sources of motion for fictive-motion blends. In our terms, this taxonomy presents a range of different motion inputs and a range of different projections to the blend. The mountain range blend has an input with a trajector moving along a linear path, but another of Talmy's examples—"The field spreads out in all directions from the granary"—has a motion input in which a material substance (such as oil or wine) distends or diffuses from an initial spot.

The formal blending in fictive-motion expressions is particularly noticeable in what Talmy calls "Access Path" expressions, such as "The bakery is across the street from the bank." The static input could be expressed by "The bakery is on the street." The motion input has something departing from one point, traversing some surface, and arriving at another point. The words "across" and "from" come from the motion input. The expression for the blended space combines grammatical elements from the two inputs, so we can say, "The bakery is *across* the street *from* the bank." The mountain range example and the bakery example have the same motion input. In the bakery example, the motion input has a surface that is traversed. Its counterpart in the static input is the street. In the blend, the surface traversed is fused with the street; we can use the label "the street" to pick out that fused element in the blend and we can put "the street" in the grammatical position for the surface traversed ("across *the street*"). Similarly, the endpoints of the trajectory in the motion input have counterparts in the static input: the bank and the bakery. In the blend, we fuse the endpoints of the trajectory with their counterparts in the static input, and we can use the labels from the static input ("the bank" and "the bakery") to pick out the fused elements in the blend and then place those labels in the grammatical positions for the endpoints. In the mountain range example, the surface traversed is left implicit: Although the United States will typically show up in the conceptual blend as the surface traversed, it is assigned no expression. But one could say "The mountain range goes across the United States from Mexico to Canada." The kinds of fictive-motion blends that can be achieved are driven by the governing principles of conceptual integration and the availability of motion inputs. Principles that play an essential role include Topology, Integration, Unpacking, Maximization of Vital Relations, and Intensification of Vital Relations. They contribute to satisfying the overarching goals of achieving human scale and strengthening vital relations.

Under "Advent Paths," Talmy gives the example "The palm trees clustered together around the oasis." Here, the motion input has multiple trajectors and trajectories converging on a location. This time, the trajectors in the motion input have counterparts in the static scene—namely, the fixed palm trees. The trajectories, on the other hand, have no counterpart in the mental space of the static scene. Blends of this type incorporate the static scene as the end state of elaborate motion in the blended space.

In the palm tree example, there is no motion and no change in the static input. The motion is projected from a different input. Similarly, in another of Talmy's examples, "As I painted the ceiling, paint spots slowly progressed across the floor," the motion that is relevant in the fictive-motion blend is not available from the input with the paint spots. The motion of the paint drops falling from ceiling to floor has nothing to do with the progress of the paint spots across the floor in the blended space. In this case, too, the trajectors have counterparts—namely, the paint spots—but the mapping is one-to-many, because each trajector in the motion space with multiple positions is mapped onto each of many distinct paint spots, each with a single position. The positions of the trajector are mapped one-to-one onto the positions of the distinct but analogical paint spots, but the single trajector is mapped one-to-many onto the distinct paint spots. In the blend, many paint spots become one paint spot that has achieved many positions. Analogy between the paint spots has been compressed into Identity of single paint spots that progress across the floor. This example illustrates the nondeterministic nature of blending; one can evoke different input spaces to achieve the blend. For example, we might see the paint spots in the blend as caravans moving across the floor, where the "lead" paint spot always has the same identity and subsequent paint spots have the identity given by their order. In that case, the "new" paint spot on the floor is analogous in the input space of painting to the immediately previous spot and, thus, in the blended space can take over its identity, as it takes over the identity of the one behind it, all the way back to the wall, where a new paint spot has just emerged onto the scene.

Under "Shadow Paths," Talmy gives the example "The tree threw its shadow down into the valley." In this case, the end state of the motion (in the motion input) has as its counterpart the perceptible shadow in the static scene. Here, fictive motion creates motion by exploiting the already available blend in which a shadow is a changing and moveable object and not just an area of diminished lighting. That blend is already lexicalized in English by the word "shadow."

It turns out, then, that fictive-motion blends are strongly double-scope. They blend an essentially static scene with an essentially dynamic one to create a blend with emergent properties that draws on the organizing frames of both inputs. Often, the path of the motion in the blend is not available to real trajectors in the real world, but part of the emergent meaning in the blend is the possibility of this motion. Fictive-motion blends are double-scope at the conceptual level, and also at the formal level. They draw grammatical elements from both inputs to create double-scope syntactic blends to express the conceptual structure in the blended space.

CULTURAL EVOLUTION OF LANGUAGE

Language changes over centuries. Latin gave rise to French. It is in fact a remarkable universal feature of all languages that they change over cultural time. For the most part, deep changes in the structure of a language take so long that we do not see them happening in a single lifespan, although we sometimes see lesser changes taking place in the space of a few years, as when new words come into the language, slang and idioms are created, or existing words acquire new extensions, such as "virus" for the nefarious program that ruins your computer. Linguists agree that this change is not a matter of either improvement or deterioration of a language. Languages do not change because they are deficient or unstable. Perfectly fine systems in a language routinely evolve into different perfectly fine systems. Why should this be?

We suggest that the central role of conceptual blending, compression, and double-scope creativity in grammar and grammatical constructions induces language change very naturally and, in fact, unavoidably. One reason for this is the natural emergence of new syntax under pressure from borrowed compressions, as we saw in the case of the French double-verb causatives. Another reason for progressive change in languages is that conceptual blending networks are underspecified. Because such networks prompt for mapping schemes without specifying the mappings and projections exhaustively, grammatical constructions leave users, singly and collectively, some leeway in the actual implementation of the mapping schemes. This kind of language change stems from variation of the underspecified aspects of the selective projections and mappings within the network. Suzanne Kemmer and Michael Israel have shown conclusively, in their extensive studies of the "way" construction, that within a relatively stable blending scheme over several centuries, usage during a certain period will emphasize certain projection patterns and not others, and, moreover, that this usage will change over time.

Modern examples of the "way" construction are "He found his way to the market," "He made his way home," "He elbowed his way through the crowd," "He jogged his way along the road," "He talked his way into the job," and "He whistled his way through the graveyard." This usage developed from a Middle English *go-your-path* construction, which accepted almost any noun that meant something like "way," as in "He lape one horse and passit his way" (1375) and "Tho wente he his strete, tho flewe I doun" (1481). Later, a new syntax developed that allowed for a complement indicating direction, as in "He went his way home." The construction developed by allowing ever more kinds of verbs indicating something about the manner of the movement. Israel remarks that these verbs tend to cluster around certain well-defined semantic prototypes. Between 1826 and 1875, a large number of verbs coding difficult motion or tortuous paths became acceptable. They include *plod, totter, shamble, grope, flounder, fumble, wend, wind, thread, corkscrew,* and *serpentine*. Dickens writes in 1837: "Mr. Bantam corkscrewed his way through the crowd." It is not until the end of the nineteenth century that we begin to find verbs like *crunch, crush, sing, toot,* and *pipe* in the construction. Israel explains that they encode "not motion per se, but rather the noise that inevitably accompanies certain forms of motion." A different development of the construction, the means thread, comes up fairly late at the end of the sixteenth century. In this thread, verbs for creating a path become acceptable, as in "Arminius paved his way" (1647). A familiar example would be "Every step that he takes he must battle his way" (1794). Israel writes: "By 1875, examples include uses with *push, struggle, jostle, elbow, shoulder, knee, beat* and *shoot*. In the nineteenth century, as the manner thread experiences a rapid expansion, the means thread begins to allow verbs encoding increasingly indirect ways of reaching a goal," as in "Not one man in five hundred could have spelled his way through a psalm" or "He smirked his way to a pedagogal desk." The role of blending is crucial:

> It is useful, in this light, to consider the ["way"] construction . . . as an example of a syntactic blend—that is, as a specialized grammatical pattern serving to combine disparate conceptual contents in a single, compact linguistic form. Essentially, the modern construction provides a way to blend the conceptual content of an activity verb with the basic idea of motion along a path. The trend toward verbs coding activities which are increasingly marginal to the achievement of motion thus reflects the construction's gradually increasing power to blend different types of events into a single conceptual package.

Israel has succeeded in showing that some underspecified projection patterns for the integration network associated with the "way" construction were extended over time. He also shows that another underspecified projection pattern for the construction diminished over time: "The third thread . . . involved usages with verbs like *keep, hold, take, snatch,* and *find* for coding the acquisition or maintenance of possession of a path. These usages were very common in early stages of the construction. But unlike the other two threads, this usage shrank rather than expanded over time, so that now only *find* . . . and a few other verbs remain to represent it."

CONCLUSION

All along, we have stressed creativity and novelty as consequences of conceptual integration. But creativity and novelty depend on a background of firmly anchored and mastered mental structures. Human culture and human thought are fundamentally conservative. They work from the mental constructs and material objects that are already available. Conceptual integration, too, has strong conservative aspects: It often uses input spaces, blending templates, and generic spaces that are anchored in existing conceptual structure; it has governing principles that drive blends in the direction of familiar, human-scale structures; and it readily anchors itself on existing material objects. Emergent structure—both conceptual and formal—can arise through conceptual integration within basically conservative integration networks.

This general pattern of development in culture and thought has the evolution of grammar as a special case. We have seen sample evidence of the central role of blending and compression both in superficially simple constructions such as Noun-Noun and in constructions that are acknowledged as highly complex such as French double-verb causatives. Advanced conceptual integration operates simultaneously for both conceptual and formal structure. It requires double-scope capacity and, by its very nature, promotes both continuity and change at the conceptual and formal levels. Indeed, for conceptual blending to happen at all, continuity is essential. We have shown how—thanks to conceptual integration—novel constructions or variants of constructions draw on deeply entrenched constructions, conceptualizations, and blending templates. But because blending involves selective projection, composition, completion, and elaboration under a set of governing principles, it can produce new, well-anchored conceptual and formal structures. In his article on "way" constructions, Israel points to just these features of simultaneous conservatism and novelty. He writes, "Utterances should sound like things the speaker has heard before." There are also, he notes, forces for innovation. The world is rich, both physically and culturally, and it evolves, and this places pressures on us to create new conceptions and expressions. We do so through double-scope integration, but the products are not wholly novel. Constitutive and governing principles ensure that the network is in many ways deeply familiar, not least in using familiar frames, a canonical set of vital relations, an easily accessible initial cross-space mapping, and human-scale organization and compression in the blended space. For grammar, all this delivers slightly new expressions that, however novel, are intelligible precisely because they are for the most part strongly anchored to existing constructions. When we hear an expression, we try to construct an integration network; but to do so we have to do some selective projection, composition, completion, and elaboration that is not specified by what we hear, so there is yet more room for creativity and novelty. We do as much blending as we need to do to make sense of the utterance, and this work is simultaneously conservative and innovative.

As we have argued in some detail in this chapter, blending turns out to be a central feature of grammar. Far from being an independently specified set of forms, grammar is an aspect of conceptual structure and its evolution. Grammar has something in common with metaphor ("You are digging your own grave"), counterfactuals ("If I were you, I would quit my job, . . . "), category extensions ("same-sex marriage," "computer virus"), and material culture (watches as descendants of sundials). All are products of integration and compression. And so, not surprisingly, all show degrees of entrenchment, novelty, and change over time. They depend on simultaneous continuity and creativity. "If I were you, I would do X . . ." is a very conventional counterfactual construction, and the implication "you should do X" is typical. But as we discussed in Chapter 12, "If I were you, I would quit my job," however anchored in familiar patterns, has ample room for variable selective projection and emergent meaning. Counterfactuals like the Woman in a Coma, in Chapter 11, require blending that is unconventional but nevertheless routinely available to newspaper readers. Just so, a noun compound like "divorce fixer" (meaning a fixer-upper house you could buy cheap because it was being sold as part of a divorce settlement) may look bizarre, but it is still a noun and still has inside it two very conventional nouns, and in fact the entire blending template that subtends it is the extremely conventional pattern we saw for "land yacht" and "boat house." The general point here is that speakers and hearers are continually called upon to construct blends. There is no reason to think that "novel" blends are inherently more costly or less desirable than totally conventional ones. In fact, presumably because people are highly competent double-scopers, finding the network that best fits a particular conceived situation is something that people enjoy and are proud of. Finding optimal networks has always been a highly valued skill, for which writers, poets, statesmen, teachers, scientists, and lawyers are highly regarded. The central blending at play in grammatical constructions, metaphor, and counterfactuals means that language users are perpetually constructing blends with varying degrees of novelty of which they are seldom aware. In this way, we suggest, diachrony is built into synchrony: The very way in which language is used synchronously (at a given time by a given community) is also the way language will change.

CHAPTER 17 ZOOM OUT

RECURSION IN GRAMMAR

When we do the kinds of blends where nouns are put together, the grammatical category in the blend is Noun. That is, putting the noun "jail" and the noun "bait" together gives us the noun "jail bait."

Question:

- Since the output grammatical forms are different from the input forms, why aren't special grammatical categories created for them?

Our answer:

As we saw in Chapter 16, the overarching goal of blending is to Achieve Human Scale. A corollary of this overarching goal is that human-scale blends serve as inputs to further blending. And as we saw in the case of personification, the category *person* organizes both one of the inputs and the output, so the output becomes a candidate for any blending template that takes *person* as an input, including the personification template itself. We called this process "recursion" and showed how it created the possibility for multiple successive blends, as in the elaboration of the concept of *number* in mathematics, which changed dramatically over time while always preserving the same human-scale category *number*.

Grammatical categories are also human-scale elements, grounded in human-scale conceptual structures such as Objects, Events, Processes, Force Dynamics, and Visual Focus and Perspective. Noun is an especially basic grammatical category, with a prototypical meaning of Thing. The disposition to achieve human scale constrains blending networks in the direction of having grammatical forms that fit existing grammatical categories. This achievement is not required by any logical or computational or linguistic consideration. It would be acceptable on all these grounds, perhaps even preferable, for the new structure in the blend to be a new grammatical category. But that would violate the overarching goal that the blend be at human scale. It is remarkable, in fact, from a merely formal point of view, that in the many thousands of languages studied by linguists we find a very small number of basic grammatical categories occurring over and over again. This formally surprising observation is explained by the principles, power, and goals of blending. In particular, it is explained by these three things: (1) Grammatical categories are human-scale elements. (2) Novel grammatical constructions are created through blending from existing grammatical constructions. (3) The Human Scale/Recursion Principle places value on having human-scale blended spaces.

In grammatical constructions, the blended space achieves human scale by borrowing its compression from an input with existing grammatical categories, which are already at human scale. So the compound of Noun and Noun in the blend takes Noun as its grammatical category from the inputs. The grammatical form for the blended space is now able to fit into other patterns in the language that employ that existing grammatical form. This central feature of preserving category from input to blend favors recursion at the formal level, just as it favors recursion at the conceptual level, because the two forms of recursion are the same thing. So "ballet school" achieves the human scale of being a Noun by borrowing that category from the inputs. So does "girl scout." Since they are both Nouns, they are suitable inputs for the Nominal Compounding template, so we can form "girl scout ballet school," which is perfectly grammatical, and also "ballet school girl scout." They, too, are Nouns, so we can form "lace stocking girl scout ballet school."

And yet, just as The Grim Reaper fits the category *person* but, when looked at on the inside, is a strange sort of person who remains connected up to the input spaces out of which he was formed, so "lace stocking girl scout ballet school" fits the grammatical category *noun* but has unusual internal structure. Preserving an organizing category like *number, person,* or *noun* favors recursion through the steps of successive blending networks, but the internal organization of the category changes. The Grim Reaper is not a standard person: It would be odd to think of him undergoing bypass surgery, for example. Complex numbers are certainly challenging sorts of numbers, palpably different from 1, 2, and 3. And a noun that consists of six nouns is indeed a noun, but it looks odd on the inside and prompts for exceptionally heavy amounts of decompression. Although recursion in theory allows unlimited iterations, it is widely known that human beings typically top out after a handful of repetitions. We say "The scarf my aunt bought" and "The scarf my aunt my uncle married bought," but it gets hard at "The scarf my aunt my uncle my father disliked married bought." This difficulty is commonly attributed to "performance"—perhaps involving limitations of working memory, cognitive load, or conceptual complexity. We now see that the ability for recursion and the limitations on that ability do not come from different considerations: The preservation of the category under projection from input to blend and the changing of the internal structure of that category as it is built in the blend are parts of the same process of conceptual integration. The preservation part preserves human scale while the alteration part moves away from human scale. The preservation part favors recursion while the alteration part impedes recursion. A sequence of too many iterations of the recursive blending gives us something that, while still fitting the human-scale category, now has content too far removed from human scale to be suitable for yet further recursive blending. That content, through successive blending, comes to involve more and more levels of decompression involving levels of "upstairs" input spaces. "Lace stocking girl scout ballet school," like "the scarf my aunt my uncle my father dislikes married bought," forces decompression. By contrast, there is the capacity of language to come up with a single word for the same content. This is called *chunking*, and it removes the forced decompression. So if "lipple" comes to mean "lace stocking girl scout ballet school," or, in a family, "scad" comes to mean "the scarf my aunt my uncle my father dislikes married bought," then it becomes easy to put them through further recursions, because the internal content of many levels of decompression is no longer forced upon us. As we saw in the Zoom Out section of Chapter 8, terms like "great-grandfather," meaning "father of parent of parent" (which can be further chunked to "Gramps," for example), offer such relief from forced decompression while preserving the human-scale grammatical category.

Eighteen

THE WAY WE LIVE

YOUNG CHILDREN DELIGHT AND frustrate us by spending hours working out connections that we find obvious. What is money? How many quarters is a dollar worth? What is this letter? Is it an M or a W? They turn pages in a book at random, sometimes going from back to front, sometimes "reading" the book upside down. The clock shows them nothing at first, and even later their ability to read it has strange gaps. Much later, they will struggle to understand what 1/3 could be, and to see that 2/3 + 1/3 = 1.

We adults, on the other hand, have complete mastery of the conceptual blends of money, writing, and timepieces. We live directly in these blends: We manipulate these elaborate networks with no conscious attention to the topologies and projections across the network. Our attention is focused on running the blend itself and attending to the relevant material anchors—the dollar, the book, the watch. Although it took us a long time to master the complex blends linked to a cultural activity like writing, once we have them, we have the greatest difficulty escaping them even when we want to: Look at this page and try not to see words and letters, but only black shapes on white paper; that is, try to see only the original input that you had when you were a two-year-old. (It's not fair to unfocus your vision so the page is a blur—that wasn't your problem when you were a two-year-old.)

The story that we have been telling in this book runs along the following lines. Double-scope conceptual integration is crucial to the activities that make us what we are. Around 50,000 years ago, this level of blending was achieved, presumably through neurological evolution, although that final evolutionary step need not have been a great biological leap. The biological evolution of conceptual blending took evolutionary time, but once double-scope blending was achieved, culture as we know it emerged. Cultures could create specific double-scope integration networks that then quickly showed up in languages, number systems, rituals and sacraments, art forms, representation systems, technologies (from advanced stone scrapers to computer interfaces), table manners, games, money, and sexual fantasies.

Before the appearance of double-scope integration, elaborate and powerful integrations were set up by biological evolution. When we see a blue cup, we are not and cannot be aware of the extremely complex neurobiological causes of our perception. The neurophysiologist might explain them, but in those explanations we will see nothing that resembles a blue cup. Our very biology locks us into perceptual integrations of this sort. Such biologically given integrations exist throughout the animal world: Mere lizards are born with many very impressive integration patterns into which every lizard is locked.

Of course, human children are animals, and they too have biology that locks them into many integrations: They quickly develop visual acuity, they have color vision and develop color constancy, they turn their heads to locate the origin of sounds they hear, and they see blue cups if there are blue cups to be seen. But given their powerful capacity for double-scope blending, children are born into a world richly structured by complex, entrenched, cultural conceptual blends, many of which they must master to function in society. Such blends are products of cultural—not biological—evolution, and they undergo continual change. Languages, for example, always change, even though their degree of formal complexity stays relatively constant. The same holds for cultural models in general.

During the first three years of life, a child constructs seemingly impossibly complex system of elaborate blending networks. The child's biological capacity for double-scope integration meets the culture's offering of specific integration networks, and the two combine to impressive effect.

Once a fundamental integration network motivated by culture is in place, its compressed, human-scale blend seems as obvious and inevitable as the perception of a blue cup. Communicating with language, making and recognizing representations, and using a fork and a spoon come to seem straightforward both for the child and the adults who interact with the child. That combinations of vocal sounds mean what they mean, that the marks on the paper are a dog, and that the fork is an instrument for eating are then taken as direct and inevitable. We can no longer see the marks without seeing the dog, or the fork without seeing its purpose and its etiquette. This learning is exceptionally difficult: A young human brain is the quickest, most plastic, and most capacious complex system in the universe, but it takes at least three years of constant working to bring it to mastery of all these culturally motivated blends. This development is almost entirely unconscious for the child and mostly unobservable by adults, who see only superficial signs of the child's mental activity and take them for granted because adults do not remember the difficulty. Like biology, culture and learning give us entrenched integrations that we can manipulate directly. In both cases, once we have the integration it is hard or impossible to escape it. We construe the physical, mental, and social worlds we live in by virtue of the integrations we achieve through biology and culture. There is no other way for us to apprehend the world. Blending is not something we do in addition to living in the world; it is our means of living in the world. Living in the human world is "living in the blend" or, rather, living in many coordinated blends. Even remembering the world and our activity in it seems to depend upon the existence of the kinds of blends three-year-olds have developed. We retain only fragmentary and unorganized conscious memories from before that stage.

The story is no different for the learning of numbers, writing, history, social patterns, or any other integration, except that after about three years of age we can remember the work we went through to acquire the blend. We know when we look at writing that we are living directly in the blend and cannot escape it, but most of us can also remember the time when writing was only marks on a page. We can probably remember learning fractions, although as adults we manipulate the blend directly and cannot help seeing the relationship between 1/2 and 1. Playing the piano, understanding sacraments, interpreting adult social actions, and using complex numbers all show this pattern. It is the pattern of human cultural learning worldwide.

It is also the apparent pattern of human consciousness and memory. Consciousness is dedicated to the products of blending. Once we have the blend of money or the watch or social action or ritual, we are not consciously aware of the different input spaces and the projections across the network. In the blend, the money has its value and the watch shows the time, and this is what we are aware of. It is also what we have conscious memory of: We remember that someone said something or that the watch showed a certain hour; we do not remember the integration networks.

The capacity for double-scope integration confers several related aptitudes. Humans can build new integration networks that are not biologically given. We can operate by directly manipulating the blends of those networks. We can also focus attention on escaping the blends, investigating the connections, changing the network, and rebuilding the blend.

If we look at the course of a human life, we see that until about age three, a child is engaged in building complex and difficult blends that count culturally as obvious and natural. After that, the child begins to build more and more blends that are recognized culturally as requiring work and learning—playing piano, reading, doing crossword puzzles, doing carpentry, learning a second language, learning world history, government, biology, and computer systems. But also, beginning at about three years of age, the child focuses attention on decompressing, on learning to manipulate input spaces and projections independently. Perhaps one of the earliest domains of decompressing is intentionality: At first, objects that appear to be self-moving and that also appear to react appropriately to the right sorts of things (such as the baby's cry, other noises, foods, smells, voices) count as intentional beings in a primary and unanalyzed sense. But the child begins to decompress: Although the wind-up puppy can be regarded as intentional in the blend, once the child develops access to the input spaces she is able to regard it as purely a dynamic physical object.

Beginning early, blending and deblending, compression and decompression, go hand in hand. The early infant sees the voice-activated crib mobile as intentional. Later, it can decompress. But the adult who has no confusion about the status of the mobile can still achieve the intentional blend, and in fact create and understand many blends that bestow intentionality upon objects like computers, buildings, clouds, and flowers. The child is considered immature for not seeing that the mobile is not alive, but the adult is considered smart for using metaphor to create intentionality where it is not. The culture requires the child to achieve the right coordinated complex blends. Only then will the child be able to live in the blend as a member of the culture. But, paradoxically, it is also essential for the culture to support decompression, so that the child can learn further skills and achieve greater overall flexibility. To learn to read and write, the child must not only blend but also deblend, not only compress but also decompress. Specifically, the child must decompress spoken language in order to manipulate the input space of grammatical forms with some independence of the blend in which they are fused with vocal sounds or gestures. To learn fractions, one must decompress the concept of number, in which for each number there is a next number ("and what comes next, Betsy?"), each number has a name, and each number is associated with a countable quantity of objects. The child has to be able to make a selective projection that leaves those properties behind. In the new blend of rational numbers, numbers have an order but there is never a "next" number, every number has an infinite number of names, and "most" numbers (such as 17/232) do not correspond to a countable set of objects. The new integration network uses blending to create rational numbers but at the same time, as part of that blending, decompresses the previous number input by projecting only selectively from it. Once the student learns rational numbers, she can work directly in that blend, but she is also able to work in the old space of what are now called "whole" numbers. The student has the blends but can also manipulate the input spaces independently. The effect of blending and deblending, compressing and decompressing, is to create much richer networks, with the greater flexibility of going from outer space to inner space and back; but this does not incapacitate or overwhelm the mind because we can still work in one blend at a time. The lizard is trapped in the integrations it has and cannot make new ones, but those integrations work very efficiently. The human being, like the lizard, has some biologically given blends it cannot escape, but it also has powerful culturally motivated blends that it has at least the chance of escaping, to decompress and reblend. It may also create new culturally motivated blends out of inputs not previously integrated. Strikingly, the human being has for these flexible culturally motivated blends some of the power the lizard has only for its biologically given blends: Once we have built a blend, we can inhabit it and manipulate it directly.

Why do we not teach the child rational numbers directly at age two? Because the child can learn these advanced blends only by going through processes of simultaneous blending and deblending, to create new blends out of old. Blending creates emergent structure, but it is also conservative, working from inputs that it has. In this way, conceptual knowledge develops step by step, through the cascade of blends. So does cultural and scientific knowledge. Firm intermediate blends are needed before the advanced ones can be created. Creating the advanced blends typically requires decompressing the intermediate ones.

Culture elaborates blends that are complex and hard to discover but relatively easy to manipulate and learn. The cultural search for an optimal blend, which can last for many years or even centuries, explores huge numbers of possibilities and retains only those that fit the governing principles optimally for the purpose at hand. This ensures that conceptual blends transmitted culturally to a new generation have excellent design, from the culture's point of view, but also that the culture will regard them as difficult, since they were a long time coming and still run up against the difficulty of the decompression that children must perform in order to learn them. Complex numbers, for instance, would typically be viewed as "hard" for children, because they are a notion developed late in the history of mathematics and one that requires severe decompression of previous conceptions of numbers such as whole, rational, and real. But the blend that needs to be manipulated for the use of complex numbers (vectors in a plane) is not formally of any obviously higher magnitude of difficulty than the blend for fractions. Writing also developed relatively late in human history, and was in fact at one time regarded as an exceptionally difficult and specialized skill (the domain of expertise of scribes or highly revered priests). But in many countries today, most people learn to write when they are children, and so they live in the blend, where written words and sentences are apprehended automatically and effortlessly. The history is remote, so it looks to us as if writing has always been around. The impression of complexity is lost in this case, and adults find it much harder to see why children might have trouble in getting from the spoken to the written word. And yet, the decompression and remapping required to learn writing are possibly more extensive than what is needed for numbers.

If constructing a useful network takes such complex activations, projections, and blendings and decompressions, how does the child do it?

Part of the answer is that we have powerful brains; part of the answer is that we are embedded in a rich world and exploit what it provides; part of the answer is that culture provides every assistance. But there are two other crucial parts of the answer.

First, children are built to trust and imitate. A two-year-old trying to open the locked door of a store hears "It's Sunday" and, without knowing what Sunday is, remarks "It's Sunday" when he discovers that the next store is locked up. He picks up a pocket calculator and walks around talking into it as if it were a mobile telephone. As everyone since Aristotle has noticed, human beings are distinguished from other animals in having fantastic and equipotential powers of mimicry. We pick up many pieces from our culture without knowing their useful connections, but then, once we have picked them up, there they are, ready to fall into place as the networks are achieved. This is the story, too, of grammar.

Second, we do not learn the domains of human life separately. We do not learn meaning, language, and reasoning separately. The child picks them up and puts them together all at once. From the beginning, she is building vital relations and compressions under governing principles. At any stage, the child is building integration networks that simultaneously serve form, meaning, emotion, and reasoning. Just as cultures naturally build up further blends on existing blends, the child builds up further blends from the ones she has already set up. The child's brain is not attempting to capture regularities of form in the adult way of speaking and then find out under what conditions they count as true, and then use them for logical inference, and then suffuse them with emotion. On the contrary, mappings, projections, and dynamic emergent structure in integration networks operate without regard for such distinctions. This puts the child's learning in a very different light from what is usually assumed by nativists and non-nativists alike.

For the nativists, what needs to be learned is just too hard and, therefore, must be largely inborn. This view seriously underestimates the extraordinary double-scope creativity of the child, the richness of the social and physical world, and the richness of the child's own mental world. Aspects (e.g., syntax or truth-conditions) that adult experts have isolated (through decompressions of their own) are not learned separately. Formal aspects of syntax or logic might indeed be unlearnable in isolation, but they are emergent in a system of conceptual integration networks that does not separate conceptual and formal blending. This is true of learning in general—a fact that severely undermines the "poverty of stimulus" argument that a child's experiences are too thin to account for learning. Far from being deprived of adequate stimulus, the child is under heavy bombardment all the time from the surrounding world, which is itself a massively blended physical, mental, and social complex. Data keep pouring in for the construction and dynamic elaboration of conceptual integration networks that the child is actively engaged in.

For non-nativists, what needs to be learned is indeed very hard to learn but can be learned through sufficiently powerful mental capabilities of statistical extraction. This view does not take account of the species-specific capacity for double-scope integration, which drives both human ontogeny and human culture. This human capacity includes, in our view, the governing principles for achieving human-scale blends through operations on the inner-space and outer-space topology of mental space configurations.

Our mental world is not an amorphous collection of relations and formal structures. Rather, it is inherently shaped by the rich topology of conceptual integration networks and mental spaces, including vital relations and their compressions. This landscape of conceptual integration networks is not something that gets superimposed onto the child's apprehension of basic meaning and form. It is the very way in which meaning and form, and hence the mental world itself, are constructed: The child with double-scope capacity latches onto the blending networks of its culture, which are themselves mental constructions built through double-scope creativity.

Consider a child building something with a Lego construction set. When the child is finished playing, we can see all the places where something new can be added. So can she. In fact, she knows all the building that went into it, all the fracture lines and components. She can instantly add something new, pick up a tower and move it from one place to another, or separate the building into two and join the halves with a bridge. There are clear constraints on the set and the construction: The physical objects accept only some actions. They cannot be divided, they cannot be put together just any way, they will not stay together if they are unbalanced and gravity pulls them apart. These constraints are inherent in the objects and their manipulation; it is the design of the pieces, not Mom and Dad or a set of rules, that imposes these limitations. No one has to tell the child that the Legos will not go together in the ways they will not go together. She ends up building a very wide variety of objects that all satisfy the Lego constraints, even though she may previously have seen almost no Lego constructions. The child has plenty of stimulus to make all these objects, because she is integrating the Lego construction set with objects she sees, vehicles she rides in, animals from stories, houses and rooms she knows, bridges she has been told about. Things made out of Legos are not the major stimulus for making things out of Legos, any more than linguistic expressions the child hears are the major stimulus for her making other linguistic expressions.

If the only constraint were "making something out of this amount of plastic," and the child had at her disposal every method of melting and molding plastic, the Lego constructions themselves would be only the tiniest fraction of the possible products, and constraints that would limit the child's productions to such Lego constructions would look factitious and unmotivated. But evolution is impervious to such accusations. Just as the culture gives the child the Lego set, so evolution gives the child double-scope integration, with its very strong construction principles.

The theory we have presented of the way we think and live has many intricate technical details in complex interactions, and many deployments over the ranges of human thought and action. The massive and repeated application of blending in diverse areas of human endeavor has led to realms of distinctive richness and complexity. But the story that this theory tells is, in hindsight, a simple one. From mammals to primates to hominids, there was a biological development of increasing capacity for conceptual integration. Once that biological development reached the stage of double-scope integration, cognitively modern human beings were born. The distinguishing behaviors of humans are products of that biological ability for double-scope conceptual integration.

Conceptual integration is strongly conservative: It always works from stable inputs and under the constitutive and governing principles. But conceptual integration is also creative, delivering new emergent structure that is intelligible because it is tied to stable structures. The bubble chamber of the brain runs constantly, making and unmaking integration networks. Cultures, too, running a bubble chamber over the collection of their members' brains, develop integration networks that can be disseminated because the members of the culture all have the capacity for double-scope integration. Very few of the networks tried out in these bubble chambers of brain and culture actually survive. A network that does survive takes its place in individual or collective memory and knowledge.

From weaponry to ideology, language to science, art to religion, fantasy to mathematics, human beings and their cultures have, step by step, made blends, unmade them, reblended them, and made new blends, always arriving at human-scale blends that they can manipulate directly. This progression from blends to newer blends, blending and deblending, compressing and decompressing, is the pattern of child learning, too.

The story of human beings—50,000 years ago, now, for the infant, the child, the adult, the novice, the expert, for the many different cultures we have developed—is always the same story, with the same operations and principles. That is the story we have tried to tell in this book.

NOTES

PREFACE

vi "cognitive fluidity": Mithen 1996.
vi "roughly 50,000 years ago": Mithen (ed.) 1998.

CHAPTER 1

3 "by looking inside ourselves": Donald 1994, p. 538.
5 "Eliza": See Weizenbaum 1966. Eliza Doolittle is the woman in *Pygmalion* and *My Fair Lady* who pretends to be more than she is.
9 "Mathematics may be defined": Russell 1918, ch. 4.
9 "a science of algebra": Kline 1972, p. 259.
10 "sum of the two surd roots": Translated by Kline 1972, p. 186.
10 "this quasi-symbolic style": Kline 1972, p. 186.
10 "Cardan wrote": Kline 1972, p. 260.
11 "Smolensky and Shastri": Smolensky 1990, Shastri 1996.
14 "analogical thinking": Gentner, Holyoak, & Kokinov 2001, Gentner 1983, Gentner 1989, Hofstadter 1995, Mitchell 1993.
14 "mental images": Shepard and Cooper 1982, Kosslyn 1980.
15 "metaphor": Pepper 1942, Black 1962, Sacks 1979, Ortony 1979, Lakoff & Johnson 1980, McNeill 1992.

CHAPTER 2

17 "How can two ideas be merged?": Boden 1994, p. 525.
17 "Conceptual framing": Goffman 1974, Bateson 1972. Special cases of what we refer to here as the general notion of conceptual blending have been discussed insightfully by Koestler 1964, Goffman 1974, Talmy 1977, Fong 1988, Moser & Hofstadter n.d., Hofstadter and Moser 1989, Hofstadler et al. 1989, and Kunda, Miller, & Claire 1990. Fauconnier 1990 and Turner 1991 also provide analyses of such phenomena. All of these authors, however, take blends to be somewhat exotic, marginal manifestations of meaning.

22 "an enchanted loom": Sherrington 1940, p. 225.

25 "There is no fixed property of 'safe'": Fillmore & Atkins 1992 presents an analysis of "risk" that can be reinterpreted as a blending analysis. On "safe," see Turner & Fauconnier 1995, Fauconnier & Turner 1998, and Sweetser 1999.

26 "the variety of possible roles": Turner & Fauconnier 1995, Turner & Fauconnier 1998, Sweetser 1999.

27 "color adjectives": See the discussion of "black kettle" in Travis 1981 and the sections on "active zones" in Langacker 1987/1991 and 1990.

27 "racism": Mithen 1996, pp. 196–197.

28 "image club": The newspaper chose to report the facts in a winking style of titillation: "TOKYO, April 1—With a shy, quivering glance that she had practiced a thousand times, eyes slightly downcast but luminous with innocence and apprehension, Kaori sat at her desk in her prim school uniform, framed by a blackboard, waiting for a 'teacher' to walk into the classroom and rip her clothes off.

"It looks like a real classroom, and baby-faced Kaori looks just like a real Japanese schoolgirl. But it is all make-believe, for here in Japan the best way for a prostitute to recruit clients is to put on a school uniform and adopt the naive anxiety of a frightened schoolgirl. 'Japanese men tend to be obsessed by schoolgirls,' said Kaori, who would not give her last name but cheerfully conceded she is really 26. 'The men who come here are looking for submissive schoolgirls.'

"This is an 'image club,' one of several hundred in Tokyo where Japanese men pay about $150 an hour to live out their fantasies about schoolgirls. In this club, customers can choose from 11 rooms, including classrooms, a school gym changing room, and a couple of imitation railroad cars where to the recorded roar of a commuter train, men can molest straphangers in school uniforms" (Nicholas D. Kristof, April 1, 1997).

30 "the sexiest vehicle on the road": *Parade*, October 3, 1999, p. 18.

31 "no fussy little grammatical exercise": Goodman 1947, p. 113.

32 "Most of the analytic claims in the social sciences": Tetlock & Belkin 1996, Roese & Olson 1995.

33 "metaphor in the design of interfaces": See, for example, Hutchins 1989 and Wozny 1989.

33 "every child can master grammar": See also Sweetser 1990, end of ch. 2.

34 "yield the wrong results": Norman 1981, Sellen & Norman 1992.

36 "Nay, shame, O Philomela": See section 1406b.

37 "special cases": See Goffman 1974, Talmy 1977, Fong 1988, Moser & Hofstadter n.d., Hofstadter 1989, and Kunda, Miller, and Claire, 1990. (We, too, in earlier works, gave analyses of such local and seemingly exceptional phenomena; see Fauconnier 1990 and Turner 1991.)

38 "particular individuals in particular circumstances": Mithen 1998, pp. 168–169.

CHAPTER 3

39 "the two separate journeys?": A version of this riddle appears in Koestler 1964, pp. 183–189; Koestler attributes its invention to Carl Dunker.

40 "dynamic mappings in thought and language": Fauconnier 1994, Fauconnier 1997, Fauconnier & Sweetser 1996.

49 "entrenched mappings and frames": Something is entrenched if it is routinely manipulated and resists displacement. An example of an entrenched frame is *going to a restaurant*, with its places for menu, server, payment, food, and so on. Knowing this frame allows us to dine out in foreign cities. An entrenched frame can also be elaborated in novel ways, as when we encounter, say, a drive-through sushi restaurant in Westwood. Similarly, when the American astronauts dock at the Russian space station and enter it to eat, they can refer to it as "dining out," and we all know they are prompting us to construct a highly novel elaboration of the *restaurant* frame. They can then make remarks like "Send the bill to Uncle Sam," "My compliments to the chef," "The best restaurants are always off the beaten path," and so on. An example of an entrenched cross-space mapping is the identity mapping of persons or objects across temporally distinct mental spaces. When we say, "Paul was a teenager in the 1960s," we automatically map Paul today onto the teenager in the 1960s. When we say, "The book you were reading has gone back to the library," we automatically map the book in the library onto the book that you were reading.

51 "Micronesian navigators": Hutchins 1995a.

51 "ETAK and Ngatik": Lewis 1972.

51 "mysterious Micronesian system": Hutchins & Hinton 1984.

52 "meet yourself coming down the street": Rodriguez 1998.

52 "meet myself coming down the street": Ibid.

52 "Spanish short story": Zarraluki 1993, pp. 299–314. We thank Mikkel Hollaender Jensen for bringing this passage to our attention.

56 "metaphor is not part of the study of meaning": Davidson 1978.

CHAPTER 4

59 "Syntax": Harris 1957, 1968; Chomsky 1957.

59 "invisible blending becomes visible": We thank Gerald Graff for bringing such examples to our attention.

71 "une issue heureusc": "Besides," as José Fort points out, "you never deal with murderers; you fight them, you isolate them; you stop them from doing harm. But these accords are doing just the opposite." "No, France has no reason to blame itself," Alan Danjou answers back in *Courrier de l'Ouest*. "France has supported the return to a type of government that is more attentive to human rights, and the negotiations that were started could have led to a good outcome."

72 "The method of loci": Hutchins (in preparation).
72 "the speaker's own house": See Yeats 1966 and Carruthers 1990.

CHAPTER 5

75 "one global idea": Hadamard 1945, p. 65.
80 "the Baby's Ascent": Sweetser 2001.

CHAPTER 6

91 "explained by so few assumptions": Dawkins 1995, p. ix.
93 "Identity and Change are compressed into Uniqueness": We capitalize Identity, Change, and so on when they are used in a strict technical sense to refer to vital relations in a network.
94 "My tax bill gets bigger every year": Sweetser 1996, 1997.
99 "Disanalogy is grounded on Analogy": Gentner & Markman 1994, 1997.
104 "scales, force-dynamic patterns, image-schemas, and vital relations": On scales, see Coulson 2001; on force-dynamics, see Talmy 2000; and on image-schemas, see Johnson 1987.
105 "the case of analogy": Hofstadter 1995.
109 "back to our starting point": Dawkins 1996, p. 36.

CHAPTER 7

113 "I see my life": Excerpt from Yeats's "Fergus and the Druid."
120 "captured formally in predicate calculus notation": See also Sweetser 1987 and 1999.
131 "You are digging your own grave": See Coulson 2001.

CHAPTER 8

139 "the Father of some Strategem": *Henry the Fourth, Part Two,* 1.1.
160 "Milton's portrayal of Satan as father": Discussed in Turner 1987.
161 "These yelling Monsters": *Paradise Lost*, II.795–802.
161 "all my nether shape thus grew Transform'd": *Paradise Lost*, II.781–785.
166 "the bathwater of traditional religion": Charles Fillmore (personal communication).

CHAPTER 9

171 "we are symbol-using, networked creatures": Donald 1991, p. 382.
172 "each step small enough to have been produced by a random mutation or recombination": Pinker & Bloom 1990.

173 "statistical inferencing": See Elman et al. 1996.
173 "the unusual nature of symbolic learning": Deacon 1997, p. 142.
173 "One line of thinking": Calvin & Bickerton 2000.
173 "preadaptations for language": Wilson 1999.
173 "co-evolutionary proposals": Deacon 1997.
174 "Art makes a dramatic appearance":Mithen 1998, p. 165.
178 "the resultative construction in English": See Goldberg 1994.

186 "these neurologically advanced human beings": "Thus," says Klein, "the text argues that after an initial human dispersal from Africa by 1 million years ago, at least three geographically distinct human lineages emerged. These culminated in three separate species: *Homo sapiens* in Africa, *Homo neanderthalensis* in Europe, and *Homo erectus* in eastern Asia. *Homo sapiens* then spread from Africa, beginning perhaps 50,000 years ago to extinguish or swamp its archaic Eurasian contemporaries. The spread was prompted by the development of the uniquely modern ability to innovate and to manipulate culture in adaptation. This ability may have followed on a neural transformation or on social and technological changes among Africans who already had modern brains. Whichever alternative is favored, the fossil, archeological, and genetic data now show that African *H. sapiens* largely or wholly replaced European *H. neanderthalensis*" (1999, p. xxiv). Klein further argues that "only fully modern humans after 50 ky [kiloyears] ago possessed fully modern language ability, and that the development of this ability may underlie their modernity" (1999, p. 348). He also writes: "But even if important details remain to be fixed, the significance of modern human origins cannot be overstated. Before the emergence of modern people, the human form and human behavior evolved together slowly, hand in hand. Afterward, fundamental evolutionary change in body form ceased, while behavioral (cultural) evolution accelerated dramatically. The most likely explanation is that the modern human form—or more precisely the modern human brain—permitted the full development of culture in the modern sense and that culture then became the primary means by which people responded to natural selective pressures. As an adaptive mechanism, not only is culture far more malleable than the body, but cultural innovations can accumulate far more rapidly than genetic ones, and this explains how, in a remarkably short time, the human species has transformed itself from a relatively rare, even insignificant large mammal to the dominant life form on the planet" (1999, p. 494). In an earlier work, Klein similarly notes that "[t]he archeological record is geographically uneven, but where it is most complete and best-dated, it implies that a radical transformation in human behavior occurred 50,000 to 40,000 years ago, the exact time perhaps depending on the place. Arguably, barring the development of those typically human traits that produced the oldest known archeological sites between 2.5 and 2 million years ago, this transformation represents the most dramatic behavioral shift that archaeologists will ever detect" (1992, p. 5). "Thus, while both Mousterians and Upper Paleolithic people buried their dead, Upper Paleolithic graves tend to be significantly more elaborate. These graves are the first to suggest a burial ritual or ceremony, with its

obvious implications of religion or ideology in the ethnographic sense of the term" (1992, p. 7). "The list of contrasts can be extended, and in each case the conclusion is not just that Upper Paleolithic people were qualitatively different, but also that they were behaviorally more advanced than Mousterians and earlier people in the same way that living people are. The evidence does not demonstrate that every known Upper Paleolithic trait was present from the very beginning. It is, in fact, only logical that many features, particularly those involving advances in technology, took time to accumulate. What the evidence does show is that, compared to their antecedents, Upper Paleolithic people were remarkably innovative and inventive; this characteristic, more than any other, is their hallmark. In the broad sweep of European prehistory, they were the first people for whom archeology clearly implies the presence of both 'Culture' and 'cultures' (or ethnicity) in the classic anthropological sense" (1992, p. 7).

186 "advanced form of vocalization": Mithen 1996, p. 142.

187 "mitochondrial DNA": We quote from "Out of Africa: Part 2," a website press release from *Nature Genetics* dated November 29, 1999: "Fossil evidence indicates that modern humans originated in Africa and then expanded from North Africa into the Middle East about 100,000 years ago. Silvana Santachiara-Benerecetti (of the University of Pavia) and colleagues now provide evidence that supports a second route of exit from Africa, whereby ancient peoples dispersed from eastern Africa and migrated along the coast to South Asia.

"Mitochondria are tiny intracellular bodies that generate the energy needed to drive the activities of a cell. They have their own DNA, distinct and independent from nuclear DNA. Mitochondrial DNA can be 'fingerprinted' according to small variations in sequence and, because mitochondria are only inherited from the mother, used to trace maternal ancestry. Closely related mitochondrial DNA sequences fall within the same 'haplogroup', and insinuate—but do not prove—a close genetic relationship between the people who carry them. People in Asia and Ethiopia carry the 'M' mitochondrial haplogroup, which raises the question: how has this come about? Have their mitochondrial DNAs evolved independently, but, through coincidence, converged onto the same haplotype? Or does the similarity reflect a genetic relationship?

"On scrutinizing the region of mitochondrial sequence in Africans and Indians, Santachiara-Benerecetti and coworkers ruled out the possibility that the M haplogroups in eastern-African and Asian populations arose independently—rather, they have a common African origin. These findings, together with the observation that the M haplogroup is virtually absent in Middle-Eastern populations, support the idea that there was a second route of migration out of Africa, approximately 60,000 years ago, exiting from eastern Africa along the coast towards Southeast Asia, Australia, and the Pacific Islands (p. 437).

187 "Y chromosomes": "We focused on estimating the expected time to the most recent common ancestor and the expected ages of certain mutations with

interesting geographic distributions. Although the geographic structure of the inferred haplotype tree is reminiscent of that obtained for other loci (the root is in Africa, and most of the oldest non-African lineages are Asian), the expected time to the most recent common ancestor is remarkably short, on the order of 50,000 years. Thus, although previous studies have noted that Y chromosome variation shows extreme geographic structure, we estimate that the spread of Y chromosomes out of Africa is much more recent than previously was thought" (Thomson et al. 2000, p. 736).

187 "language as an invention of behaviorally modern human beings": Cavalli-Sforza 2000.

187 "weaving and cord-making": As reported in Angier 1999.

CHAPTER 10

195 "everyday objects as material anchors": Hutchins (in preparation).

199 "aviation dials": Hutchins 1995b, p. 274.

201 "blue hockey stick": Holder 2000.

203 "what do we mean by worth?": Gordon 2000.

207 "the idea of the Gothic cathedral": Scott (in press).

207 "creation of a habitat for the sacred": Ibid.

207 "method of loci": Ibid. See also Hutchins (in preparation).

208 "emergent property of this compound blended space": Hutchins (in preparation).

208 "long narratives": Ibid.

209 "imaginary spaces": Scott (in press); emphasis added.

209 "a space in which to meditate": Scott (in press), quoting Carruthers 1990, p. 244.

210 "Automated Teller Machine": Holder 1999.

212 "the modality of sign": Liddell 1998, Van Hoek 1996, Poulin 1996.

212 "the blended Garfield": Liddell 1998, pp. 293–294.

215 "How to build a baby": Mandler 1992.

CHAPTER 11

218 "no fussy little grammatical exercise": Goodman 1947, p. 113.

218 "no form of causal inference": King, Keohane, and Verba 1994, pp. 77–79.

218 "fundamental problem of causal inference": Ibid., p. 79.

218 "everything remains the same": Ibid., p. 77; emphasis added.

219 "all counterfactual conditions are causal assertions": Roese & Olson 1995, p. 11.

219 "Would she be pro-life?": Law professor Goldberg, quoted in *the Los Angeles Times*.

221 "abortion debates": Coulson 2001.
223 "Bishop Berkeley would have preferred": Penrose 1994, p. 417.
224 "wonderful and absurd counterfactual": Moser, 1988.
232 "anosognosia": Ramachandran 1998, pp. 127–130.
232 "Mrs. Macken": Ibid., pp. 143–146.
232 "her arm was fine": Ibid., pp. 149–150.
233 "lottery depression": Kahneman, Slovic, & Tversky 1982.
234 "non-Euclidean geometry": Kline 1972 and Bonola 1912.
236 "most credit must be accorded to Saccheri": Kline 1972, p. 869.
244 "the absence of number": Ibid., p. 184.
244 "blending in the development of mathematics": Lakoff & Núñez 2000.

CHAPTER 12

250 "condemned him to be, magnificently, only a man": From "La Tour de France Comme Épopée" in Barthes 1957, pp. 125, 129.

256 "I would have liked to have met me": Michel Charolles (personal communication), Eve Sweetser (personal communication).

256 "His Ancient Philosophy professor": Jeff Pelletier (personal communication).

256 "I would have been his widow": Nili Mandelblit (personal communication).

257 "if she were not my wife": "First Lady" 1999, p. 18A.

257 "Consider a dialogue from the movie": See Fauconnier 1997, pp. 120–126.

265 "Todd Oakley discusses": Oakley 1995, Spiegelman 1991.

266 "Erving Goffman observes": Goffman 1974. In particular, see the chapter on theatrical framing.

CHAPTER 13

272 "Wallis provided geometric constructions": See Kline 1980.

274 "just as alive as biological viruses": John Markoff, "Beyond Artificial Intelligence, a Search for Artificial Life." *New York Times,* Week in Review Section, p. 5, February 25, 1990.

CHAPTER 14

279 "then death had died today": *King Henry the Sixth, Part One*, 4.7.

280 "The Dracula crowd will scream":Reeves 1993. (This example was pointed out to us by Bill Gleim.)

280 "These health care vampires": The interpretation we are analyzing here is the one most people get in isolation, but it has been reported to us that at the time

the newspaper editorial was published, there was a television commercial paid for by the health care industry in which two actors play a couple who are afraid that Clinton's proposed changes to the health care system will deprive them of what they need.

289 "the following excerpt from a letter": Coulson 2001, pp. 238–241.

292 "the Grim Reaper": Some points of our analysis are prompted by the analysis of the Grim Reaper in Lakoff & Turner 1989. These authors, in discussing "Death-in-general" as the abstract cause behind all individual deaths, implicitly detect the causal tautology involved in the blend of the Grim Reaper. They also detect and analyze the pattern of conception whereby Death is personified. They name this pattern EVENTS ARE ACTIONS, and it consists, in part, of a cross-space mapping between an actor in an action and a nonactor in an event: The nonactor in the event is understood metaphorically by projection from the actor in the action. Finally, Lakoff and Turner discuss constraints on such projections, which we build on in the following analysis.

293 "withering into old age": Analyzed in Lakoff & Johnson 1980.

293 "computer desktop": See Fauconnier 2001.

CHAPTER 15

299 "folk models of anger": Lakoff [and Kövecses] 1987.

303 "buildings, companies, and people": Nunberg 1979.

306 "college dormitory with a wastebasket": This example is based on Coulson 2001.

308 "evil magician": Lakoff & Turner 1989.

CHAPTER 16

310 "consider football and language": Searle 1969.

316 "the Bastille": Mailer 1968, p. 113. (We thank Jennifer Harding for bringing this passage to our attention.)

318 "likely to give an interview": Sweetser 1999.

326 "not identical to them": Ramachandran & Blakeslee 1998.

333 "already active in the context of communication": Douglas Hofstadter (personal communication).

333 "make their communications relevant": Grice 1975, Sperber & Wilson 1986.

340 "trashcan essentially never fills up": Ricardo Maldonado (personal communication).

343 "exist only in the imagination": Kline 1972, p. 594.

343 "irrational numbers had no existence": Kline 1980, p. 114.

344 "imaginary quantities": Kline 1972, p. 252.
344 "greater than infinity": Kline 1980, p. 116.
345 "a general process": Lansing (personal communication).
345 "remarkable analogies": Hofstadter 2000.
349 "EVENTS ARE ACTIONS": "Lakoff & Turner 1989.
349 "Fictive motion": Fictive motion is studied in Talmy 1995 and 2000.

CHAPTER 17

353 "Linguistics is arguably the most hotly contested property": Rymer 1992, p. 48.
355 "Following Charles Travis": Travis 1981.
359 "fire of love": Brooke-Rose 1958.
359 "the bathwater of traditional religion": Charles Fillmore (personal communication).
361 "likely candidate": Sweetser 1999.
361 "fake": Coulson 1997, Coulson & Fauconnier 1999.
365 "The most sophisticated analysis of size terms": Langacker 1987, p. 118; Langacker 1988, p. 70.
367 "Vallejo 94": These pictures appear on pages 116 and 117 of the June 1994 issue.
370 "The caused-motion construction": Goldberg 1994.
374 "French offers three complex blends": Fauconnier & Turner 1996.
375 "the verbal system of Hebrew": Mandelblit 1997, 2000.
377 "basic reference is to motion": Talmy 2000, p. 104.
379 "The bakery is across the street from the bank": "Ibid., p. 136.
379 "The palm trees clustered": Ibid., p. 134.
380 "paint spots": Ibid., p. 128.
380 "shadow": Ibid., p. 114.
381 "way construction": Kemmer & Israel 1994, Israel 1996.
382 "single conceptual package": Ibid., p. 226.
382 "this usage shrank": Ibid., p. 221.

REFERENCES

Angier, Natalie. 1999. "Furs for Evening, But Cloth Was the Stone Age Standby." *The New York Times on the Web,* December 14.

Barsalou, Lawrence W. 1999. "Perceptual Symbol Systems." *Behavioral and Brain Sciences,* 22, pp. 577–609.

Barthes, Roland. 1957. *Mythologies.* Paris: Éditions du Seuil.

Bateson, Gregory. 1972. *Steps to an Ecology of Mind.* New York: Ballantine Books.

Black, Max. 1962. *Models and Metaphors.* Ithaca, N.Y.: Cornell University Press 1962.

Boden, Margaret. 1994. "Précis of *The Creative Mind: Myths and Mechanisms.*" *Behavioral and Brain Sciences,* 17, pp. 519–570.

______. 1990. *The Creative Mind: Myths and Mechanisms.* London: Weidenfeld & Nicholson.

Boden, Margaret (ed.). 1994. *Dimensions of Creativity.* Cambridge, Mass.: MIT Press.

Bonola, Roberto. 1912. *Non-Euclidean Geometry: A Critical and Historical Study of Its Development.* Translated by H. S. Carslaw. Chicago: Open Court Publishing Company. [Reprinted in 1955.]

Brooke-Rose, Christine, 1958. *A Grammar of Metaphor.* London: Secker & Warburg.

Brugman, Claudia. 1990. "What Is the Invariance Hypothesis?" *Cognitive Linguistics,* 1:2, pp. 257–266.

Calvin, William, and Derek Bickerton. 2000. *Lingua ex Machina: Reconciling Darwin and Chomsky with the Human Brain.* Cambridge, Mass.: MIT Press.

Carruthers, Mary. 1990. *The Book of Memory: A Study of Memory in Medieval Culture.* Cambridge, U.K.: Cambridge University Press.

Cavalli-Sforza, Luigi Luca. 2000. *Genes, Peoples, and Languages.* New York: Farrar, Straus, and Giroux.

Chomsky, Noam. 1957. *Syntactic Structures.* Berlin/New York: Mouton de Gruyter.

Coulson, Seana. 2001. *Semantic Leaps: Frame-Shifting and Conceptual Blending in Meaning Construction.* New York/Cambridge: Cambridge University Press.

______. 1997. "Semantic Leaps: Frame-Shifting and Conceptual Blending." Ph.D. dissertation, University of California, San Diego.

______. 1995. "Analogic and Metaphoric Mapping in Blended Spaces." *Center for Research in Language Newsletter,* 9:1, pp. 2–12.

Coulson, Seana, and Gilles Fauconnier. 1999. "Fake Guns and Stone Lions: Conceptual Blending and Privative Adjectives." In B. Fox, D. Jurafsky, and L. Michaelis (eds.), *Cognition and Function in Language*. Stanford: Center for the Study of Language and Information.

Csikszentmihalyi, Mihaly. 1991. *Flow: The Psychology of Optimal Experience.* New York: HarperCollins.

Davidson, Donald. 1978. "What Metaphors Mean." *Critical Inquiry*, 5:1.

Dawkins, Richard. 1996. *Climbing Mount Improbable.* London/New York: Norton.

______. 1995. *River Out of Eden.* New York: Basic Books.

Deacon, Terrence. 1997. *The Symbolic Species: The Co-Evolution of Language and the Brain.* New York: W. W. Norton.

Donald, Merlin. 1994. "Computation: Part of the Problem of Creativity." *Behavioral and Brain Sciences* 17:3, pp. 537–538.

______. 1991. *Origins of the Modern Mind: Three Stages in the Evolution of Culture and Cognition.* Cambridge, Mass.: Harvard University Press.

Education Excellence Partnership website: www.edex.org.

Elman, Jeffrey L., Elizabeth A. Bates, Mark H. Johnson, Annette Karmiloff-Smith, Domenico Parisi, and Kim Plunkett. 1996. *Rethinking Innateness: A Connectionist Perspective on Development.* Cambridge, Mass.: MIT Press.

Fauconnier, Gilles. 2001. "Conceptual Blending and Analogy." In Dedre Gentner, Keith Holyoak, and Boicho Kokinov (eds.), *The Analogical Mind: Perspectives from Cognitive Science* (pp. 255–286). Cambridge, Mass.: MIT Press.

______. 1997. *Mappings in Thought and Language.* Cambridge, U.K.: Cambridge University Press.

______. 1994. *Mental Spaces.* New York: Cambridge University Press. [Originally published in 1985 by MIT Press.]

______. 1990. "Domains and Connections." *Cognitive Linguistics* 1:1.

Fauconnier, Gilles, and Eve Sweetser (eds.). 1996. *Spaces, Worlds, and Grammar.* Chicago: University of Chicago Press.

Fauconnier, Gilles, and Mark Turner. 1998. "Principles of Conceptual Integration." In Jean-Pierre Koenig (ed.), *Discourse and Cognition* (pp. 269–283). Stanford: Center for the Study of Language and Information.

______. 1996. "Blending as a Central Process of Grammar." In Adele Goldberg (ed.), *Conceptual Structure, Discourse, and Language*. Stanford: Center for the Study of Language and Information.

______. 1994. *Conceptual Projection and Middle Spaces.* San Diego: University of California, Department of Cognitive Science Technical Report 9401 (available on-line at *blending.stanford.edu* and *mentalspace.net*).

Fillmore, Charles J., and Beryl T. Atkins. 1992. "Toward a Frame-Based Lexicon: The Semantics of RISK and Its Neighbors." In Adrienne Lehrer and Eva Feder Kittay (eds.), *Frames, Fields, and Contrasts: New Essays in Semantic and Lexical Organization* (pp. 75–102). Hillsdale, N.J.: Lawrence Erlbaum Associates.

Fillmore, Charles J., and Paul Kay. n.d. "On Grammatical Constructions." Unpublished ms., University of California at Berkeley.

"First Lady Looks Closer to Senate Run." 1999. Reuters, *San Jose Mercury News*, October 24.

Fong, Heatherbill. 1988. "The Stony Idiom of the Brain: A Study in the Syntax and Semantics of Metaphors." Ph.D. dissertation, University of California, San Diego.

Forbus, Kenneth D., Dedre Gentner, and Keith Law. 1994. "MAC/FAC: A Model of Similarity-Based Retrieval." *Cognitive Science,* 19, pp. 141–205.

Frazer, Sir James George. 1922. *The Golden Bough: A Study in Magic and Religion.* New York: Macmillan.

French, Robert Matthew. 1995. *The Subtlety of Sameness: A Theory and Computer Model of Analogy-Making*. Cambridge, Mass.: MIT Press.

Gentner, Dedre. 1989. "The Mechanisms of Analogical Reasoning." In S. Vosniadou and A. Ortony (eds.), *Similarity and Analogical Reasoning*. Cambridge, U.K.: Cambridge University Press.

______. 1983. "Structure-Mapping: A Theoretical Framework for Analogy." *Cognitive Science,* 7, pp. 155–170.

Gentner, Dedre, Keith Holyoak, and Boicho Kokinov (eds.). 2001. *The Analogical Mind: Perspectives from Cognitive Science.* Cambridge, Mass.: MIT Press.

Gentner, D., and A. B. Markman. 1997. "Structure Mapping in Analogy and Similarity." *American Psychologist,* 52, pp. 45–56.

______. 1994. "Structural Alignment in Comparison: No Difference Without Similarity." *Psychological Science*, 5:3, pp. 152–158.

Gibbs, R. W., Jr. 1994. *The Poetics of Mind: Figurative Thought, Language, and Understanding.* Cambridge, U.K.: Cambridge University Press.

Goffman, Erving. 1974. *Frame Analysis: An Essay on the Organization of Experience.* New York: Harper & Row.

Goguen, Joseph. 1999. "An Introduction to Algebraic Semiotics, with Application to User Interface Design." In Chrystopher Nehaniv (ed.), *Computation for Metaphor, Analogy, and Agents* (pp. 242–291). Berlin: Springer-Verlag. [A volume in the series Lecture Notes in Artificial Intelligence.]

Goldberg, Adele. 1994. *Constructions: A Construction Grammar Approach to Argument Structure.* Chicago: University of Chicago Press.

Goldberg, Adele (ed.). 1996. *Conceptual Structure, Discourse, and Language.* Stanford: Center for the Study of Language and Information.

Gombrich, E. H. 1965. "The Use of Art for the Study of Symbols." *American Psychologist,* 20, pp. 34–50.

Goodman, Nelson. 1955. *Fact, Fiction, and Forecast.* Indianapolis: Bobbs-Merrill.

______. 1947. "The Problem of Counterfactual Conditionals," *Journal of Philosophy*, 44, pp. 113–128.

Gordon, Mary. 2000. *The New York Times on the Web,* June 4.

Grice, H. P. 1975. "Logic and Conversation." In Peter Cole and Jerry L. Morgan (eds.), *Syntax and Semantics.* Vol. 3: *Speech Acts* (pp. 41–58). New York: Academic Press.

Hadamard, Jacques. 1945. *The Psychology of Invention in the Mathematical Field.* Princeton: Princeton University Press.

Harris, Zellig. 1968. *Mathematical Structures of Language* (*Interscience Tracts in Pure and Applied Mathematics,* Vol. 21). New York: Interscience Publishers/John Wiley & Sons.

______. 1957. "Co-Occurrence and Transformation in Linguistic Structure." *Language,* 33:3, pp. 283–340.

Hodder, Ian. 1998. "Creative Thought: A Long-Term Perspective." In Steven Mithen (ed.), *Creativity in Human Evolution and Prehistory* (pp. 61–77). London/New York: Routledge.

Hofstadter, Douglas R. 2000. "The Ubiquity and Power of Analogy in Discovery in Physics." The Ninth Annual Hofstadter Lecture (in memory of Richard Hofstadter), Stanford University, February 29.

______. 1995. *Fluid Concepts and Creative Analogies.* New York: Basic Books.

______. 1985. "Analogies and Roles in Human and Machine Thinking." In Douglas R. Hofstadter, *Metamagical Themas.* New York: Bantam Books.

Hofstadter, Douglas R., Liane Gabora, Salvatore Attardo, and Victor Raskin. 1989. "Synopsis of the Workshop on Humor and Cognition." *Humor: International Journal of Humor Research,* 2:4, pp. 417–440.

Hofstadter, Douglas R., and David J. Moser. 1989. "To Err is Human; To Study Error-Making Is Cognitive Science." *Michigan Quarterly Review,* 28:2, pp. 185–215.

______. n.d. "Errors: The Royal Road." Unpublished manuscript.

Holder, Barbara. 2000. "Conceptual Blending in Airbus A320 Displays." Sensation and Perception Seminar, Department of Cognitive Sciences, University of California, Irvine, November 15.

______. 1999. "Blending and Your Bank Account: Conceptual Blending in ATM Design." *Newsletter of the Center for Research in Language,* 11:6 (available online at *crl.ucsd.edu*).

Holland, J. H., Keith James Holyoak, R. E. Nisbett, and Paul Thagard. 1986. *Induction: Processes of Inference, Learning, and Discovery.* Cambridge, Mass.: Bradford Books/MIT Press.

Holland, Paul. 1986. "Statistics and Causal Inference." *Journal of the American Statistical Association,* 81, pp. 945–960.

Holyoak, Keith James, and Paul Thagard. 1995. *Mental Leaps: Analogy in Creative Thought.* Cambridge, Mass.: MIT Press.

______. 1989. "Analogical Mapping by Constraint Satisfaction." *Cognitive Science,* 13, pp. 295–355.

Hummel, John, and Keith Holyoak. 1997. "Distributed Representations of Structure: A Theory of Analogical Access and Mapping." *Psychological Review,* 104:3, pp. 427–466.

Hutchins, Edwin (in preparation). "Material Anchors for Conceptual Blends."

______. 1995a. *Cognition in the Wild.* Cambridge, Mass.: MIT Press.

______. 1995b. "How a Cockpit Remembers Its Speeds." *Cognitive Science,* 19:3, pp. 265–288.

______. 1989. "Metaphors for Interface Design." In M. M. Taylor, F. Néel, and D. G. Bouwhuis (eds.), *The Structure of Multimodal Dialogue* (pp. 11–28). Amsterdam: Elsevier Science Publishers.

Hutchins, Edwin, and Geoffrey Hinton. 1984. "Why the Islands Move." *Perception,* 13, pp. 629–632.

Indurkhya, Bipin. 1992. *Metaphor and Cognition: An Interactionist Approach.* Dordrecht: Kluwer.

Israel, Michael. 1996. "The *Way* Constructions Grow." In Adele Goldberg (ed.), *Conceptual Structure, Discourse, and Language* (pp. 217–230). Stanford: Center for the Study of Language and Information.

Johnson, Mark. 1987. *The Body in the Mind.* Chicago: University of Chicago Press.

Kahneman, Daniel. 1995. "Varieties of Counterfactual Thinking." In Neal J. Roese and James M. Olson (eds.), *What Might Have Been: The Social Psychology of Counterfactual Thinking.* Mahwah, N.J.: Lawrence Erlbaum Associates.

Kahneman, Daniel, Paul Slovic, and Amos Tversky. 1982. *Judgment Under Uncertainty: Heuristics and Biases.* Cambridge/New York: Cambridge University Press.

Kay, Paul. 1995. "Construction Grammar." In Jef Verschueren, Jan-Ola Ostman, and Jan Blommaert (eds.), *Handbook of Pragmatics.* Amsterdam/Philadelphia: John Benjamins.

Kay, Paul, and Charles Fillmore. 1995. "Grammatical Constructions and Linguistic Generalizations: The *What's X doing Y?* Construction." Unpublished ms., Department of Linguistics, University of California at Berkeley.

Keane, Mark T., Tim Ledgeway, and Stuart Duff. 1994. "Constraints on Analogical Mapping: A Comparison of Three Models." *Cognitive Science,* 18, pp. 387–438.

Kemmer, Suzanne, and Michael Israel. 1994. "Variation and the Usage-Based Model." In Katie Beals et al. (eds.), *Papers from the Parasession on Variation and Linguistic Theory.* Chicago: Chicago Linguistic Society.

King, Gary, Robert O. Keohane, and Sidney Verba. 1994. *Designing Social Inquiry: Scientific Inference in Qualitative Research.* Princeton: Princeton University Press.

Klein, Richard G. 1999. *The Human Career: Human Biological and Cultural Origins,* 2nd ed. Chicago: University of Chicago Press, 1999.

______. 1992. "The Archeology of Modern Human Origins." *Evolutionary Anthropology*, 1:1, pp. 5–14.

Kline, Morris. 1980. *Mathematics: The Loss of Certainty.* Oxford: Oxford University Press.

______. 1972. *Mathematical Thought from Ancient to Modern Times.* New York: Oxford University Press.

Koestler, Arthur. 1964. *The Act of Creation.* New York: Macmillan.

Kosslyn, Stephen M. 1980. *Image and Mind.* Cambridge, Mass.: Harvard University Press.

Kunda, Ziva, Dale T. Miller, and Theresa Clare. 1990. "Combining Social Concepts: The Role of Causal Reasoning." *Cognitive Science,* 14, pp. 551–577.

Lakoff, George. 1993. "The Contemporary Theory of Metaphor." In Andrew Ortony (ed.), *Metaphor and Thought,* 2nd ed. (pp. 202–251) (Cambridge, U.K.: Cambridge University Press).

______. 1990. "The Invariance Hypothesis." *Cognitive Linguistics,* 1:1, pp. 39–74.

Lakoff, George, and Mark Johnson. 1980. *Metaphors We Live By.* Chicago: University of Chicago Press.

Lakoff, George [and Zoltán Kövecses]. 1987. *Women, Fire, and Dangerous Things.* Chicago: University of Chicago Press.

Lakoff, George, and Rafael Núñez. 2000. *Where Mathematics Comes From: How the Embodied Mind Brings Mathematics into Being.* New York: Basic Books.

Lakoff, George, and Mark Turner. 1989. *More Than Cool Reason: A Field Guide to Poetic Metaphor.* Chicago: University of Chicago Press.

Langacker, Ronald. 1990. *Concept, Image, and Symbol: The Cognitive Basis of Grammar.* Berlin/New York: Mouton de Gruyter.

______. 1988. "A View of Linguistic Semantics." In Brygida Rudzka-Ostyn (ed.), *Topics in Cognitive Linguistics* (pp. 49–90). Amsterdam/Philadelphia: John Benjamins.

______. 1987–1991. *Foundations of Cognitive Grammar,* Vols. 1 and 2. Stanford: Stanford University Press.

Lewis, David. 1973. *Counterfactuals.* Cambridge, Mass.: Harvard University Press.

______. 1972. *We, the Navigators. The Ancient Art of Landfinding in the Pacific.* Honolulu: The University Press of Hawaii.

Liddell, Scott. 1998. "Grounded Blends, Gestures, and Conceptual Shifts." *Cognitive Linguistics,* 9:3, pp. 283–314.

Mailer, Norman. 1968. *Armies of the Night.* New York: New American Library.

Mandelblit, Nili. 1997. "Grammatical Blending: Creative and Schematic Aspects in Sentence Processing and Translation." Ph.D. dissertation, University of California, San Diego.

______. 2000. "The Grammatical Marking of Conceptual Integration: From Syntax to Morphology." *Cognitive Linguistics* 11:314, pp. 197–252.

Mandelblit, Nili, and Oron Zachar. 1998. "The Notion of Dynamic Unit: Conceptual Developments in Cognitive Science." *Cognitive Science* 22:2, pp. 229–268.

Mandler, Jean. 1992. "How to Build a Baby." *Psychological Review,* 99:4, pp. 587–604.

Maslow, Abraham. 1968. *Toward a Psychology of Being.* New York: Van Nostrand Reinhold.

McNeill, David. 1992. *Hand and Mind: What Gestures Reveal About Thought.* Chicago: University of Chicago Press.

Mitchell, M. 1993. *Analogy-Making as Perception.* Cambridge, Mass.: MIT Press.

Mithen, Steven. 1998. "A Creative Explosion? Theory of Mind, Language, and the Disembodied Mind of the Upper Paleolithic." In Steven Mithen (ed.), *Creativity in Human Evolution and Prehistory* (pp. 165–191). London/New York: Routledge.

______. 1996. *The Prehistory of the Mind: A Search for the Origins of Art, Science and Religion.* London/New York: Thames & Hudson.

Mithen, Steven (ed.). 1998. *Creativity in Human Evolution and Prehistory.* London/New York: Routledge.

Moser, David. 1988. "If This Paper Were in Chinese, Would Chinese People Understand the Title?" Unpublished ms., Center for Research on Concepts and Cognition, Indiana University.

Moser, David, and Douglas Hofstadter. n.d. "Errors: A Royal Road to the Mind." Unpublished ms., Center for Research on Concepts and Cognition, Indiana University.

Norman, D. A. 1981. "Categorization of Action Slips." *Psychological Review,* 88, pp. 1–15.

Nunberg, G. 1979. "The Non-Uniqueness of Semantic Solutions: Polysemy." *Linguistics and Philosophy,* 3:2.

Oakley, Todd. 1995. "Presence: The Conceptual Basis of Rhetorical Effect." Ph.D. dissertation, University of Maryland.

Ortony, Andrew (ed.). 1979. *Metaphor and Thought.* Cambridge/New York: Cambridge University Press.

Pavel, Thomas. 1986. *Fictional Worlds.* Cambridge, Mass.: Harvard University Press.

Penrose, Roger. 1994. *Shadows of the Mind: A Search for the Missing Science of Consciousness.* New York: Oxford University Press.

Pepper, Stephen. 1942. "Root Metaphors." In Stephen Pepper, *World Hypotheses.* Berkeley: University of California Press.

Perkins, D. 1994. "Creativity: Beyond the Darwinian Paradigm." In Margaret Boden (ed.), *Dimensions of Creativity* (pp. 119–142). Cambridge, Mass.: MIT Press.

Pinker, Steven, and Paul Bloom. 1990. "Natural Language and Natural Selection." *Behavioral and Brain Sciences,* 13, pp. 707–784.

Poulin, Christine. 1996. "Manipulation of Discourse Spaces in ASL." In Adele Goldberg (ed.), *Conceptual Structure, Discourse, and Language* (pp. 421–433). Stanford: Center for the Study of Language and Information.

Ramachandran, V. S., and Sandra Blakeslee. 1998. *Phantoms in the Brain.* New York: Morrow.

Reeves, Richard. 1993. "Best Performance by a Politician." *Los Angeles Times.*

Robert, Adrian. 1998. "Blending in the Interpretation of Mathematical Proofs." In Jean-Pierre Koenig (ed.), *Discourse and Cognition.* Stanford: Center for the Study of Language and Information.

Rodriguez, Paul. 1998. "Identity Blends and Discourse Context." Unpublished ms., University of California, San Diego.

Roese, Neal, and James Olson (eds.). 1995. *The Social Psychology of Counterfactual Thinking.* Hillsdale, N.J.: Lawrence Erlbaum Associates.

Russell, Bertrand. 1918. *Mysticism and Logic and Other Essays.* London/New York: Longmans.

Rymer, Russ. 1992. "Annals of Science: A Silent Childhood-I." *New Yorker*, April 13.

Sacks, Sheldon (ed.). 1979. *On Metaphor.* Chicago: University of Chicago Press.

Santachiara-Benerecetti, Silvana. 1999. "Out of Africa: Part 2," website press release from *Nature Genetics,* November 29.

Schwartz, Daniel L., and John B. Black. 1996. "Shuttling Between Depictive Models and Abstract Rules: Induction and Fallback." *Cognitive Science,* 20, pp. 457–497.

Scott, Robert A. (in press). *The Gothic Enterprise: The Idea of the Cathedral and Its Uses in Medieval Europe, 1134–1550.* Berkeley/Los Angeles: University of California Press.

Searle, John. 1969. *Speech Acts: An Essay in the Philosophy of Language.* London: Cambridge University Press.

Sellen, A. J., and D. A. Norman. 1992. "The Psychology of Slips." In B. J. Baars (ed.), *Experimental Slips and Human Error: Exploring the Architecture of Volition.* New York: Plenum Press.

Shastri, Lokendra. 1996. "Temporal Synchrony, Dynamic Bindings, and SHRUTI—A Representational But Non-Classical Model of Reflexive Reasoning." *Behavioral and Brain Sciences,* 19:2, pp. 331–337.

Shepard, Roger, and Lynn A. Cooper. 1982. *Mental Images and Their Transformations.* Cambridge, Mass.: MIT Press.

Smolensky, Paul. 1990. "Tensor Product Variable Binding and the Representation of Symbolic Structures in Connectionist Systems." *Artificial Intelligence,* 46:1–2, pp. 159–216.

Sperber, Dan, and Deirdre Wilson. 1986. *Relevance: Communication and Cognition.* Cambridge, Mass.: Harvard University Press.

Spiegelman, Art. 1991. *Maus II, a Survivor's Tale.* New York: Pantheon Books.

Sweetser, Eve. 2001. "Blended Spaces and Performativity." *Cognitive Linguistics*, 11:3/4, pp. 305–334.

______. 1999. "Compositionality and Blending: Semantic Composition in a Cognitively Realistic Framework." In Theo Janssen and Gisela Redeker (eds.). *Cognitive Linguistics: Foundations, Scope and Methodology* (pp. 129–162). Berlin/New York: Mouton de Gruyter.

______. 1997. "Role and Individual Readings of Change Predicates." In Jan Nuyts and Eric Pederson (eds.), *Language and Conceptualization*. Oxford University Press.

______. 1996. "Changes in Figures and Changes in Grounds: A Note on Change Predicates, Mental Spaces, and Scalar Norms." *Cognitive Studies: Bulletin of the Japanese Cognitive Science Society*, 3:3 (September), pp. 75–86. [Special issue on cognitive linguistics.]

______. 1990. *From Etymology to Pragmatics: Metaphorical and Cultural Aspects of Semantic Structure.* Cambridge, U.K.: Cambridge University Press.

______. 1987. "The Definition of *Lie:* An Examination of the Folk Theories Underlying a Semantic Prototype." In Dorothy Holland and Naomi Quinn (eds.), *Cultural Models in Language and Thought* (pp. 43–66). Cambridge, U.K.: Cambridge University Press.

Talmy, Len. 2000. *Toward a Cognitive Semantics.* Vol. 1: *Concept Structuring Systems*; Vol. 2: *Typology and Process.* Cambridge, Mass.: MIT Press.

______. 1995. "Fictive Motion in Language and 'Ception.'" In Paul Bloom, Mary Peterson, Lynn Nadel, and Merrill Garrett (eds.), *Language and Space* (pp. 307–384). Cambridge, Mass.: MIT Press.

______. 1977. "Rubber-Sheet Cognition in Language." *Proceedings of the 13th Regional Meeting of the Chicago Linguistic Society.*

Tetlock, Philip, and Aaron Belkin (eds.). 1996. *Counterfactual Thought Experiments in World Politics.* Princeton: Princeton University Press.

Thomson, Russell, Jonathan Pritchard, Peidong Shen, Peter Oefner, and Marcus Feldman. 2000. "Recent Common Ancestry of Human Y Chromosomes: Evidence from DNA Sequence Data." *Proceedings of the National Academy of Sciences*, 97:13 (June), pp. 7360–7365.

Travis, Charles, 1981. *The True and the False: The Domain of the Pragmatic.* Amsterdam: John Benjamins.

Turner, Mark. 1996a. "Conceptual Blending and Counterfactual Argument in the Social and Behavioral Sciences." In Philip Tetlock and Aaron Belkin (eds.), *Counterfactual Thought Experiments in World Politics.* Princeton: Princeton University Press.

______. 1996b. *The Literary Mind.* New York: Oxford University Press.

______. 1991. *Reading Minds: The Study of English in the Age of Cognitive Science.* Princeton: Princeton University Press.

______. 1990. "Aspects of the Invariance Hypothesis." *Cognitive Linguistics,* 1:2, pp. 247–255.

______. 1987. *Death Is the Mother of Beauty: Mind, Metaphor, Criticism.* Chicago: University of Chicago Press.

Turner, Mark, and Gilles Fauconnier. 1998. "Conceptual Integration in Counterfactuals." In Jean-Pierre Koenig (ed.), *Discourse and Cognition* (pp. 285–296). Stanford: Center for the Study of Language and Information.

______. 1995. "Conceptual Integration and Formal Expression." *Journal of Metaphor and Symbolic Activity,* 10:3, pp. 183–204.

Van Hoek, Karen. 1996. "Conceptual Locations for Reference in American Sign Language." In Gilles Fauconnier and Eve Sweetser (eds.), *Spaces, Worlds, and Grammar* (pp. 336–350). Chicago: University of Chicago Press.

Weisberg, R. W. 1993. *Creativity: Beyond the Myth of Genius.* New York: W. H. Freeman.

Weizenbaum, Joseph. 1966. "ELIZA—A Computer Program for the Study of Natural Language Communication Between Man and Machine." *Communications of the ACM,* 9:1, pp. 36–45.

Westney, D. Eleanor. 1987. *Imitation and Innovation.* Cambridge, Mass.: Harvard University Press.

Wexo, John Bennett. 1992. *Dinosaurs.* A volume of *Zoobooks.* San Diego: Wildlife Education, Limited.

Wilson, Frank R. 1999. *The Hand.* New York: Vintage.

Wozny, Lucy Anne. 1989. "The Application of Metaphor, Analogy, and Conceptual Models in Computer Systems." *Interacting with Computers,* 1:3, pp. 273–283.

Yeats, Frances. 1966. *The Art of Memory.* Chicago: University of Chicago Press.

Zarraluki, Pedro. 1993. "Páginas Inglesas." In *Cuento Español Contemporáneo.* Edicion de Ma. Ángeles Encinar y Anthony Percival. Madrid: Ediciones Cátedra.

FURTHER IMPORTANT WORK ON CONCEPTUAL BLENDING

THE LAST FIVE YEARS HAVE seen an explosion of remarkable work on conceptual blending, with fascinating applications to art, literature, poetics, mathematics, political science, musicology, linguistics, theology, psychoanalysis, and film.

It would have taken a book ten times the size of this one to do justice to the rich new avenues of investigation that have been opened in all these domains, and we regret that the present book, focused as it is on theory, cannot carry a review of this stimulating work. Below is a partial list of additional recent and ongoing work on blending. Some additional work is presented on the web at *blending.stanford.edu*.

BOOKS

Turner, Mark. 2001. *Cognitive Dimensions of Social Science*. Oxford: Oxford University Press.

Zbikowski, Lawrence. 2001. *Conceptualizing Music: Cognitive Structure, Theory, and Analysis*. New York: Oxford University Press.

ARTICLES AND TALKS

Bizup, Joseph. 1998. "Blending in Ruskin." Paper presented at the Annual Meeting of the Modern Language Association.

Brandt, Per Aage. 1998. "Cats in Space." *The Roman Jakobson Centennial Symposium: International Journal of Linguistics Acta Linguistica Hafniensia* Volume 29. C. A. Reitzel: Copenhagen. [Jakobson and Lévi-Strauss's structuralist reading of Baudelaire's "Les Chats" is reconsidered in light of cognitive rhetoric and conceptual blending theory.]

Bundgård, Peer F. 1999. "Cognition and Event Structure." *Almen Semiotik*, 15, 78–106. [A review of conceptual integration theory.]

Casonato, Marco M. 2000. "Scolarette sexy: processi cognitivi standard nella scena della perversione." *Psicoterapia: clinica, epistemologia, ricerca*, 20–21, Spring. [An analysis of the role of blending in sexual imagination and realized fantasy, including but not restricted to "perverse" scenes.]

Casonato, Marco, Gilles Fauconnier, and Mark Turner. 2001. "L'immaginazione e il cosiddetto 'conflitto' psichico." *Annuario di Itinerari Filosofici*, Vol. 5 (*Strutture dell'esperienza*), No. 3 (*Mente, linguaggio, espressione*). Milano: Mimesis.

Chen, Melinda. 2000. "A Cognitive-Linguistic View of Linguistic (Human) Objectification." Paper presented at the 5th Conference on Conceptual Structure, Discourse, and Language. [A discussion of blends in objectifying human beings.]

Cienki, Alan, and Deanne Swan. 1999. "Constructions, Blending, and Metaphors: Integrating Multiple Meanings." Paper presented at the 6th International Cognitive Linguistics Conference.

Collier, David, and Stephen Levitsky. 1997. "Democracy with Adjectives: Conceptual Innovation in Comparative Research." *World Politics,* 49:3 (April), pp. 430–451.

Coulson, Seana. 1999. "Conceptual Integration and Discourse Irony." Paper presented at Beyond Babel: 18th Annual Conference of the Western Humanities Alliance.

Csabi, Szilvia. 1997. "The Concept of America in the Puritan Mind." Paper presented at the 5th Conference of the International Cognitive Linguistics Association.

Evans, Vyvyan. 1999. "The Cognitive Model for Time." Paper presented at Beyond Babel: 18th Annual Conference of the Western Humanities Alliance.

Fauconnier, Gilles. 2000. "Methods and Generalizations." In T. Janssen and G. Redeker (eds.), *Cognitive Linguistics: Foundations, Scope, and Methodology* (pp. 95–127). The Hague: Mouton de Gruyter. [A volume in the Cognitive Linguistics Research Series.]

______. 1999. "Embodied Integration." Paper presented at the 6th International Cognitive Linguistics Conference.

Fauconnier, Gilles, and Mark Turner. 2000. "Compression and Global Insight." *Cognitive Linguistics*, 11:3–4, pp. 283–304.

______. 1999a. "Metonymy and Conceptual Integration." In Klaus-Uwe Panther and Günter Radden (eds.), *Metonymy in Language and Thought* (pp. 77–90). Amsterdam: John Benjamins. [A volume in the series Human Cognitive Processing.]

______. 1999b. "Polysemy and Conceptual Blending" (available on-line at *blending.stanford.edu*).

______. 1998. "Conceptual Integration Networks." *Cognitive Science,* 22:2 (April-June), pp. 133–187.

Forceville, Charles. 2001. "Blends and Metaphors in Multimodal Representations." Paper presented at the 7th International Cognitive Linguistics Conference.

Freeman, Donald. 1999. "'Speak of me as I am': The Blended Space of Shakespeare's *Othello*." Paper presented at Beyond Babel: 18th Annual Conference of the Western Humanities Alliance.

Freeman, Margaret. 1999a. "The Role of Blending in an Empirical Study of Literary Analysis." Paper presented at the 6th International Cognitive Linguistics Conference.

______. 1999b. "Sound Echoing Sense: The Evocation of Emotion Through Sound in Conceptual Mapping Integration of Cognitive Processes." Paper presented at Beyond Babel: 18th Annual Conference of the Western Humanities Alliance.

______. 1998. "'Mak[ing] new stock from the salt': Poetic Metaphor as Conceptual Blend in Sylvia Plath's 'The Applicant.'" Paper presented at the Annual Meeting of the Modern Language Association.

______. 1997. "Grounded Spaces: Deictic-Self Anaphors in the Poetry of Emily Dickinson." *Language and Literature*, 6:1, pp. 7–28. [Contains a blended-space analysis of Dickinson's "Me from Myself—to banish—."]

Grady, Joseph., Todd Oakley, and Seana Coulson. 1999. "Conceptual Blending and Metaphor." In G. Steen and R. Gibbs (eds.), *Metaphor in Cognitive Linguistics.* Amsterdam/Philadelphia: John Benjamins.

Gréa, Philippe. 2001. "La théorie de l'intégration conceptuelle appliquée à la métaphore et la métaphore filée." Doctoral dissertation, Université de Paris.

Grush, Rick, and Nili Mandelblit. 1998. "Blending in Language, Conceptual Structure, and the Cerebral Cortex." In Per Aage Brandt, Frans Gregersen, Frederik Stjernfelt, and Martin Skov (eds.), *The Roman Jakobson Centennial Symposium: International Journal of Linguistics Acta Linguistica Hafniensia,* Vol. 29 (pp. 221–237). Copenhagen: C. A. Reitzel.

Herman, Vimala. 1999. "Deictic Projection and Conceptual Blending in Epistolarity." *Poetics Today,* 20:3, pp. 523–542.

Hiraga, Masako. 1999. "Blending and an Interpretation of Haiku." *Poetics Today*, 20:3, pp. 461–482.

______. 1999. "Rough Sea and the Milky Way: 'Blending' in a Haiku Text." In Chrystopher Nehaniv (ed.), *Computation for Metaphor, Analogy, and Agents* (pp. 27–36). Berlin: Springer-Verlag. [A volume in the series Lecture Notes in Artificial Intelligence.]

______. 1998. "Metaphor-Icon Links in Poetic Texts: A Cognitive Approach to Iconicity." *Journal of the University of the Air,* 16. ["The model of 'blending' . . . provides an effective instrument to clarify the complexity of the metaphor-icon link."]

Hofstadter, Douglas. 1999. "Human Cognition as a Blur of Analogy and Blending." Paper presented at Beyond Babel: 18th Annual Conference of the Western Humanities Alliance.

Holder, Barbara, and Seana Coulson. 2000. "Hints on How to Drink from a Fire Hose: Conceptual Blending in the Wild Blue Yonder." Paper presented at the 5th Conference on Conceptual Structure, Discourse, and Language.

Jappy, Tony. 1999. "Blends, Metaphor, and the Medium." Paper presented at the 6th International Cognitive Linguistics Conference.

Kim, Esther. 2000. "Analogy as Discourse Process." Paper presented at the 5th Conference on Conceptual Structure, Discourse, and Language. [Includes a discussion of blending in discourse.]

Lakoff, George, and Rafael E. Núñez. 1997. "The Metaphorical Structure of Mathematics: Sketching Out Cognitive Foundations for a Mind-Based Mathematics." In Lyn English (ed.), *Mathematical Reasoning: Analogies, Metaphors, and Images*. Hillsdale, N.J.: Erlbaum. [Analyzes blending in the invention of various mathematical structures.]

Lee, Mark, and John Barnden. 2000. "Metaphor, Pretence, and Counterfactuals." Paper presented at the 5th Conference on Conceptual Structure, Discourse, and Language. [Includes a discussion of blending in counterfactuals.]

Maglio, Paul P., and Teenie Matlock. 1999. "The Conceptual Structure of Information Space." In A. Munro, D. Benyon, D., and K. Hook (eds.), *Personal and Social Navigation of Information Space*. Berlin: Springer-Verlag. [Includes a section on "Conceptual Blends in Information Space."]

Maldonado, Ricardo. 1999. "Spanish Causatives and the Blend." Paper presented at the 6th International Cognitive Linguistics Conference.

Mandelblit, Nili. 1995. "Beyond Lexical Semantics: Mapping and Blending of Conceptual and Linguistic Structures in Machine Translation." *Proceedings of the Fourth International Conference on the Cognitive Science of Natural Language Processing*. Dublin.

Mandelblit, Nili, and Gilles Fauconnier. 2000. "Underspecificity in Grammatical Blends as a Source for Constructional Ambiguity." In A. Foolen and F. van der Leek (eds.), *Constructions*. Amsterdam: John Benjamins.

Narayan, Shweta. 2000. "Mappings in Art and Language: Conceptual Mappings in Neil Gaiman's *Sandman*." Honors thesis, University of California Berkeley.

Oakley, Todd. 1998. "Conceptual Blending, Narrative Discourse, and Rhetoric." *Cognitive Linguistics*, 9, pp. 321–360.

Olive, Esther Pascual. 2001. "Why Bother to Ask Rhetorical Questions (If They Are Already Answered)? A Conceptual Blending Account of Argumentation in Legal Settings." Paper presented at the 7th International Cognitive Linguistics Conference.

Ramey, Lauri. 1998. "'His Story's Impossible to Read': Creative Blends in Michael Palmer's *Books Against Understanding*." Paper presented at the Twentieth Century Literature Conference, University of Louisville.

______. 1997a. "A Film Is/Is Not A Novel: Blended Spaces in Sense and Sensibility." Paper presented at the Popular Culture Association/American Culture Association in the South Conference.

______. 1997b. "What n'er was Thought and cannot be Expres't: Michael Palmer and Postmodern Allusion." Paper presented at the 9th Annual Conference on Linguistics and Literature.

______. 1996. "The Poetics of Resistance: A Critical Introduction to Michael Palmer." Ph.D. dissertation, University of Chicago. [See especially Chapter 4.]

______. 1995. "Blended Spaces in Thurber and Welty." Marian College Humanities Series.

Ramey, Martin. 2000. "Cognitive Blends and Pauline Metaphors in 1 Thessalonians." In *Proceedings of the 2000 World Congress on Religion*, organized by the Society of Biblical Literature.

______. 1997. "Eschatology and Ethics," Chapter 4 of "The Problem of the Body: The Conflict Between Soteriology and Ethics in Paul." Doctoral dissertation, Chicago Theological Seminary. [Contains a discussion of blending in 1 Thessalonians.]

Récanati, François. 1995. "Le présent épistolaire: une perspective cognitive." *L'information grammaticale*, 66 (June), pp. 38–45. [Récanati applies the earliest work on blended spaces to problems of tense. He translates "blended space" as "espace mixte."]

Rohrer, Tim. "The Embodiment of Blending." Paper presented at the 6th International Cognitive Linguistics Conference.

Sinding, Michael. 2001. "Assembling Spaces: The Conceptual Structure of Allegory." Paper presented at the Annual American Comparative Literature Association Conference.

Sondergaard, Morten. 1999. "Blended Spaces in Contemporary Art." Paper presented at Beyond Babel: 18th Annual Conference of the Western Humanities Alliance.

Sovran, Tamar. 1999. "Generic Level Versus Creativity in Metaphorical Blends." Paper presented at the 6th International Cognitive Linguistics Conference.

Steen, Francis. 1998. "Wordsworth's Autobiography of the Imagination." *Auto/Biography Studies*, Spring. [Includes a discussion of blending in, e.g., memory, perception, dreaming, and pretend play, and consequences for literary invention.]

Sun, Douglas. 1994. "Thurber's Fables for Our Time: A Case Study in Satirical Use of the Great Chain Metaphor." *Studies in American Humor*, 3:1 (new series), pp. 51–61.

Swan, Deanne, and Alan Cienki. 1999. "Constructions, Blending, and Metaphors: The Influence of Structure." Paper presented at the 6th International Cognitive Linguistics Conference.

Sweetser, Eve. 1999. "Subjectivity and Viewpoint as Blended Spaces." Paper presented at the 6th International Cognitive Linguistics Conference.

Sweetser, Eve, and Barbara Dancygier. 1999. "Semantic Overlap and Space-Blending." Paper presented at the 6th International Cognitive Linguistics Conference.

Tobin, Vera. 2001. "Texts That Pretend to Be Talk: Frame-Shifting and Frame-Blending Across Frames of Utterance in Mystery Science Theater 3000." Paper presented at the 7th International Cognitive Linguistics Conference.

Turner, Mark. 2002. "The Cognitive Study of Art, Language, and Literature." *Poetics Today.*

______. 2000. "Backstage Cognition in Reason and Choice." In Arthur Lupia, Mathew McCubbins, and Samuel L. Popkin (eds.), *Elements of Reason: Cognition, Choice, and the Bounds of Rationality* (pp. 264–286). Cambridge, U.K.: Cambridge University Press.

______. 1999a. "Forbidden Fruit." Paper presented at the 6th International Cognitive Linguistics Conference.

______. 1999b. "Forging Connections." In Chrystopher Nehaniv (ed.), *Computation for Metaphor, Analogy, and Agents* (pp. 11–26). Berlin: Springer-Verlag. [A volume in the series Lecture Notes in Artificial Intelligence.]

Turner, Mark, and Gilles Fauconnier. 2000. "Metaphor, Metonymy, and Binding." In Antonio Barcelona (ed.), *Metonymy and Metaphor at the Crossroads* (pp. 264–286). Berlin/New York: Mouton de Gruyter. [A volume in the series Topics in English Linguistics.]

______. 1999a. "A Mechanism of Creativity". *Poetics Today,* 20:3, pp. 397–418. Reprinted as "Life on Mars: Language and the Instruments of Invention." In Rebecca Wheeler (ed.), *The Workings of Language* (pp. 181–200). Westport, Conn.: Praeger.

______. 1999b. "Miscele e metafore." *Pluriverso: Biblioteca delle idee per la civiltà planetaria,* 3:3 (September), pp. 92–106. [Translation by Anna Maria Thornton.]

Veale, Tony. 1999. "Pragmatic Forces in Metaphor Use: The Mechanics of Blend Recruitment in Visual Metaphors." In Chrystopher Nehaniv (ed.), *Computation for Metaphor, Analogy, and Agents* (pp. 37–51). Berlin: Springer-Verlag. [A volume in the series Lecture Notes in Artificial Intelligence.]

______. 1996. "Pastiche: A Metaphor-Centred Computational Model of Conceptual Blending, with Special Reference to Cinematic Borrowing." Unpublished ms.

Veale, Tony, and Diarmuid O'Donogue. 1999. "Computational Models of Conceptual Integration." Paper presented at the 6th International Cognitive Linguistics Conference.

Vorobyova, Olga. "Conceptual Blending in Narrative Suspense: Making the Pain of Anxiety Sweet." Paper presented at the 7th International Cognitive Linguistics Conference.

Zbikowski, Lawrence. 1999. "The Blossoms of 'Trockne Blumen': Music and Text in the Early Nineteenth Century." *Music Analysis,* 18:3 (October 1999), pp. 307–345.

______. 1997. "Conceptual Blending and Song." Unpublished ms.

Zunshine, Lisa. "Domain Specificity and Conceptual Blending in A. L. Barbauld's *Hymns.*" Unpublished ms.

INDEX